Konstantin Meskouris

Structural Dynamics

Models, Methods, Examples

Konstantin Meskouris

Structural Dynamics

Models, Methods, Examples

Univ.-Prof. Dr.-Ing. Konstantin Meskouris
Rhein.-Westf. Technische Hochschule Aachen (RWTH)
Lehrstuhl für Baustatik und Baudynamik
Mies-van-der-Rohe-Straße 1
D-52074 Aachen

Die Deutsche Bibliothek – CIP-Einheitsaufnahme
A catalogue record for this book is available from Die Deutsche Bibliothek

ISBN 3-433-01327-6

Cover design: grappa blotto design, Berlin
Printing: betz-druck GmbH, Darmstadt
Binding: Wilh. Osswald + Co., Neustadt
Printed in Germany

Dedicated to

Professor emeritus Dr.-Ing. Dr.-Ing. h.c. Christian Petersen

Preface

Principles there are,
but even they remain unreal
until you actually apply them

Forman S. Acton

An explanation is surely in order when someone undertakes the task of adding another volume to the already impressive number of the available textbooks on structural dynamics, with special emphasis on earthquake engineering. The peculiarity of this book is that many standard computer programs used to solve everyday problems in structural dynamics are actually included in a ready-to-use form on CD-ROM, which makes it easy for the reader to embrace learning by doing, with a hands-on approach. I believe that actually applying these tools to solving one's own structural dynamics problems is the best way to achieve a certain degree of mastery in the subject, even if making some "fruitful mistakes" on the way cannot always be avoided. The book itself, which evolved from lecture notes over the years, should be particularly useful as an introductory text to undergraduate students and also to practising engineers.

It is my pleasure to thank my present and past co-workers for their help in preparing this book. Ms. Marion Brüggemann has contributed Chapter 9 on the seismic behaviour of liquid-filled tanks, Mr. Uwe Weitkemper and Mr. Falko Schube have worked out some of the practical examples and Mr. Hamid Sadegh-Azar has developed the program shell for the CD-ROM. My very best thanks also to Mrs. Anke Madej for typing the manuscript and Miss Julia Thomas for correcting my English!

Aachen, February 2000 Konstantin Meskouris

Table of Contents

1 Introduction and outline

It may be safely stated that today every structural engineer needs at least a nodding acquaintance of the principles and methods of structural dynamics in order to be able to properly fulfil his duties. This has not been always so; the massive buildings and bridges erected in past centuries with their large in-built damping capacity were normally quite insensitive to dynamic loads (with the possible exception of earthquakes). Contemporary structures are, on the contrary, mostly light-weight with relatively low damping, so that their susceptibility to vibration is much larger. In addition to dynamic excitations of natural origin, such as wind or earthquake loads, today's structure are also subject to a whole series of man-made vibration sources, such as traffic-induced vibrations, periodic loads from machinery, impulsive loads due to car accidents, blast or explosion and many more. Since many structures are potentially vibration sensitive, in the sense that their overall safety and performance (including the occupant's comfort) might be seriously impaired in the presence of dynamic loads, the pertinent checks must be included in the structure's general stability investigation. Typical vibration-sensitive structures include high-rise buildings, large-span bridges (especially suspension and cable-stayed bridges), offshore structures, natural draught cooling towers, high masts and towers, and also large multi-span sheds, machinery supports, bell towers and also cultural heritage monuments.

It is the civil engineer's task to produce a safe and economically optimal design for a given structure, considering all pertinent loading conditions. Fig. 1-1 shows the theoretical outline of this process.

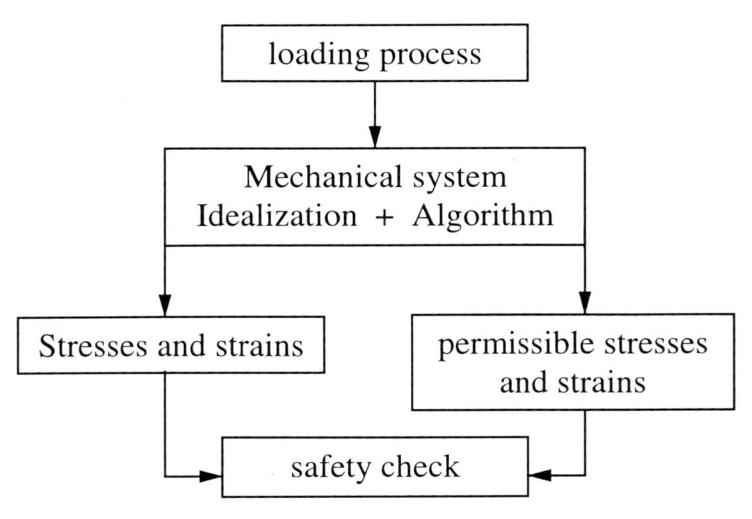

Fig. 1-1: Theoretical outline of structural safety checks

Stability checks for structures subject to time-dependent loads differ from those for static, time-independent loads in three respects:

- The description and modelling of time-dependent loads is certainly more complicated than for static loads.

- The mathematical model of the structure which is used for the determination of structural stresses and strains becomes also more complex than for static analysis alone. In the latter case, only the stiffness properties of the structure must be modelled, while for a dynamic analysis its mass and damping properties must also be considered.
- The allowable stress and strain levels in the structure are also different for static and for dynamic loads, e.g. in the cases of fatigue, low-cycle fatigue, progressive damage etc.

It is evident that not every time-dependent loading of a structure merits a full dynamic analysis. The latter is the case only for excitations which give rise to medium or large inertia forces as products of mass and acceleration, and in order for this to happen, the structure must be able to siphon off energy from the loading process (e.g. seismic event or strong wind) and convert it into kinetic energy.

This book discusses the well-established methods and models of structural dynamics and presents ready-to-use tools for the solution of many practical problems in form of 52 computer programs. Chapter 2 is devoted to some basic relationships from mechanics and mathematics which play an important role in structural dynamics; Chapter 3 deals with the single-degree-of-freedom oscillator, which, because of its simplicity, is a very popular model. Furthermore, many important concepts and methods, both in the time and in the frequency domain, are best explained in the context of a single-degree-of-freedom model. Some aspects of vibration isolation and the treatment of non-linear single-degree-of-freedom systems are also dealt with in this chapter.

Chapter 4 deals with discrete multi-degree-of-freedom systems. After introducing sub-structuring and condensation techniques, the presentation concentrates on modal analysis, the solution of the eigenvalue problem, direct integration methods and frequency domain methods.

Chapter 5 is devoted to plane frame structures with distributed mass and stiffness under harmonic excitation; both the EULER/BERNOULLI- and the TIMOSHENKO theory are employed.

Chapter 6 discusses the practical problem of bell-induced vibrations of church towers, both theoretically and experimentally, with some practical examples included.

Chapters 7, 8 and 9 deal with seismic loads. Chapter 7 discusses the geophysical background and the mathematical tools used by engineers for describing seismic loads as well as the structural models mostly in use. Chapter 8 deals with practical examples concerning the seismic behaviour of high-rise buildings, a seismic investigation of the Cologne Cathedral and the determination of the natural frequencies and mode shapes of a complex refinery container.

Chapter 9 presents methods for investigating the seismic response of liquid-filled tanks and finally, in the Appendix, all 52 computer programs are described with their input and output files.

2 Introduction

2.1 Fundamental quantities and their units

The metric system (SI, „Système international") will be used throughout this book. In the context of structural dynamics, we shall be dealing with the following three fundamental variables:

Length l	measured in meters (m)
Mass m, M	measured in kilograms (kg)
Time t	measured in seconds (s)

Their physical dimension is often given in brackets, e.g.

Length l:	[L]
Mass m, M:	[M]
Time t:	[T]

The following variables also appear regularly in structural dynamics applications; their customary symbols are given in braces followed by dimensions and units:

- Displacement: {u, w}, dimension [L], unit m.
- Velocity: $\{\dot{u}, \dot{w}\}$, defined as the derivative of displacement with respect to time, dimension [L T^{-1}], unit m/s.
- Acceleration: $\{\ddot{u}, \ddot{w}\}$, defined as the second temporal derivative of displacement, dimension [L T^{-2}], unit m/s^2.
- Force: According to NEWTON's second law, the force acting on a body equals its inertial mass m times the acceleration $\underline{a}$, $\underline{F} = m\,\underline{a}$. Here, as well as in the following vectors and matrices are designated by underlining the corresponding symbol. The force {F} has the dimension [M L T^{-2}] and its unit is 1 $kgms^{-2}$ or 1 N (Newton).
- Stress: It is defined as force per unit area, with the dimension [M L^{-1} T^{-2}] and is measured in N/m^2 or Pa (Pascal).
- Work: {W}, equals the vector product (force · incremental displacement) with [M L^2 T^{-2}] as dimension and 1 Nm = 1 J (Joule) as unit.
- Power: {P}, is equal to the work per unit time or the time integral of force times velocity. Its dimension is [M L^2 T^{-3}] with 1 Nm/s = 1 W (Watt) as unit.
- Density: {ρ}, is equal to mass per unit volume with [M L^{-3}] as dimension and 1 kg/m³ as unit.

Some useful conversion factors with other units follow:

Stress, pressure:	1 Millibar (mbar), equals 0.1 kPa or 0.1 kN/m^2
Work:	1 J = 1 Nm = 2.78 10^{-7} kWh
Power:	1 W = 1 J/s = 1.36 10^{-3} PS

The adoption of a consistent unit system is very important in structural dynamics, since it dispenses with the need for error-prone force-mass conversions. This can be achieved by measuring masses in 1000-kg units = 1 (continental) ton, forces in kN, lengths in m and time in s. In the standard equation describing the action of the earth's gravity upon a body

$$G = mg \tag{2.1.1}$$

G is the weight of the mass m and g is the gravity acceleration ($g \approx 9.81\ m/s^2$). The latter may be set equal to 10 m/s^2 (2% error), so that a mass of 1 ton weighs 10 kN.

2.2 D´ALEMBERT's principle, law of conservation of momentum

Introducing into NEWTON's second law

$$\underline{F} = m\,\underline{a} \tag{2.2.1}$$

the inertia force vector $\underline{I}$ according to

$$\underline{I} = -m\underline{a} \tag{2.2.2}$$

leads to the „pseudo-static" formulation of the equilibrium condition:

$$\underline{F} + \underline{I} = 0 \tag{2.2.3}$$

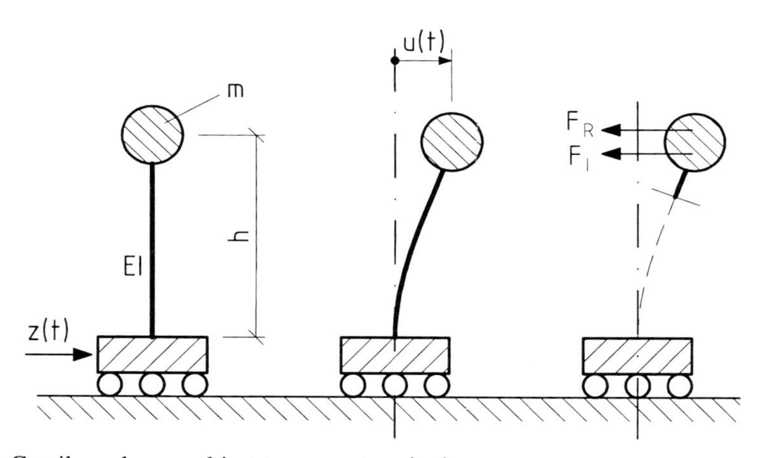

Fig. 2.2-1: Cantilever beam subject to support excitation

This is the well-known principle of D'ALEMBERT, which states that dynamic equilibrium can be expressed formally by considering inertia forces $\underline{I}$ as equivalent to, and in addition to, the acting external loads $\underline{F}$. It should be noted that in defining the inertia force vector it has been implicitly assumed that the frame of reference employed (reference co-ordinate system) is at rest or moving with constant velocity (NEWTONIAN frame of reference). If a co-ordinate system fixed to an accelerating body is used instead, the body itself is seemingly at rest relative to this reference frame; however, inertia forces are present which are formally indistinguishable from external loads. Through the introduction of inertia forces as equivalent external loads, the well-known static equilibrium equations can be applied to dynamic problems. Of course, the resulting expressions are now differential equations rather than algebraic equations as is the case in statics, because of the appearance of second derivatives of the displacements as accelerations. The setting up of such a differential equation by means of D'ALEMBERT's principle is shown in the following example Fig. 2.2-1 shows a massless cantilever beam (with length h, bending stiffness EI and a point or lumped mass m at its free end), the base of which is subjected to a time-varying displacement z(t); the determination of its equation of motion is required. A single independent co-ordinate u(t), in this case the horizontal displacement of the mass with respect to its support, is sufficient for describing the system's configuration. The system can therefore be termed as a single-degree-of-freedom oscillator, or SDOF system. If the lumped mass is cut free from the rest of the structure, the restoring force F_R acting opposite to the positive displacement u(t) must be introduced; additionally, the inertia force F_I also opposes the displacement u(t). The force equilibrium condition in the direction of u as the only kinematic degree of freedom present states that the sum of all horizontal forces must equal zero:

$$F_R + F_I = 0 \tag{2. 2. 4}$$

Here

$$F_R = (\text{spring stiffness k}) \cdot (\text{displacement u}) = k \cdot u = \frac{3EI}{h^3} \cdot u$$

and

$$F_I = \text{mass} \cdot \text{total acceleration} = m \cdot (\ddot{z} + \ddot{u})$$

This leads to the differential equation of motion

$$m \cdot \ddot{u} + \frac{3EI}{h^3} \cdot u = -m \cdot \ddot{z} \tag{2. 2. 5}$$

As already mentioned, consistent units should be used throughout (e.g. kN, m, s and tons).

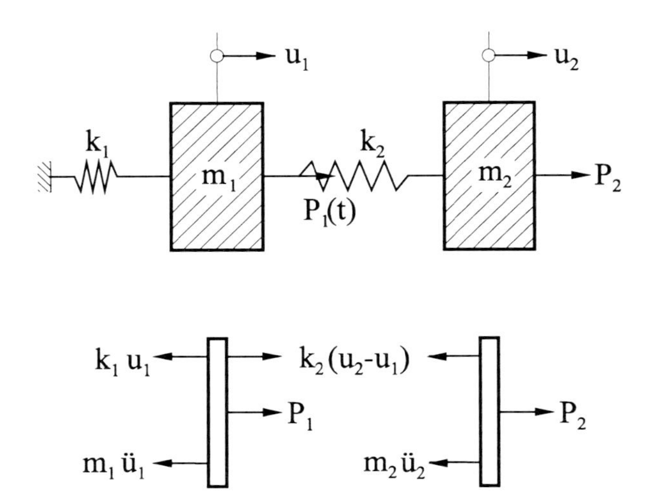

Fig. 2.2-2: Two-degree-of-freedom system

The two-degree-of-freedom system of Fig. 2.2-2 will be considered next. By "cutting free" both bodies and writing down the equilibrium equations in the horizontal direction, the following expressions emerge:

$$\begin{aligned} m_1\ddot{u}_1 + k_1u_1 - k_2(u_2 - u_1) &= P_1 \\ m_2\ddot{u}_2 + k_2(u_2 - u_1) &= P_2 \end{aligned} \tag{2. 2. 6}$$

They can be written down more succinctly in matrix form as follows:

$$\begin{bmatrix} m_1 & 0 \\ 0 & m_2 \end{bmatrix}\begin{bmatrix} \ddot{u}_1 \\ \ddot{u}_2 \end{bmatrix} + \begin{bmatrix} k_1+k_2 & -k_2 \\ -k_2 & k_2 \end{bmatrix}\begin{bmatrix} u_1 \\ u_2 \end{bmatrix} = \begin{bmatrix} P_1 \\ P_2 \end{bmatrix} \tag{2. 2. 7}$$

or

$$\underline{M}\,\underline{\ddot{u}} + \underline{K}\,\underline{u} = \underline{F} \tag{2. 2. 8}$$

with the mass matrix $\underline{M}$, the stiffness matrix $\underline{K}$, the displacement vector $\underline{u}$, the acceleration vector $\underline{\ddot{u}}$ and the load vector $\underline{F}$.

Another example is shown in Fig. 2.2-3. A rigid beam with total mass m and moment of inertia about the centre of gravity S equal to Θ is supported by longitudinal springs at its end sections:

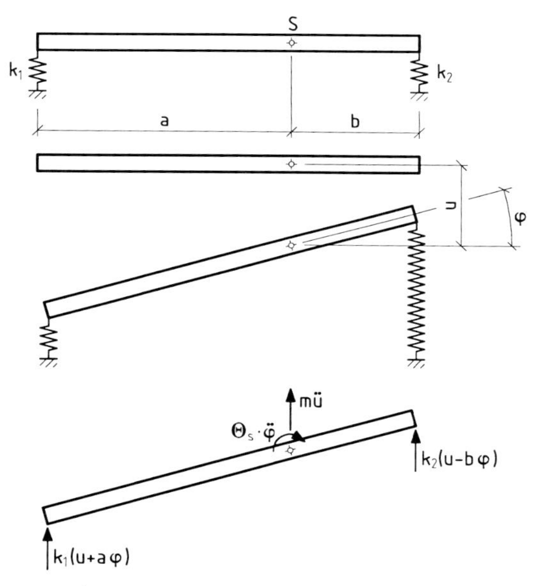

Fig. 2.2-3: Rigid beam on spring supports

The equilibrium equation in the vertical direction ($\Sigma V = 0$) yields

$$m\ddot{u} + (k_1 + k_2)u + (ak_1 - bk_2)\varphi = 0 \tag{2. 2. 9}$$

The moment equation $\Sigma M = 0$ about the centre of gravity leads to

$$\Theta\ddot{\varphi} + (ak_1 - bk_2)u + (a^2k_1 + b^2k_2)\varphi = 0 \tag{2. 2. 10}$$

In matrix form, these expressions can be written down as:

$$\begin{bmatrix} m & 0 \\ 0 & \Theta \end{bmatrix}\begin{bmatrix} \ddot{u} \\ \ddot{\varphi} \end{bmatrix} + \begin{bmatrix} k_1+k_2 & ak_1-bk_2 \\ ak_1-bk_2 & a^2k_1+b^2k_2 \end{bmatrix}\begin{bmatrix} u \\ \varphi \end{bmatrix} = \begin{bmatrix} 0 \\ 0 \end{bmatrix} \qquad (2.2.11)$$

Lastly, the railway carriage shown in Fig. 2.2-4 is considered with its wheel bodies subjected to support excitations y_{L1}, y_{L2}, y_{R1} and y_{R2} . Here damping forces, modelled by viscous dashpots, are also being considered. They are set equal to the products of velocities with the corresponding viscous damping coefficients and they oppose positive displacements in the same way as restoring and inertia forces.

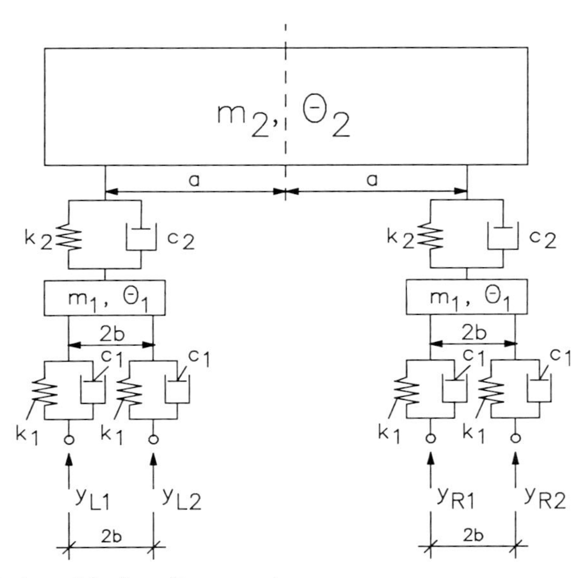

Fig. 2.2-4: Simplified model of a railway carriage

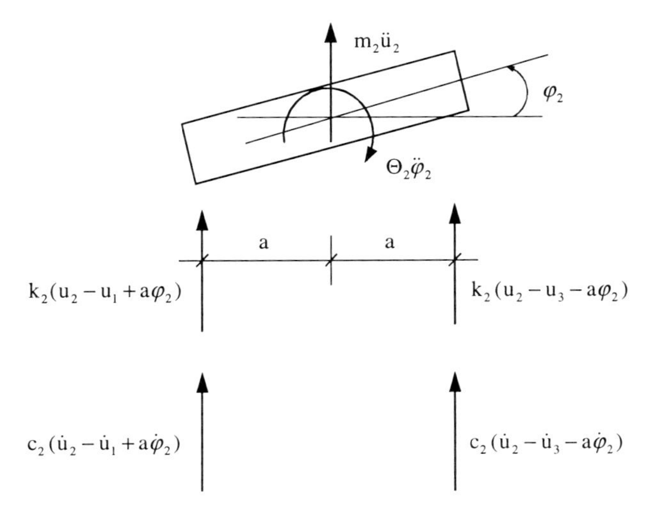

Fig. 2.2-5: Upper part of the carriage

As before, "cutting free" the single rigid bodies and writing down the equilibrium conditions furnishes two equations ($\Sigma V = 0$, $\Sigma M = 0$) for each body. For the upper part of the carriage, Fig. 2.2-5 shows the forces involved.

The equilibrium condition for the sum of the vertical force components $\Sigma V = 0$ yields

$$m_2\ddot{u}_2 - c_2\dot{u}_1 + 2c_2\dot{u}_2 - c_2\dot{u}_3 - k_2u_1 + 2k_2u_2 - k_2u_3 = 0 \qquad (2.2.12)$$

and analogously for the moment sum $\Sigma M = 0$:

$$\Theta_2\ddot{\varphi}_2 - ac_2\dot{u}_1 + ac_2\dot{u}_3 + 2a^2c_2\dot{\varphi}_2 - ak_2u_1 + ak_2u_3 + 2a^2k_2\varphi_2 = 0 \qquad (2.2.13)$$

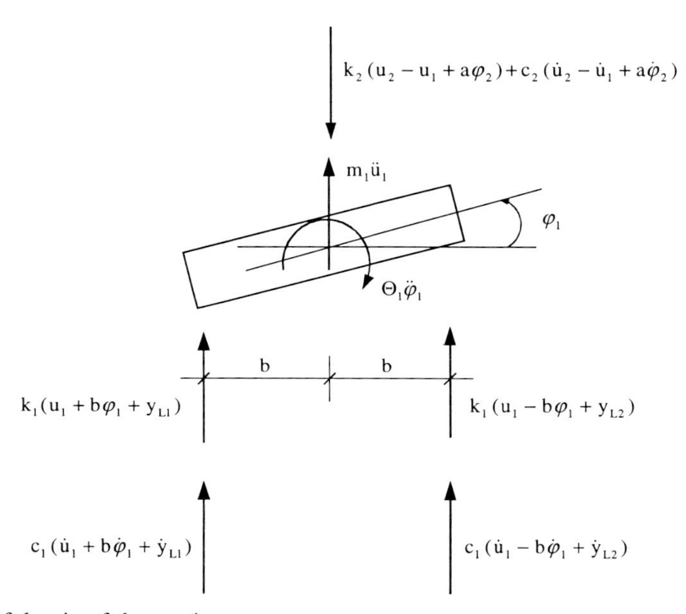

Fig. 2.2-6: Left bogie of the carriage

Fig. 2.2-6 shows the left bogie of the carriage; the equilibrium condition $\Sigma V = 0$ leads to

$$\begin{aligned} & m_1\ddot{u}_1 + (2c_1 + c_2)\dot{u}_1 - c_2\dot{u}_2 - ac_2\dot{\varphi}_2 + (2k_1 + k_2)u_1 - k_2u_2 - ak_2\varphi_2 = \\ & = -c_1(\dot{y}_{L1} + \dot{y}_{L2}) - k_1(y_{L1} + y_{L2}) \end{aligned} \qquad (2.2.14)$$

and $\Sigma M = 0$ as before

$$\Theta_1\ddot{\varphi}_1 + 2b^2c_1\dot{\varphi}_1 + 2b^2k_1\varphi_1 = bc_1(-\dot{y}_{L1} + \dot{y}_{L2}) + bk_1(-y_{L1} + y_{L2}) \qquad (2.2.15)$$

For the right bogie (Fig. 2.2-7) the condition $\Sigma V = 0$ leads to

$$\begin{aligned} & m_1\ddot{u}_3 - c_2\dot{u}_2 + (2c_1 + c_2)\dot{u}_3 + ac_2\dot{\varphi}_2 - k_2u_2 + (2k_1 + k_2)u_3 + ak_2\varphi_2 = \\ & = -c_1(\dot{y}_{R1} + \dot{y}_{R2}) - k_1(y_{R1} + y_{R2}) \end{aligned} \qquad (2.2.16)$$

and $\Sigma M = 0$ to:

$$\Theta_1\ddot{\varphi}_3 + 2b^2c_1\dot{\varphi}_3 + 2b^2k_1\varphi_3 = bc_1(-\dot{y}_{R1} + \dot{y}_{R2}) + bk_1(-y_{R1} + y_{R2}) \qquad (2.2.17)$$

$$\underline{M}\,\underline{\ddot{V}}+\underline{C}\,\underline{\dot{V}}+\underline{K}\,\underline{V}=\underline{F} \tag{2. 2. 18}$$

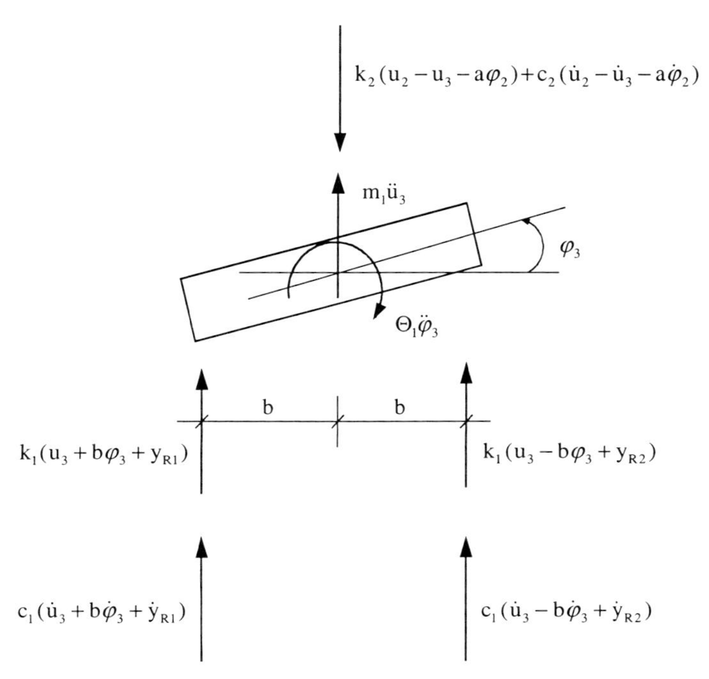

Fig. 2.2-7: Right bogie of the carriage

All six equilibrium equations derived so far can be expressed by the matrix equation with the matrices and vectors

$$\underline{M}=\begin{bmatrix} m_1 & & & & & \\ & m_2 & & & 0 & \\ & & m_1 & & & \\ & & & \Theta_1 & & \\ & 0 & & & \Theta_2 & \\ & & & & & \Theta_1 \end{bmatrix}; \quad \underline{\ddot{V}}=\begin{bmatrix} \ddot{u}_1 \\ \ddot{u}_2 \\ \ddot{u}_3 \\ \ddot{\varphi}_1 \\ \ddot{\varphi}_2 \\ \ddot{\varphi}_3 \end{bmatrix} \tag{2. 2. 19}$$

$$\underline{C}=\begin{bmatrix} 2c_1+c_2 & -c_2 & 0 & 0 & -ac_2 & 0 \\ -c_2 & 2c_2 & -c_2 & 0 & 0 & 0 \\ 0 & -c_2 & 2c_1+c_2 & 0 & ac_2 & 0 \\ 0 & 0 & 0 & 2b^2c_1 & 0 & 0 \\ -ac_2 & 0 & ac_2 & 0 & 2a^2c_2 & 0 \\ 0 & 0 & 0 & 0 & 0 & 2b^2c_1 \end{bmatrix}, \quad \underline{\dot{V}}=\begin{bmatrix} \dot{u}_1 \\ \dot{u}_2 \\ \dot{u}_3 \\ \dot{\varphi}_1 \\ \dot{\varphi}_2 \\ \dot{\varphi}_3 \end{bmatrix} \tag{2. 2. 20}$$

$$\underline{K}=\begin{bmatrix} 2k_1+k_2 & -k_2 & 0 & 0 & -ak_2 & 0 \\ -k_2 & 2k_2 & -k_2 & 0 & 0 & 0 \\ 0 & -k_2 & 2k_1+k_2 & 0 & ak_2 & 0 \\ 0 & 0 & 0 & 2b^2k_1 & 0 & 0 \\ -ak_2 & 0 & ak_2 & 0 & 2a^2k_2 & 0 \\ 0 & 0 & 0 & 0 & 0 & 2b^2k_1 \end{bmatrix}, \quad \underline{V}=\begin{bmatrix} u_1 \\ u_2 \\ u_3 \\ \varphi_1 \\ \varphi_2 \\ \varphi_3 \end{bmatrix} \qquad (2.\ 2.\ 21)$$

The load vector on the right hand side equals

$$\underline{F}=\begin{bmatrix} -c_1(\dot{y}_{L1}+\dot{y}_{L2})-k_1(y_{L1}+y_{L2}) \\ 0 \\ -c_1(\dot{y}_{R1}+\dot{y}_{R2})-k_1(y_{R1}+y_{R2}) \\ -bc_1(\dot{y}_{L1}-\dot{y}_{L2})-bk_1(y_{L1}-y_{L2}) \\ 0 \\ -bc_1(\dot{y}_{R1}-\dot{y}_{R2})-bk_1(y_{R1}-y_{R2}) \end{bmatrix} \qquad (2.\ 2.\ 22)$$

Now to the momentum conservation law, applied to purely translational motions (Fig. 2.2-8). It states that the time integral of the force $\underline{F}$ acting on the mass m from t_1 to t_2 is equal to the difference in momentum at times t_1 and t_2 . Also, the momentum difference $\Delta\underline{I}$ is equal to the mass m times the velocity difference $\Delta\underline{v}$ between t_1 and t_2:

$$\int_{t_1}^{t_2} \underline{F}(t)\,dt = m\,(\underline{v}_2-\underline{v}_1) = m\,\Delta\underline{v} = \underline{I}_2-\underline{I}_1 = \Delta\underline{I} \qquad (2.\ 2.\ 23)$$

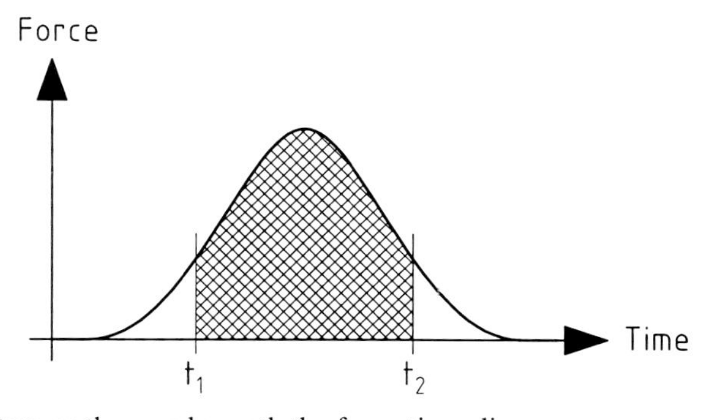

Fig. 2.2-8: Momentum as the area beneath the force-time-diagram

The integral in Eq. (2.2.23) is equal to the area beneath the force-time diagram of Fig. 2.2-8; it represents a force impulse. An important concept in this context is the so-called DIRAC delta function (or DIRAC unit impulse), which can be generated by letting the difference t_2-t_1 approach zero while keeping the magnitude of the impulse as a unit value. The DIRAC delta function acting at time $t = t_1$ is denoted by $\delta(t-t_1)$ and is characterised by the following features:

- For all $t \neq t_1$ its value equals zero, for $t = t_1$ it approaches infinity.
- The time integral of δ over any time range including t_l yields a unit value (unit impulse).

2.3 Energy conservation

A distinction is made between "energy of position" ("potential energy" after J.M. RANKINE) and "energy of motion" or "kinetic energy" (after LORD KELVIN). A body with mass m moving in a straight line at a velocity v contains the kinetic energy

$$E_K = \frac{m \cdot v^2}{2} \qquad (2.3.1)$$

Analogously, a body rotating about a point O (Fig. 2.3-1) with an circular frequency ω in rad/s contains an amount of kinetic energy equal to

$$E_K = \frac{\Theta \cdot \omega^2}{2} \qquad (2.3.2)$$

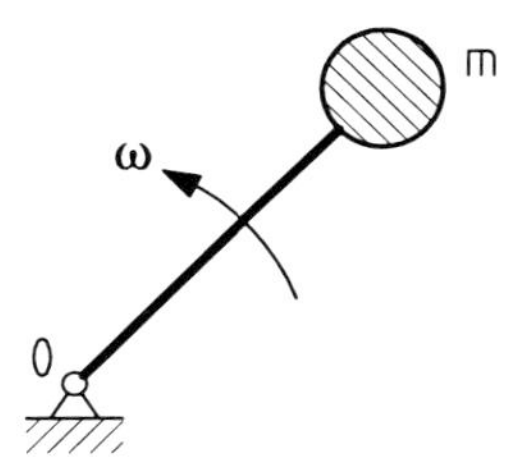

Fig. 2.3-1: A body rotating about a fixed point

Here, Θ is the moment of inertia of the body about O, expressed in kgm^2 or in (1000 kg) tons times m^2 if the consistent units recommended earlier are used. For a body of mass m and negligible dimensions in respect to the radius r rotating about O (Fig. 2.3-1), the moment of inertia is given by

$$\Theta = m \cdot r^2 \qquad (2.3.3)$$

Some expressions for mass moments of inertia of geometrically simple rigid bodies are presented in Section 2.4.

The potential energy of a body with mass m positioned at a height h over a reference level is equal to

$$E_P = G \cdot h = m \cdot g \cdot h \qquad (2.3.4)$$

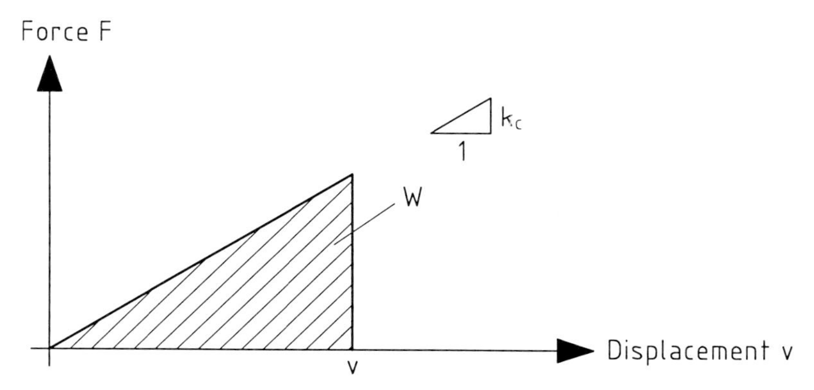

Fig. 2.3-2: Work stored in an elastic longitudinal spring

with G as weight of the body, G = mg. A special form of potential energy is the strain energy W which is stored in an elastic body in the course of its deformation process. In the special case of a longitudinal spring with a spring constant k_c in kN/m, k_c being the axial force in kN necessary to produce a spring deformation of 1 m, a force F (in kN) causes a deformation of $v = \frac{F}{k_c}$ (in m) and the strain energy stored in the spring becomes (Fig. 2.3-2):

$$W = \frac{1}{2} \cdot F \cdot \frac{F}{k_c} = \frac{1}{2} \frac{F^2}{k_c} \tag{2. 3. 5}$$

The factor $\frac{1}{2}$ in (2.3.5) points out that the deformation v is being caused by the force F itself, meaning that both v and F increase from zero to their final values. If, in contrast, a constant force F (in the direction of v) is present during the whole displacement process, no factor $\frac{1}{2}$ arises. Analogously, in a rotational spring with a spring constant k_φ in $\frac{kNm}{rad}$ (k_φ being the moment necessary to cause a rotation of 1 radian), the work done by a moment M and stored in the spring is equal to

$$W = \frac{1}{2} \cdot M \cdot \frac{M}{k_\varphi} = \frac{1}{2} \frac{M^2}{k_\varphi} \tag{2. 3. 6}$$

For mechanical systems into which no energy is introduced (e.g. by external forces) and in which no mechanical energy is dissipated (e.g. converted to heat by friction), the law of energy conservation can be written down as

$$E = E_K + E_P = \text{const.} \tag{2. 3. 7}$$

or

$$\frac{d}{dt}(E) = \frac{d}{dt}(E_K + E_P) = 0 \tag{2. 3. 8}$$

In this case, the amount of energy in the system stays constant. If, in an open system, external forces $\underline{F}$ introduce an amount of energy equal to W_F while the energy amount W_D is dissipated inside the system, an extended version of Eq. (2.3.8) holds:

$$\frac{d}{dt}(E)=\frac{d}{dt}(W_F+W_D) \qquad (2.3.9)$$

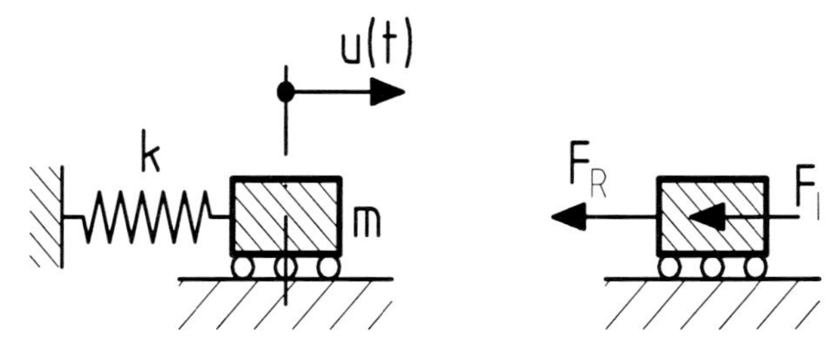

Fig. 2.3-3: Undamped single-degree-of-freedom (SDOF) system

As an example for the validity of the law of conservation of energy the undamped SDOF system shown in Fig. 2.3-3 is considered. The corresponding equation of motion is

$$F_I+F_R = m\ddot{u}+k\,u=0 \qquad (2.3.10)$$

Its solution is given by the harmonic function

$$u(t)=u_0 \sin\omega t \qquad (2.3.11)$$

Introducing Eq. (2.3.11) in (2.3.10) yields

$$k\,u_0 = m\,u_0\,\omega^2 \qquad (2.3.12)$$

and multiplying both sides by $\frac{1}{2}u_0$ leads to

$$\frac{1}{2}k\,u_0^{\,2}=\frac{1}{2}m\,(u_0\,\omega)^2 \qquad (2.3.13)$$

The left-hand side of this equation is the maximum strain energy stored in the spring, while the right-hand side is the maximum kinetic energy, as predicted by the law of energy conservation.

2.4 Area and mass moments of inertia

The evaluation of some properties of a general polygonal plane section will be presented first (Fig. 2.4-1). Its area A, its first order moments S_x and S_y and its second order moments I_x, I_y, I_{xy} can be computed by the expressions given below (according to FLESSNER [2.1]), once the section is defined by entering the (x,y) co-ordinates of all its vertices. The first point considered (node no. 1) must be entered again as the last node of the polygon, in order for it to be closed. The nodes are numbered sequentially in a counter-clockwise direction (mathematically positive sense), so that the actual section is always on the left while following the perimeter in the order of ascending node numbers. Multi-cell sections can be simply treated by introducing zero-width cuts at suitable locations, as depicted in Fig. 2.4-1.

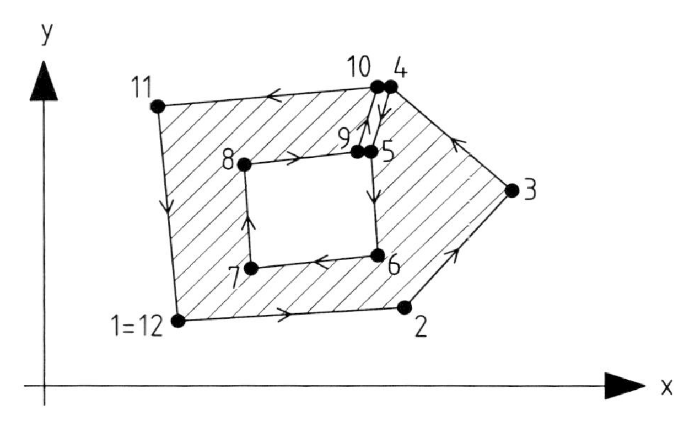

Fig. 2.4-1: Plane section and reference co-ordinate system

$$A = \frac{1}{2}\sum_{i=1}^{n-1}\left(x_i y_{i+1} - x_{i+1} y_i\right) \tag{2. 4. 1}$$

$$S_x = \frac{1}{6}\sum_{i=1}^{n-1}\left[\left(x_i y_{i+1} - x_{i+1} y_i\right)\left(y_i + y_{i+1}\right)\right] \tag{2. 4. 2}$$

$$S_y = \frac{1}{6}\sum_{i=1}^{n-1}\left[\left(x_i y_{i+1} - x_{i+1} y_i\right)\left(x_i + x_{i+1}\right)\right] \tag{2. 4. 3}$$

$$I_x = \frac{1}{12}\sum_{i=1}^{n-1}\left[\left(x_i y_{i+1} - x_{i+1} y_i\right)\left(\left[y_i + y_{i+1}\right]^2 - y_i y_{i+1}\right)\right] \tag{2. 4. 4}$$

$$I_y = \frac{1}{12}\sum_{i=1}^{n-1}\left[\left(x_i y_{i+1} - x_{i+1} y_i\right)\left(\left[x_i + x_{i+1}\right]^2 - x_i x_{i+1}\right)\right] \tag{2. 4. 5}$$

$$I_{xy} = \frac{1}{12}\sum_{i=1}^{n-1}\left[\left(x_i y_{i+1} - x_{i+1} y_i\right)\left[\left(x_i + x_{i+1}\right)\left(y_i + y_{i+1}\right) - \frac{1}{2}\left(x_i y_{i+1} + x_{i+1} y_i\right)\right]\right] \tag{2. 4. 6}$$

The determination of the section's principal axes (ξ, η) and the corresponding principal moments of inertia, I_ξ, I_η, can be carried out as follows:

- Introduction of an arbitrarily located Cartesian (x,y) co-ordinate system as shown in Fig. 2.4-1 and computation of A, S_x, S_y , I_x, I_y, I_{xy} , using the expressions (2.4.1) through (2.4.6).

- Determination of the values x_s, y_s as co-ordinates of the section's centre of gravity S in the (x,y) system:

$$x_s = \frac{S_y}{A} \tag{2. 4. 7}$$

$$y_s = \frac{S_x}{A} \tag{2.4.8}$$

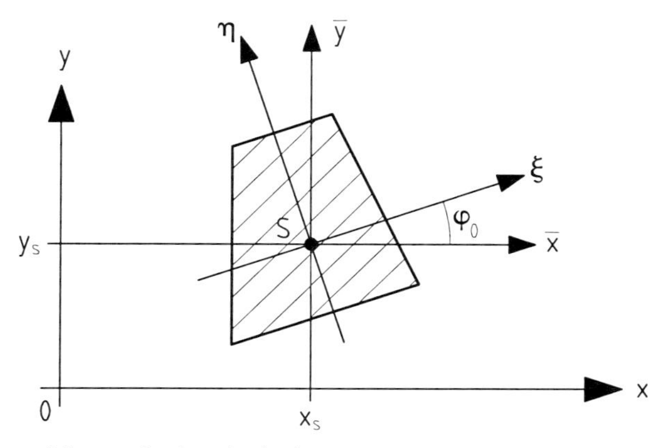

Fig. 2.4-2: Position of the section's principal axes

- The principal moments of inertia I_ξ, I_η with reference to the (ξ, η) system are given by the following expressions, in which $(\bar{x}, \bar{y})$ is the co-ordinate system parallel to (x,y) with S as its origin and φ_0 is the angle between the axes ξ and $\bar{x}$ (Fig. 2.4-2).

$$I_{\bar{x}} = I_x - Ay_s^2 \tag{2.4.9}$$

$$I_{\bar{y}} = I_y - Ax_s^2 \tag{2.4.10}$$

$$I_{\overline{xy}} = I_{xy} - Ax_s y_s \tag{2.4.11}$$

$$\tan 2\varphi_0 = \frac{2I_{\overline{xy}}}{I_{\bar{y}} - I_{\bar{x}}} \tag{2.4.12}$$

$$I_\xi = \frac{I_{\bar{x}} + I_{\bar{y}}}{2} + \left[\frac{I_{\bar{x}} - I_{\bar{y}}}{2}\cos 2\varphi_0 - I_{\overline{xy}}\sin 2\varphi_0\right] \tag{2.4.13}$$

$$I_\eta = \frac{I_{\bar{x}} + I_{\bar{y}}}{2} - \left[\frac{I_{\bar{x}} - I_{\bar{y}}}{2}\cos 2\varphi_0 - I_{\overline{xy}}\sin 2\varphi_0\right] \tag{2.4.14}$$

The actual calculations are carried out by the computer program AREMOM described in the Appendix. As an example, Fig. 2.4-3 shows the section of a two-cell canal, the section properties of which are to be determined.

The input file KOORD.ARE for the AREMOM program is in this case

```
19
        0.0, 0.0
        6.8, 0.0
        8.0, 1.2
       12.2, 1.2
       12.2, 7.4
       11.4, 7.4
       11.4, 2.0
        6.8, 2.0
        6.8, 6.6
       11.4, 6.6
       11.4, 7.4
        0.8, 7.4
        0.8, 6.6
        5.5, 6.6
        5.5, 0.8
        0.8, 0.8
        0.8, 7.4
        0.0, 7.4
        0.0, 0.0
```

and the results as they appear in the output file MOMENT.ARE are:

```
 Area        36.10000
 Sx         138.874000
 Sy         215.907000
 Ixx        767.150300
 Iyy       1805.667000
 Ixy        859.153500
 Angle (deg)  5.739
 Ixi        230.040500
 Ieta         517.242400
 xs           5.9808
 ys           3.8469
 Ixs          232.912300
 Iys          514.370500
 Ixys          28.575480
```

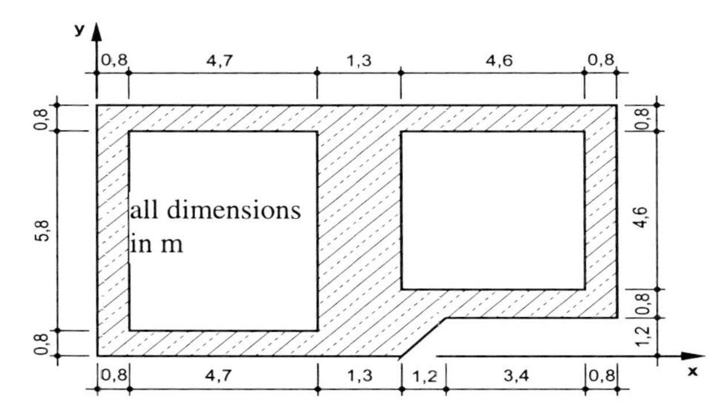

Fig. 2.4-3: Reinforced concrete section

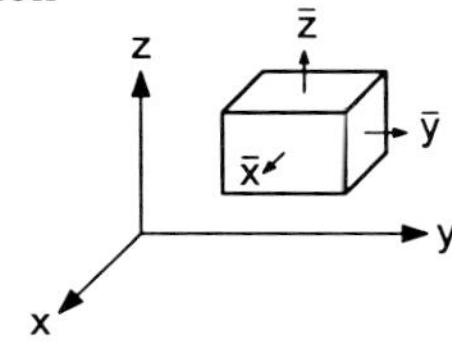

Fig. 2.4-4: Coordinate system in the 3D case

In the three-dimensional case, the second moments of inertia of a rigid body with reference to the (x,y,z) co-ordinate system of Fig. 2.4-4 are defined as follows:

$$\begin{aligned}
\Theta_{xx} &= \int_V \left(y^2 + z^2\right) dm \\
\Theta_{yy} &= \int_V \left(z^2 + x^2\right) dm \\
\Theta_{zz} &= \int_V \left(x^2 + y^2\right) dm \\
\Theta_{xy} &= \int_V xy \, dm \\
\Theta_{yz} &= \int_V yz \, dm \\
\Theta_{zx} &= \int_V zx \, dm
\end{aligned} \qquad (2.\,4.\,15)$$

In the case of homogeneous bodies with density ρ the mass differential dm is equal to $\rho\, dV$ and the integrals must be evaluated over the total volume V of the body. If the "centrifugal" moments of inertia Θ_{ij}, $i \neq j$, turn out to be zero, the corresponding co-ordinate system coincides with the principal axes of the body.

The actual computation of the principal mass moments of inertia of a rigid body consists of the following steps, in exact analogy with the two-dimensional case:

- Determination of the volume, the first and second order mass moments and also the position of the centre of gravity of the rigid body in reference to an arbitrary (x,y,z) Cartesian co-ordinate system. For homogeneous polyhedral rigid bodies PREDIGER [2.2] and PETERSEN [2.3] present closed-form expressions, which, however, are somewhat lengthy and are not reproduced here. They form the basis of the computer program BODMOM which can be used for corresponding computations.

- Using STEINER´s law, the second order mass moments of inertia $\Theta_{\bar{i}\bar{k}}$ with reference to the co-ordinate system $(\bar{x}, \bar{y}, \bar{z})$ are computed next. The $(\bar{x}, \bar{y}, \bar{z})$ system is parallel to the (x,y,z) co-ordinate system and its origin coincides with the centre of gravity. With M as the total mass of the body and x_S, y_S and z_S as co-ordinates of the centre of gravity in the (x,y,z) system, the following expressions hold:

$$\Theta_{\bar{x}\bar{x}} = \Theta_{xx} - M\,(y_s^{\,2} + z_s^{\,2}) \qquad (2.\,4.\,16)$$

$$\Theta_{\bar{y}\bar{y}} = \Theta_{yy} - M\,(z_s^{\,2} + x_s^{\,2}) \qquad (2.\,4.\,17)$$

$$\Theta_{\bar{z}\bar{z}} = \Theta_{zz} - M\,(x_s^{\,2} + y_s^{\,2}) \qquad (2.\,4.\,18)$$

$$\Theta_{\bar{x}\bar{y}} = \Theta_{xy} - M\, x_S\, y_S \qquad (2.\,4.\,19)$$

$$\Theta_{\bar{y}\bar{z}} = \Theta_{yz} - M\, y_S\, z_S \qquad (2.\,4.\,20)$$

$$\Theta_{\overline{zx}} = \Theta_{zx} - M\, z_s\, x_s \tag{2. 4. 21}$$

- The principal second order moments of inertia $\Theta_1, \Theta_2, \Theta_3$ are the solutions of the cubic equation

$$\Theta^3 - I_1\,\Theta^2 + I_2\,\Theta - I_3 = 0 \tag{2. 4. 22}$$

with the coefficients

$$I_1 = \Theta_{\overline{xx}} + \Theta_{\overline{yy}} + \Theta_{\overline{zz}} \tag{2. 4. 23}$$

$$I_2 = \Theta_{\overline{xx}} \cdot \Theta_{\overline{yy}} + \Theta_{\overline{yy}} \cdot \Theta_{\overline{zz}} + \Theta_{\overline{zz}} \cdot \Theta_{\overline{xx}} - (\Theta_{\overline{xy}} \cdot \Theta_{\overline{xy}} + \Theta_{\overline{yz}} \cdot \Theta_{\overline{yz}} + \Theta_{\overline{zx}} \cdot \Theta_{\overline{zx}}) \tag{2. 4. 24}$$

$$I_3 = \begin{vmatrix} \Theta_{\overline{xx}} & \Theta_{\overline{xy}} & \Theta_{\overline{xz}} \\ \Theta_{\overline{xy}} & \Theta_{\overline{yy}} & \Theta_{\overline{yz}} \\ \Theta_{\overline{xz}} & \Theta_{\overline{yz}} & \Theta_{\overline{zz}} \end{vmatrix} \tag{2. 4. 25}$$

For more details the reader is referred to the publications cited earlier.

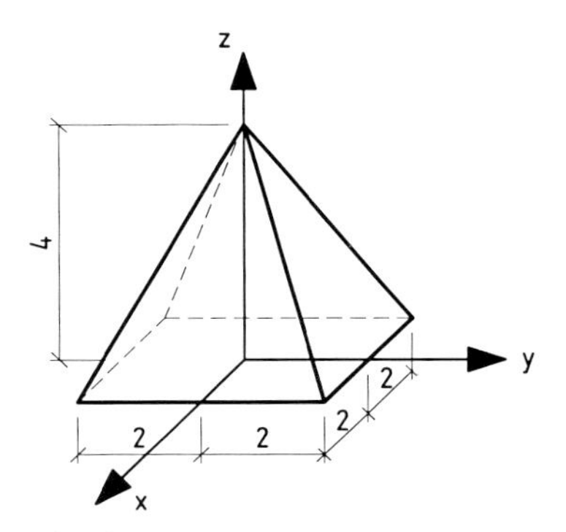

Fig. 2.4-5: Pyramid with square basis

The pyramid shown in Fig. 2.4-5 serves as an example for application of the BODMOM computer program. It is defined by a square base (with 4 nodes) and four triangular facets (3 nodes each), so that there are altogether five facets. The input file KOORD.BOD containing the co-ordinates of all nodal points of each facet is reproduced below:

```
5
3
3
3
3
4
 2.0,-2.0,0.0
 2.0, 2.0,0.0
 0.0, 0.0,4.0
 2.0, 2.0,0.0
-2.0, 2.0,0.0
 0.0, 0.0,4.0
-2.0, 2.0,0.0
-2.0,-2.0,0.0
 0.0, 0.0,4.0
-2.0,-2.0,0.0
```

```
 2.0,-2.0,0.0
 0.0, 0.0,4.0
 2.0,-2.0,0.0
 2.0, 2.0,0.0
-2.0, 2.0,0.0
-2.0,-2.0,0.0
```

The results are written in the MOMENT.BOD file and consist of the first order mass moments S_x, S_y, S_z, the co-ordinates of the centre of gravity, the body volume and the second order mass moments with reference to the (x,y,z) co-ordinate system:

```
1st order moments       Sx, Sy, Sz:          .0000E+00     .0000E+00     .2133E+02
Co-ordinates of centre of gravity:           .0000E+00     .0000E+00     .1000E+01
Volume  V:                                   .2133E+02
2nd order moments  Jxx, Jyy, Jzz:            .5120E+02     .5120E+02     .3413E+02
2nd order moments  Jxy, Jyz, Jzx:            .0000E+00     .0000E+00     .0000E+00
```

The following table presents some results for masses and second moments of inertia of simple rigid bodies.

	mass m	Θ_x	Θ_y	Θ_z
l, A, S, x, y, z Slender beam	$\rho \cdot A \cdot \ell$	0	$\frac{m \cdot \ell^2}{12}$	$\frac{m \cdot \ell^2}{12}$
a, b, c, S, x, y, z Slab	$\rho \cdot a \cdot b \cdot c$	$\frac{m}{12}(b^2 + c^2)$	$\frac{m}{12}(a^2 + c^2)$	$\frac{m}{12}(a^2 + b^2)$
2a, d, S, x, y, z Cylinder	$\rho \cdot \pi \cdot a^2 \cdot d$	$m\left(\frac{a^2}{4} + \frac{d^2}{12}\right)$	$m\left(\frac{a^2}{4} + \frac{d^2}{12}\right)$	$m\frac{a^2}{2}$
2a, t, d, S, x, y, z Thin-walled tube	$\rho \cdot 2\pi atd$	$m\left(\frac{a^2}{2} + \frac{d^2}{12}\right)$	$m\left(\frac{a^2}{2} + \frac{d^2}{12}\right)$	ma^2
2a, d, S, x, y, z Half cylinder	$\rho \cdot \frac{\pi a^2 d}{2}$	$\frac{m}{36}(3d^2 + 2{,}515\, a^2)$	$m\left(\frac{a^2}{4} + \frac{d^2}{12}\right)$	$0{,}320 \cdot ma^2$
a, S, x, y, z Sphere	$\rho \cdot \frac{4}{3} \cdot \pi \cdot a^3$	$\frac{2}{5} \cdot m \cdot a^2$	$\frac{2}{5} \cdot m \cdot a^2$	$\frac{2}{5} \cdot m \cdot a^2$

The computer programs introduced in this section are summarised below with the names of their input and output files:

Program title	Input file	Output file
AREMOM	KOORD.ARE	MOMENT.ARE
BODMOM	KOORD.BOD	MOMENT.BOD

2.5 Complex representation of harmonic vibration

Harmonic vibrations and their representation by complex variables and exponential functions are quite common in structural dynamics applications. This section summarises some basic concepts in dealing with this topic.

Firstly, a complex variable F is introduced, consisting of a real part a and an imaginary part b:

$$F = a + i \cdot b \tag{2. 5. 1}$$

In this expression, $i = \sqrt{-1}$ is the imaginary unit (Fig. 2.5-1). Multiplying a real number b by i is equivalent to rotating it counter-clockwise (from the real towards the imaginary axis) by $\frac{\pi}{2}$ radian or 90 degrees. A second multiplication by i causes another rotation

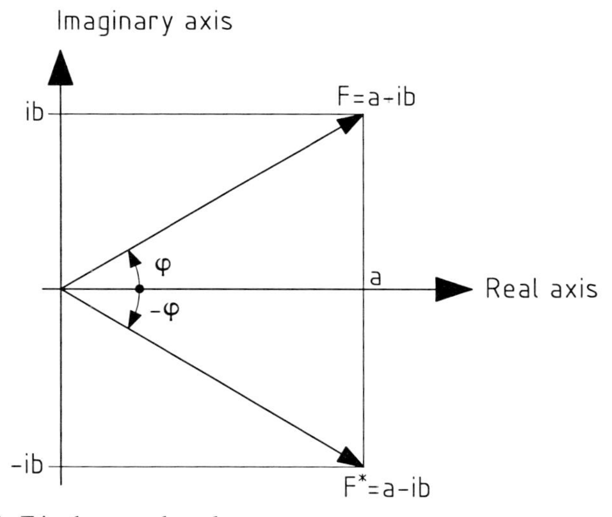

Fig. 2.5-1: Variable F in the complex plane

by $\frac{\pi}{2}$, so that b now points in the direction of the negative real axis ($i^2 = -1$). Introducing the angle φ as in Fig. 2.5-1 leads to the trigonometric form of the complex quantity F:

$$F = |F|(\cos\varphi + i \cdot \sin\varphi) \tag{2. 5. 2}$$

$|F|$ is the absolute value of F and φ (in radian) the corresponding phase angle. In terms of a and b, $|F|$ and φ are given by

$$|F|=\sqrt{a^2+b^2} \tag{2. 5. 3}$$

$$\varphi=\arctan\frac{b}{a} \tag{2. 5. 4}$$

Another representation of F is the so-called exponential form. It is based on the EULER equations for complex variables

$$\begin{aligned} e^{zi} &= \cos z + i\cdot\sin z \\ e^{-zi} &= \cos z - i\cdot\sin z \end{aligned} \tag{2. 5. 5}$$

A comparison with Eq. (2.5.2) yields:

$$F=|F|e^{i\varphi} \tag{2. 5. 6}$$

F* is the so-called complex conjugate of F; the real parts of F and F* are the same while the imaginary parts differ only in their sign (Fig. 2.5-1). Geometrically speaking, F* is the mirror image of F about the real axis.

Harmonic phenomena are often represented with the aid of vectors of equal amplitude which rotate about the origin in the complex plane with a constant angular frequency ω. Such vectors can be expressed as

$$F=|F|e^{i\theta} \tag{2. 5. 7}$$

with the phase angle θ as a simple linear function of time

$$\theta=\omega\cdot t+\varphi \tag{2. 5. 8}$$

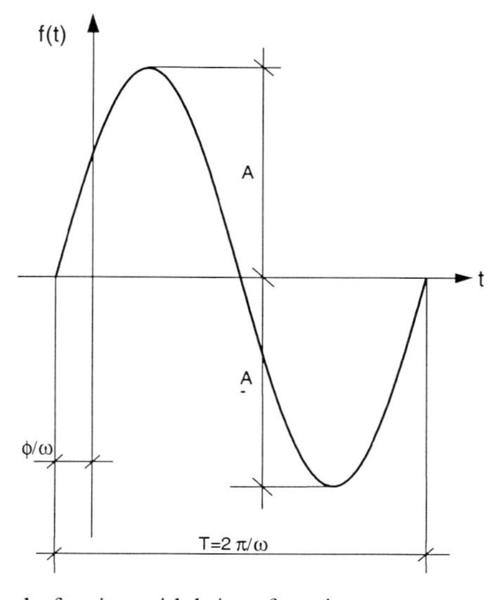

Fig. 2.5-2: A sample period of a sinusoidal time function

Any sinusoidally varying quantity f(t) can be given by

$$f(t) = A \cdot \sin(\omega \cdot t + \varphi) \qquad (2.\,5.\,9)$$

with Fig. 2.5-2 showing a sample period of f(t) with amplitude A, angular (or circular) frequency ω (in radians/s) and φ as the phase angle at time zero.

Apart from the „circular frequency" ω, measured in radians/s (rad s^{-1}), the term „frequency" f, measured in cycles/s or Hertz (Hz), is in common use. While the former is equal to the number of cycles in 2π seconds, the latter describes the number of cycles per second, so that there is a factor of 2π between ω and f:

$$\omega = 2 \cdot \pi \cdot f \qquad (2.\,5.\,10)$$

The period (of vibration) T, measured in s, describes the duration of a cycle and is correspondingly equal to the inverse of the frequency f (or 2π divided by ω):

$$T = 1/f = 2\pi/\omega. \qquad (2.\,5.\,11)$$

To recap:

The circular frequency ω in rad / s is equal to the no. of cycles occurring in 2π seconds.
The frequency f in Hz is equal to the no. of cycles per second.
The period T in s is equal to the duration of a single cycle.

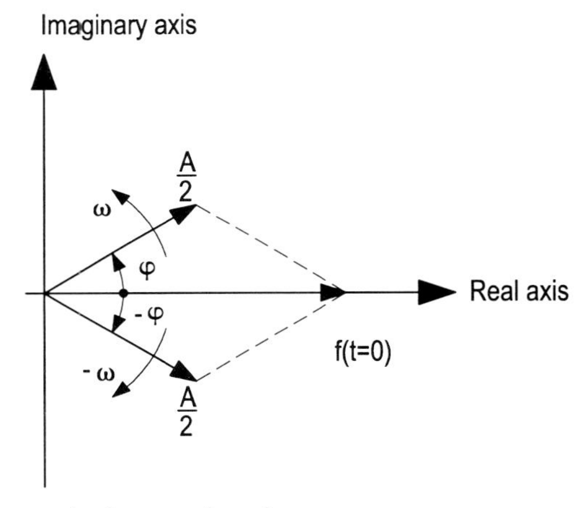

Fig. 2.5-3: Rotating vectors in the complex plane

Harmonically varying quantities of the form (2.5.9) can be visualised simply by considering two vectors of amplitude $\frac{A}{2}$ rotating counter-clockwise and clockwise respectively, with circular frequencies ω and $-\omega$ (Fig. 2.5-3). At time t=0 the vectors have angular orientations of φ and $-\varphi$, φ being the initial phase lag of the harmonic process. The (real) amplitude of the harmonically varying function is found on the horizontal axis as the vector sum of both vectors, since their imaginary parts cancel each other out.

2.6 Frequency analysis

Frequency analysis, meaning the decomposition of time functions in their harmonic components, is a much-used tool in structural dynamics applications. To begin with, a real periodic function f(t) with the period $T=\dfrac{2\pi}{\omega}$ is regarded, where

$$f(t+T)=f(t) \tag{2. 6. 1}$$

The function f(t) can be expressed as sum of complex exponentials:

$$f(t)=a_0+a_1 e^{i\omega t}+a_2 e^{2i\omega t}+\cdots+a_n e^{ni\omega t}+a_{-1} e^{-i\omega t}+a_{-2} e^{-2i\omega t}+\cdots+a_{-n} e^{-ni\omega t} \tag{2. 6. 2}$$

or, in a more compact form

$$f(t)=\sum_{n=-\infty}^{\infty} a_n e^{ni\omega t} \tag{2. 6. 3}$$

The determination of the coefficient a_0 can be carried out by integrating both sides of the equation with respect to time over a single period; due to the periodicity of the complex exponentials all terms on the right-hand side vanish with the exception of a_0 :

$$\int_0^{\frac{2\pi}{\omega}} a_n e^{ni\omega t}\,dt=\frac{a_n}{ni\omega}\left[e^{ni\omega t}\right]_0^{2\pi/\omega}=\frac{a_n}{ni\omega}\left[e^{2n\pi i}-e^0\right]=0 \tag{2. 6. 4}$$

This leads to

$$\int_0^{\frac{2\pi}{\omega}} f(t)\,dt=\frac{2\pi a_0}{\omega} \tag{2. 6. 5}$$

and accordingly

$$a_0=\frac{\omega}{2\pi}\int_0^{\frac{2\pi}{\omega}} f(t)\,dt=\frac{1}{T}\int_0^T f(t)\,dt \tag{2. 6. 6}$$

where a_0 can be seen to be simply the mean value of f(t). In order to determine the coefficient a_n, both sides of Eq. (2.6.3) are multiplied by $e^{-ni\omega t}$ and integrated over one period. On the right-hand side all terms vanish except the term containing a_n, leading to

$$a_n=\frac{\omega}{2\pi}\int_0^{\frac{2\pi}{\omega}} f(t)\,e^{-ni\omega t}dt;\quad a_{-n}=\frac{\omega}{2\pi}\int_0^{\frac{2\pi}{\omega}} f(t)\,e^{ni\omega t}dt \tag{2. 6. 7}$$

Integrating between $-\dfrac{T}{2}$ and $\dfrac{T}{2}$, instead of from zero to T, and introducing the frequency f_n of the nth harmonic component according to $f_n=\dfrac{\omega_n}{2\pi}=n\dfrac{\omega_1}{2\pi}=n f_1$ yields:

$$a_n = \frac{1}{T} \int_{-\frac{T}{2}}^{\frac{T}{2}} f(t) e^{-i2\pi f_n t} dt \qquad (2.6.8)$$

Here, n assumes all positive and negative integer values, including zero. To picture this expression simply, consider Fig. 2.5-3, which interprets sinusoidal quantities as vectors rotating with constant frequency in the complex plane. With the EULER expressions for complex variables

$$e^{zi} = \cos z + i \sin z, \quad e^{-zi} = \cos z - i \sin z \qquad (2.6.9)$$

it becomes obvious that $e^{-i2\pi f_n t}$ in Eq. (2.6.8) is simply a unit vector rotating with a frequency equal to $-f_n$ about the origin. Multiplying $e^{-i2\pi f_n t}$ by f(t) cancels the rotation of the harmonic component with the frequency f_n, so that the integral in Eq. (2.6.8) yields a finite value for a_n. The products of all other harmonic components of f(t) with $e^{-i2\pi f_n t}$ can still be seen as rotating vectors, the integrals of which vanish over a period. Renaming the complex coefficient a_n of f(t) as F_n changes Eq. (2.6.3) into

$$f(t) = \sum_{n=-\infty}^{\infty} a_n e^{i2\pi f_n t} = \sum_{n=-\infty}^{\infty} F_n e^{i2\pi f_n t} \qquad (2.6.10)$$

This is the representation of a real function f(t) as the sum of products of complex coefficients F_n with corresponding exponentials $e^{i2\pi f_n t}$, with each complex coefficient F_n mirroring the contribution of the harmonic component of frequency f_n to the function (or „signal") f(t). Sketching all coefficients F_n at the discrete frequency values f_j furnishes the function's „amplitude spectrum", which is a graphic representation of the distribution of the function's harmonic components along the frequency axis. All discrete frequencies f_j are integer multiples of $f_1 = \frac{1}{T}$ according to $f_j = j\frac{1}{T}$. Since f(t) is a real time function, for every coefficient F_n of frequency f_n there must exist a complex conjugate coefficient F^*_n with frequency $-f_n$, so that the imaginary parts may cancel. Thus, the spectrum of a real-valued function is „conjugate even":

$$F_k = F(f_k) = F^*_{-k} = F^*(-f_k) \qquad (2.6.11)$$

It follows that some special properties of symmetry are inherent in such an amplitude spectrum representation, in which both positive and negative frequencies are present („double-sided spectrum"). While the real parts of the complex coefficients F_k are symmetric about the amplitude axis, the imaginary parts of F_k are antisymmetric around the origin. The coefficient F_0 is real, as can be seen in from Eq. (2.6.6).

Introducing trigonometric functions in place of the complex exponentials and changing the summation range transforms Eq. (2.6.10) into:

$$f(t) = a_0 + \sum_{k=1}^{\infty} a_k \cos \omega_k t + \sum_{k=1}^{\infty} b_k \sin \omega_k t \qquad (2.6.12)$$

The coefficient a_0 is given by Eq. (2.6.6) and the coefficients a_k and b_k can be written as:

$$a_k = \frac{2}{T} \int_{-T/2}^{T/2} f(t) \cos \omega_k t \, dt \tag{2.6.13}$$

$$b_k = \frac{2}{T} \int_{-T/2}^{T/2} f(t) \sin \omega_k t \, dt \tag{2.6.14}$$

with the circular frequency $\omega_k = k \frac{2\pi}{T}$.

The coefficients a_k und b_k can be viewed as the real and the imaginary parts, respectively, of the harmonic component associated with the circular frequency ω_k. They can be displayed along a frequency axis at discrete points ω_k with an increment of

$$\Delta\omega = \frac{2\pi}{T} \text{ in } \frac{\text{rad}}{\text{s}} \tag{2.6.15}$$

or, alternatively

$$\Delta f = \frac{1}{T} \text{ in Hz} \tag{2.6.16}$$

Each harmonic component ($a_k + ib_k$), or $A_k \cos(2\pi f_k t + \varphi_k)$ in polar co-ordinates, can be visualised according to Fig. 2.5-3 by two vectors rotating in opposite directions with amplitudes equal to $\frac{A_k}{2}$. This amplitude, $\frac{A_k}{2}$, is the ordinate depicted at the frequency values f_k and $-f_k$. As an example, the periodic function f(t) shown in Fig. 2.6-1 for t between zero and 12s is considered. It consists of three sine waves with amplitudes of 1.0, 0.5 and 0.25 units with periods of 3, 1.5 and 0.5s respectively (the corresponding frequencies being 0.333, 0.667 and 2 Hz). Its equation is given by

$$f(t) = 1.0 \sin \frac{2\pi}{3} t + 0.50 \sin \frac{2\pi}{1.5} t + 0.25 \sin \frac{2\pi}{0.5} t \tag{2.6.17}$$

The harmonic analysis yields FOURIER coefficients with imaginary parts equal to 0.5, 0.25 and 0.125 units, corresponding to frequencies of 0.333, 0.667 and 2.0 Hz respectively, as was to be expected. The real parts of the coefficients vanish, since according to Eq. (2.6.13) they represent cosine terms.

The energy contained in any „signal“ f(t) is, generally speaking, proportional to the square of its amplitudes. This is also true for every single harmonic component, the power of which can be determined by integrating the squared function over one or more periods and dividing through the time span T or nT:

$$P = \frac{1}{T} \int_0^T [A_k \sin(2\pi f_k t)]^2 \, dt = \frac{A_k^2}{2} \tag{2.6.18}$$

This result shows that the power associated with the kth harmonic component is equal to $\frac{A_k^2}{2}$, which leads to ordinates of $\frac{A_k^2}{4}$ at the frequency values f_k and $-f_k$. Sketching these „power spectral ordinates“ over a frequency axis at the corresponding discrete frequencies furnishes the „double sided“ power spectrum ($-\infty < f < \infty$) of the periodic signal, which can, of course, also be drawn as a „one sided“ power spectrum ($0 \le f < \infty$). Fig. 2.6-2 shows an example; the „static“ (or DC) ordinate corresponding to f = 0 is the same in both cases.

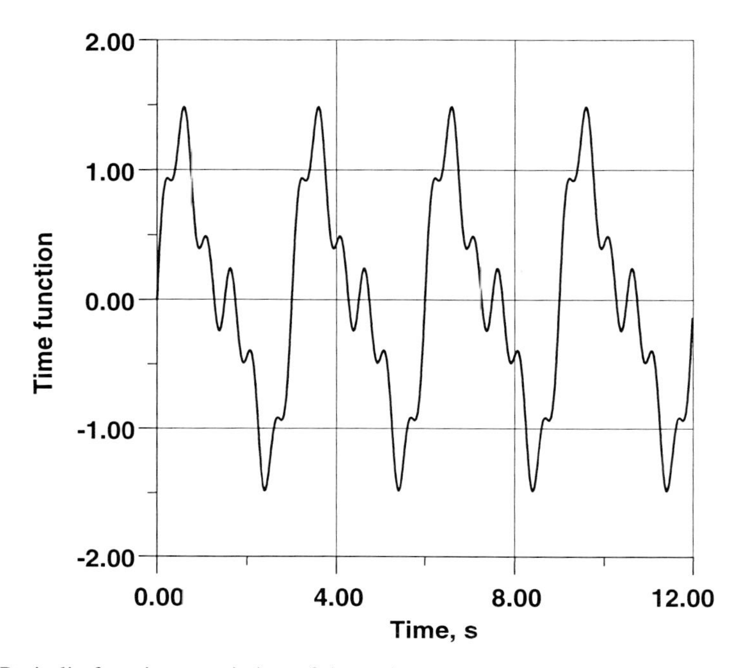

Fig. 2.6-1: Periodic function consisting of three sine waves

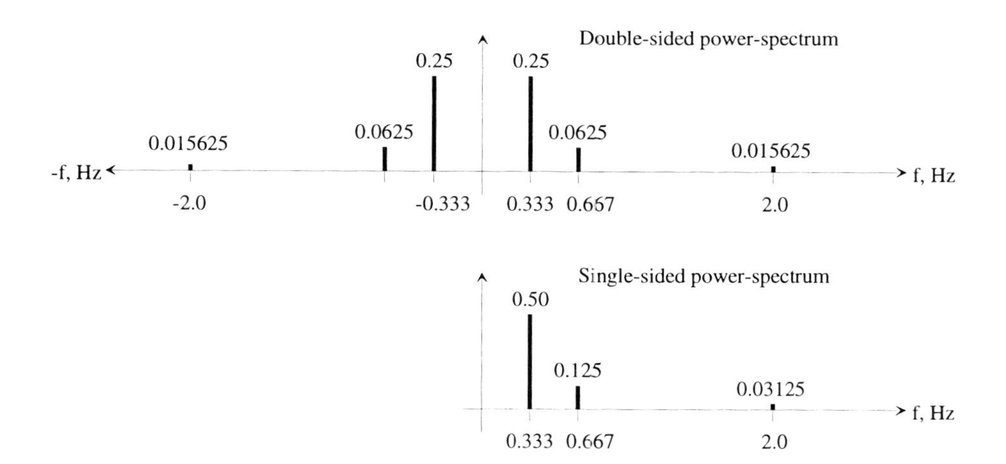

Fig. 2.6-2: Two- and one-sided power spectrum of the periodic function Eq. (2.6.17)

To illustrate: For the signal shown in Fig. 2.6-1 with its power spectrum as depicted in Fig. 2.6-2, a total power of 2 · (0.25 + 0.0625 + 0.015625) = 0.65625 (units^2) per period can be computed. It is equal to

$$P=\frac{1}{12}\int_0^{12}\left[1.0\sin\frac{2\pi}{3}t+0.50\sin\frac{2\pi}{1{,}5}t+0.25\sin\frac{2\pi}{0{,}5}t\right]^2 dt=\frac{1}{12}(6+1.5+0.375) \tag{2.6.19}$$
$$=0.65625$$

Next, aperiodic time functions or signals are considered. Formally, they are periodic functions with periods approaching infinity, $T\rightarrow\infty$, so that the frequency increment (or step) $f_1=\frac{1}{T}$, resp. $\omega_1=\frac{2\pi}{T}$ that separates the discrete spectral ordinates vanishes and a smooth curve replaces the „comb" spectrum of the periodic function. In the following, the relevant formulas for transforming a real time function f(t) in F(f) or F(ω), that is from the time to the frequency domain, will be derived.

Again, the starting point is the FOURIER series representation of a periodic function according to Eq. (2.6.12) with $a_0 = 0$:

$$f(t)=\sum_{k=1}^{\infty}\left[\frac{2}{T}\int_{-T/2}^{T/2}f(t)\cos\omega_k t\,dt\right]\cos\omega_k t+\sum_{k=1}^{\infty}\left[\frac{2}{T}\int_{-T/2}^{T/2}f(t)\sin\omega_k t\,dt\right]\sin\omega_k t \tag{2.6.20}$$

Substituting $\frac{2}{T}$ with $\frac{\Delta\omega}{\pi}$ according to Eq. (2.6. 15) leads to:

$$f(t)=\sum_{k=1}^{\infty}\left[\frac{\Delta\omega}{\pi}\int_{-T/2}^{T/2}f(t)\cos\omega_k t\,dt\right]\cos\omega_k t+\sum_{k=1}^{\infty}\left[\frac{\Delta\omega}{\pi}\int_{-T/2}^{T/2}f(t)\sin\omega_k t\,dt\right]\sin\omega_k t \tag{2.6.21}$$

and introducing $\Delta\omega\rightarrow d\omega$ while simultaneously replacing sums by integrals yields

$$f(t)=\int_{\omega=0}^{\infty}\frac{1}{\pi}\left(\int_{t=-\infty}^{\infty}f(t)\cos\omega t\,dt\right)\cos\omega t\,d\omega+\int_{\omega=0}^{\infty}\frac{1}{\pi}\left(\int_{t=-\infty}^{\infty}f(t)\sin\omega t\,dt\right)\sin\omega t\,d\omega \tag{2.6.22}$$

From the definition of the FOURIER transform X(ω) of an aperiodic time function x(t) as

$$X(\omega)=\frac{1}{2\pi}\left(\int_{-\infty}^{\infty}x(t)\cos\omega t\,dt\right)-i\,\frac{1}{2\pi}\left(\int_{-\infty}^{\infty}x(t)\sin\omega t\,dt\right) \tag{2.6.23}$$

follows

$$X(\omega)=\frac{1}{2\pi}\left(\int_{-\infty}^{\infty}x(t)\,[\cos\omega t-i\sin\omega t]\,dt\right) \tag{2.6.24}$$

or

$$X(\omega)=\frac{1}{2\pi}\int_{-\infty}^{\infty}x(t)\,e^{-i\omega t}dt \tag{2.6.25}$$

For the opposite direction, that is from the frequency to the time domain, the pertinent expression is

$$x(t)=\int_{-\infty}^{\infty} X(\omega)e^{i\omega t}d\omega \tag{2. 6. 26}$$

The factor $\frac{1}{2\pi}$ in Eq. (2.6.25) (forward transform) is often encountered, instead, in Eq. (2.6.26) for the transformation from the frequency to the time domain (inverse transform); alternatively, factors of $\frac{1}{\sqrt{2\pi}}$ may appear in Eq. (2.6.25) as well as in Eq. (2.6.26). If the frequency f in Hz is used instead of the circular frequency ω, the factor $\frac{1}{2\pi}$ disappears and the expressions for the forward and inverse transforms become more or less symmetric:

$$X(f)=\int_{-\infty}^{\infty} x(t)e^{-i2\pi ft}dt \tag{2. 6. 27}$$

$$x(t)=\int_{-\infty}^{\infty} X(f)e^{i2\pi ft}df \tag{2. 6. 28}$$

The functions x(t) on the one hand and X(ω), X(f) on the other hand form a „FOURIER Transform Pair“ .

As an example, the function f(t) shown in Fig. 2.6-3 is subjected to a forward FOURIER transformation to the frequency domain. Its counterpart F(ω) is computed by Eq. (2.6.25) as

$$F(\omega) = \frac{1}{2\pi}\int_{-\infty}^{+\infty} f(t)\cos\omega t dt - i\frac{1}{2\pi}\int_{-\infty}^{+\infty} f(t)\sin\omega t dt \tag{2. 6. 29}$$

The original time function is even, f(t) = f(-t), so that the imaginary part of Eq. (2.6.29) vanishes. The final result is:

$$F(\omega) = \frac{1}{2\pi}\int_{-t_1}^{t_1} a_0 \cos\omega t\, dt = \frac{a_0\, t_1}{\pi}\frac{\sin\omega t_1}{\omega t_1} \tag{2. 6. 30}$$

With

$$\lim_{\omega\to 0}\frac{\sin\omega t_1}{\omega t_1}=1 \tag{2. 6. 31}$$

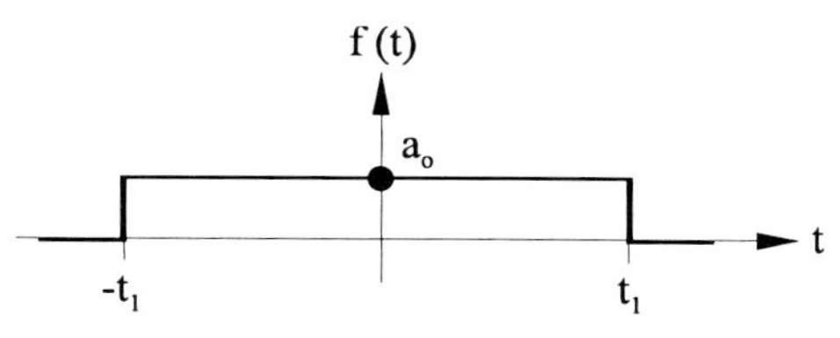

Fig. 2.6-3: Rectangular pulse in the time domain

the frequency domain function F(ω) obtains a value of $a_0 t_1 / \pi$ for ω = 0, as shown in Fig. 2.6-4. It is clear that with increasing pulse duration t_1 the frequency domain representation becomes steadily narrower, with the single line at ω = 0 as the limiting case for a static load. On the other hand, short pulses („sharp tap") exhibit a broad-band FOURIER transform.

A closed form evaluation of the transform integrals in Eqs. (2.6.25) and (2.6.26) is feasible only in special cases. Usually, the function f(t) is given in the form of a time series, which must be handled numerically by the so-called „Discrete FOURIER Transform" (DFT) algorithm. This method will be applied to a time series f_r, r = 0, 1, 2,...N-1, consisting of N points with a constant time step (sampling interval) Δt. The DFT expression for the forward transform can be given as

$$F_k = \frac{1}{N\,\Delta t}\sum_{r=0}^{N-1} f_r e^{-i\omega_k r \Delta t}\,\Delta t = \frac{1}{N}\sum_{r=0}^{N-1} f_r\, e^{-i\frac{2\pi kr}{N}} \qquad (2.\,6.\,32)$$

or, in a different form:

$$F_k = \frac{1}{N}\sum_{r=0}^{N-1} f_r\left(\cos\frac{2\pi kr}{N} - i\,\sin\frac{2\pi kr}{N}\right) \qquad (2.\,6.\,33)$$

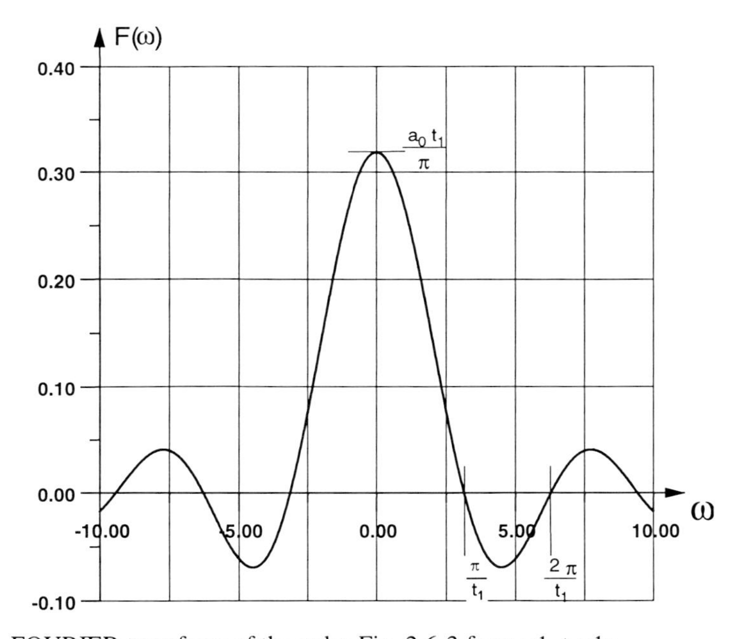

Fig. 2.6-4: FOURIER transform of the pulse Fig. 2.6-3 for a_0=1, t_1=1s

This expression furnishes N complex coefficients F_k, k = 0, 1, 2,..N-1, which correspond to the circular frequency values $\omega_k = (2\pi / T)\,k$. In this expression,

$$T = N \cdot \Delta t \qquad (2.\,6.\,34)$$

is the fictitious period of the time series, since the algorithm assumes that the time series is periodic with period T. However, this does not generally lead to serious problems, because „spill-over" phenomena can be avoided by opting for a sufficiently long period T. Of the N complex coefficients F_k only those for k = 0, 1, 2,...N/2, describing the frequency content of the time series for circular frequency values up to the so-called NYQUIST circular frequency $\frac{\pi}{\Delta t}$ are important. For k greater than N/2 the coefficients F_k are repeated; they are, in fact, mirror images of the first N/2 coefficients. The real parts of complex coefficients at the same distance from the NYQUIST frequency (also designated as „folding frequency" for obvious reasons) are the same, while the imaginary parts have the same amplitude and the opposite sign. Clearly, the NYQUIST frequency is also the highest frequency for which information can be gained from the harmonic analysis of the available signal with Δt as sampling interval.

If the time series contains harmonic components with frequencies higher than the NYQUIST frequency, so-called „aliasing" occurs during harmonic analysis, leading to more or less erroneous results for the forward transform. If the presence of unwanted high-frequency components is detected by observing that the corresponding coefficients F_k near the NYQUIST frequency are not negligibly small, it might be advisable to eliminate these high-frequency components from the original signal (e.g. by low-pass filtering) before embarking on the final harmonic analysis. This is particularly important because the results for lower frequencies can be seriously distorted through aliasing.

The inverse transform to Eq. (2.6.32) is given by

$$f_r = \sum_{k=0}^{N-1} F_k \, e^{i\frac{2\pi kr}{N}}, \quad r = 0, 1, 2, \ldots N-1 \qquad (2.\,6.\,35)$$

Eqs. (2.6.32) and (2.6.35) are very easy to program. However, they are much too time-consuming for practical purposes if the number of data points, N, equals several tens or hundreds of thousands. So-called Fast-FOURIER transform (FFT) algorithms, of which the classical COOLEY-TUKEY method [2.4] is an example, are much more efficient. For the COOLEY-TUKEY method the number of data points N must be equal to 2^n with n integer; this is usually done by „padding" the measured signal with zeros.

Two FFT-based programs, FFT1 and FFT2, are available for carrying out forward (FFT1) and inverse (FFT2) transforms. While FFT1 is used for transformations from the time domain to the frequency domain, FFT2 allows inverse transformations from the frequency domain to the time domain. Their input and output files are summarised in the following table:

Program title	Input file	Output files
FFT1	TIMSER.DAT	OMCOF.DAT OMQU.DAT
FFT2	OMCOF.DAT	RESULT.DAT

The file TIMSER.DAT contains the time series to be transformed in (2E14.7) format, with the time points (with constant time step) in the first and the ordinates in the second column. The maximum number of data points, NANZ, compatible with the chosen array sizes in the program is equal to 8192. In the output file OMCOF.DAT the computed (NANZ/2) complex FOURIER coefficients are written in three columns (3E14.7 format) with the circular frequency values in the first and the real and imaginary parts of the coefficients in the second and third column, respectively. The file OMQU.DAT contains in two columns (2E14.7 format) the circular frequencies in the first and the squares of the FOURIER coefficients in the second column.

For the inverse transform, the (max. 4096) complex FOURIER coefficients are read from the file OMCOF.DAT, formatted as above, and the computed time series (with max. 8192 values) is written in RESULT.DAT (format 2E14.7) with time points in the first and ordinates in the second column, respectively.

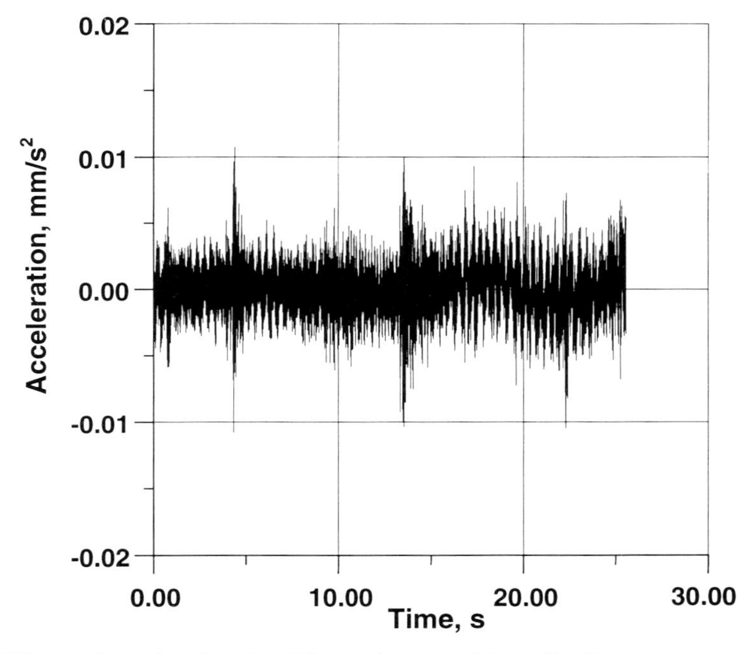

Fig. 2.6-5: Measured acceleration time history due to ambient vibration

The time series of Fig. 2.6-5, consisting of 8192 points with a time step of 0.003125 s is considered as an example. It depicts the measured acceleration time history of a church tower due to ambient vibration. The computation of the squares of the FOURIER coefficients by means of the program FFT1 furnishes the results shown in Fig. 2.6-6, from which a natural frequency of the tower at 15.4 rad/s can be inferred.

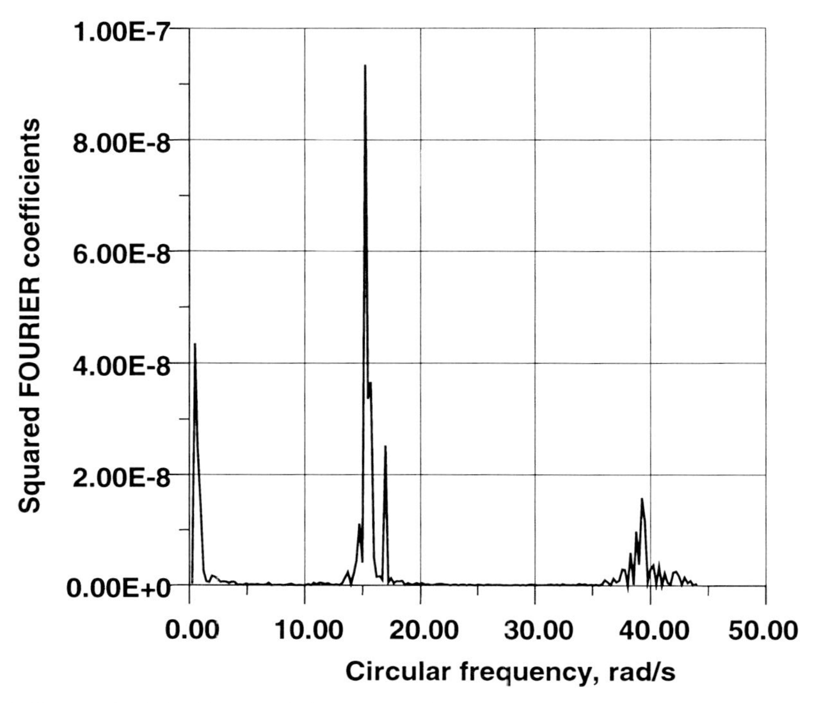

Fig. 2.6-6: Squared FOURIER coefficients over circular frequency

2.7 Classification of dynamic processes, fundamentals of random vibration theory

In structural dynamics we are dealing with time-dependant phenomena, with the physical variables involved being functions of both space and time. A meaningful classification of dynamic processes is shown in Fig. 2.7-1.

The first distinction is between deterministic and stochastic (random) processes:

- For deterministic processes the variables involved are known functions of time.
- In the case of stochastic or random processes, the time functions can obtain random values, so that predictions of future values are only possible by statistical means. Some typical examples from structural dynamics are wind loads, earthquake-induced ground accelerations or the accelerations of vehicles travelling on uneven roads.

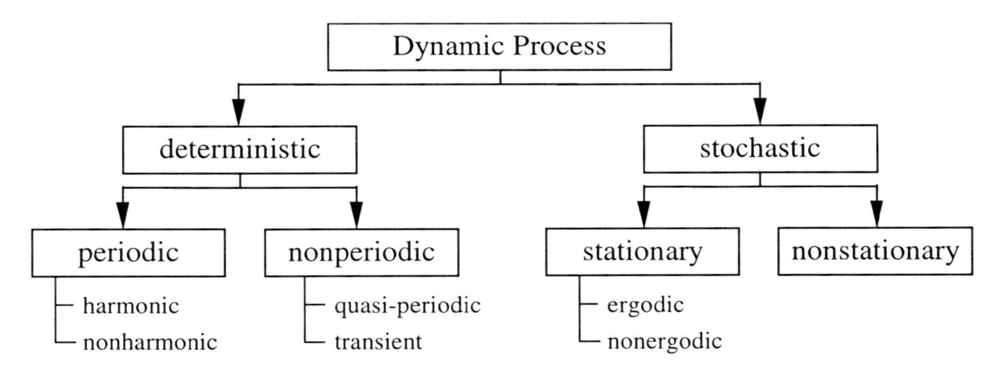

Fig. 2.7-1: Classification of dynamic processes

An example for deterministic dynamic loadings important in structural dynamics applications are machine support reactions. A special form of aperiodic processes are transients with long periods of standstill and short "bursts" of activity, where $x(t) \neq 0$.

Stochastic processes can be further classified as stationary or non-stationary. In the former, temporal averages along a sample function (e.g. mean values or standard deviations) are constant, whereas this is not the case with the latter. Fig. 2.7-2 shows a series of sample functions ${}^jx(t)$, with $x_i = x(t_i)$ for each function as a random variable. The ensemble of all random functions ${}^jx(t)$ is a random process, with the single ${}^jx(t)$ being termed as sample functions, or realisations of the process. If the statistic parameters of the sample functions taken across the ensemble are the same as the corresponding ones taken along a single realisation, we are dealing with a so-called ergodic process. Stationary and ergodic processes will be considered in the following, unless stated otherwise. Some fundamental statistical parameters of a single sample function x(t) are introduced next:

1. The mean value as a temporal average is defined as

$$\bar{x} = m = \frac{1}{T}\int_0^T x(t)\,dt \tag{2.7.1}$$

where T is assumed to be sufficiently large (theoretically $T \rightarrow \infty$). If the function x(t) is given only in the form of a time series $\{x_r\}$, $r = 1, 2, \ldots, N$ of ordinates sampled with a constant time step Δt , its mean is equal to

$$m = \frac{1}{N}\sum_{i=1}^{N} x_i \tag{2.7.2}$$

2. The mean square, that is the temporal average of the square of the sample function taken along T, is defined as

$$\overline{x^2} = \frac{1}{T}\int_0^T x^2(t)\,dt \tag{2.7.3}$$

and in the case of a discrete time series

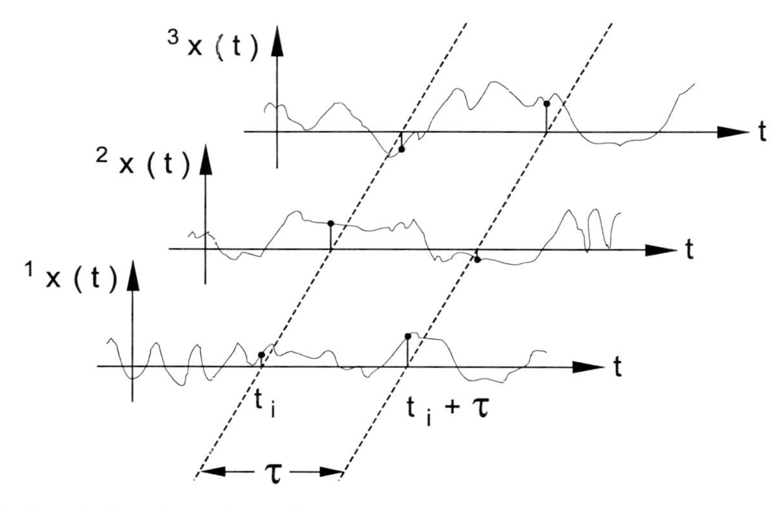

Fig. 2.7-2: Sample functions of a random process

$$\overline{x^2} = \frac{1}{N} \sum_{i=1}^{N} x_i^{\,2} \tag{2.7.4}$$

3. The variance σ^2, being the square of the standard deviation σ, is given by

$$\sigma^2 = \overline{x^2} - m^2 \tag{2.7.5}$$

A simple co-ordinate transformation can ensure that the temporal mean m is equal to zero, in which case the variance becomes equal to the square of the mean value.

4. The autocorrelation function $R_{xx}(t)$ is defined as

$$R_{xx}(\tau) = \frac{1}{T} \int_{0}^{T} x(t)\, x(t+\tau)\, dt \tag{2.7.6}$$

for the general case, and as

$$R_{xx}(\tau = (k-1) \cdot \Delta t) = \frac{1}{N-(k-1)} \sum_{i=1}^{N-(k-1)} x_i \cdot x_{i+(k-1)} \quad k = 1, 2, ... \tag{2.7.7}$$

for a time series. It is an even function, $R_{xx}(\tau) = R_{xx}(-\tau)$, its ordinates varying between $(\sigma^2 + m^2)$ and $(-\sigma^2 + m^2)$. For $t = 0$ its value is

$$R_{xx}(0) = \sigma^2 + m^2 \tag{2.7.8}$$

Cross correlation functions between two different sample functions x(t) and y(t) can be defined analogously:

$$R_{xy}(\tau) = \frac{1}{T} \int_{0}^{T} x(t)\, y(t+\tau)\, dt \tag{2.7.9}$$

If both sample functions x(t) and y(t) have zero mean, the following expression holds

$$-\sigma_x\sigma_y \leq R_{xy}(\tau) \leq \sigma_x\sigma_y \tag{2.7.10}$$

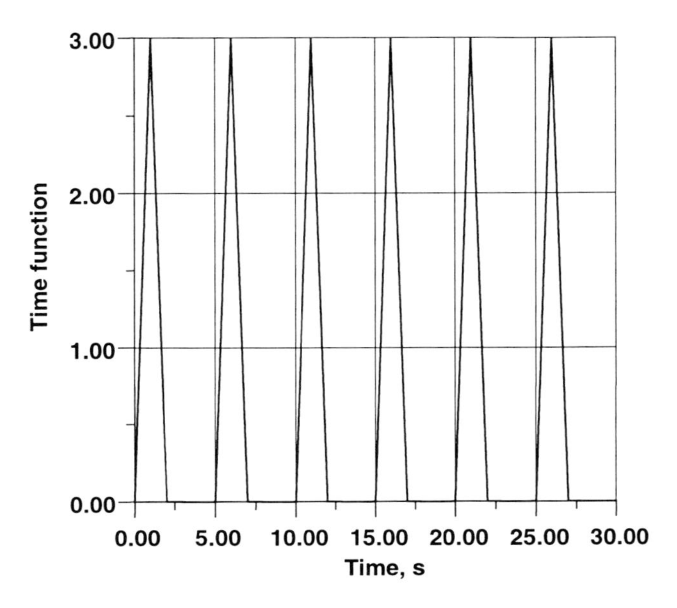

Fig. 2.7-3: Sample function consisting of triangular pulses

To illustrate, we consider Fig. 2.7-3 depicting a periodic sample function (period T = 5s) which consists of triangular pulses with an amplitude of 3 units and a duration of 2s followed by a 3s quiescent interval. Its mean value is equal to m = 0.60 units/s and its mean square equals $\frac{1}{5}\left(\frac{1}{3}\cdot 3^2\cdot 2\right) = 1.20\,(\text{units}^2)/\text{s}$. Fig. 2.7-4 shows the corresponding autocorrelation function which, for $\tau = 0$, exhibits a value of 1.20 as predicted by Eq. (2.7.8).

5. The power spectral density of a process is defined as the FOURIER transform of its autocorrelation function. Conversely, the autocorrelation function is the inverse FOURIER transform of the power spectral density function (WIENER-KHINTCHINE theorem):

$$S_{xx}(\omega) = \frac{1}{2\pi}\int_{-\infty}^{\infty} R_{xx}(\tau)\, e^{-i\omega\tau}\, d\tau \tag{2.7.11}$$

$$R_{xx}(\tau) = \int_{-\infty}^{\infty} S_{xx}(\omega)\, e^{i\omega\tau}\, d\omega \tag{2.7.12}$$

Since $R_{xx}(\tau)$ is an even function, its FOURIER transform can be given as:

$$S_{xx}(\omega) = \frac{1}{2\pi}\int_{-\infty}^{\infty} R_{xx}(\tau)\cos\omega\tau\; d\tau = \frac{1}{\pi}\int_{0}^{\infty} R_{xx}(\tau)\cos\omega\tau\; d\tau \tag{2.7.13}$$

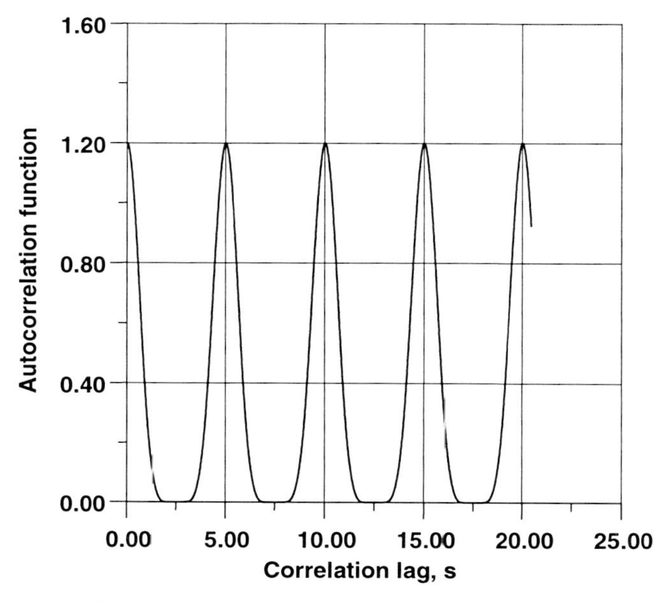

Fig. 2.7-4: Autocorrelation function for a series of triangular pulses

$S_{xx}(\omega)$ is a real and even function, as is the case with R_{xx} . An important relationship can be derived from Eq. (2.7.12) for $\tau=0$:

$$R_{xx}(0)=\int_{-\infty}^{\infty} S_{xx}(\omega)\,d\omega = \overline{x^2} \tag{2.7.14}$$

This means that the area below the power spectral density function equals the mean square of the process.

Cross spectral density functions can be introduced analogously:

$$S_{xy}(\omega)=\frac{1}{2\pi}\int_{-\infty}^{\infty} R_{xy}(\tau)e^{-i\omega\tau}\,d\tau=A(\omega)-i\,B(\omega) \tag{2.7.15}$$

However, in contrast to auto spectral density functions, they are generally complex-valued. For the complex conjugate cross spectral density $S_{xy}^{*}(\omega)$ of $S_{xy}(\omega)$ the following equality holds:

$$S_{xy}^{*}(\omega)=S_{yx}(\omega) \tag{2.7.16}$$

Power spectral densities of derived processes (e.g. $\dot{x}$, $\ddot{x}$) are given by:

$$S_{\dot{x}\dot{x}}(\omega) = \omega^2 S_{xx}(\omega) \tag{2.7.17}$$

$$S_{\ddot{x}\ddot{x}}(\omega) = \omega^4 S_{xx}(\omega) \tag{2.7.18}$$

This enables us to compute, for example, the mean square of velocity and acceleration if the power spectral density function of the corresponding displacement is known.

The unit for a power spectral density function $S_{xx}(\omega)$ is given by the square of the unit of x per circular frequency unit (rad/s). Sometimes instead of $S_{xx}(\omega)$ the single-sided power spectral density function $G_{xx}(\omega)$ is employed, which is defined only for $\omega \geq 0$. The following relationship holds:

$$G_{xx}(\omega) = 2\,S_{xx}(\omega) \tag{2. 7. 19}$$

If frequencies f in Hz are used instead of circular frequencies ω in rad/s, the ordinates of the corresponding single-sided power spectral density function $W_{xx}(f)$ are related to those of the double-sided power spectral density $S_{xx}(\omega)$ by means of

$$W_{xx}(f) = 4\pi\,S_{xx}(\omega) \tag{2. 7. 20}$$

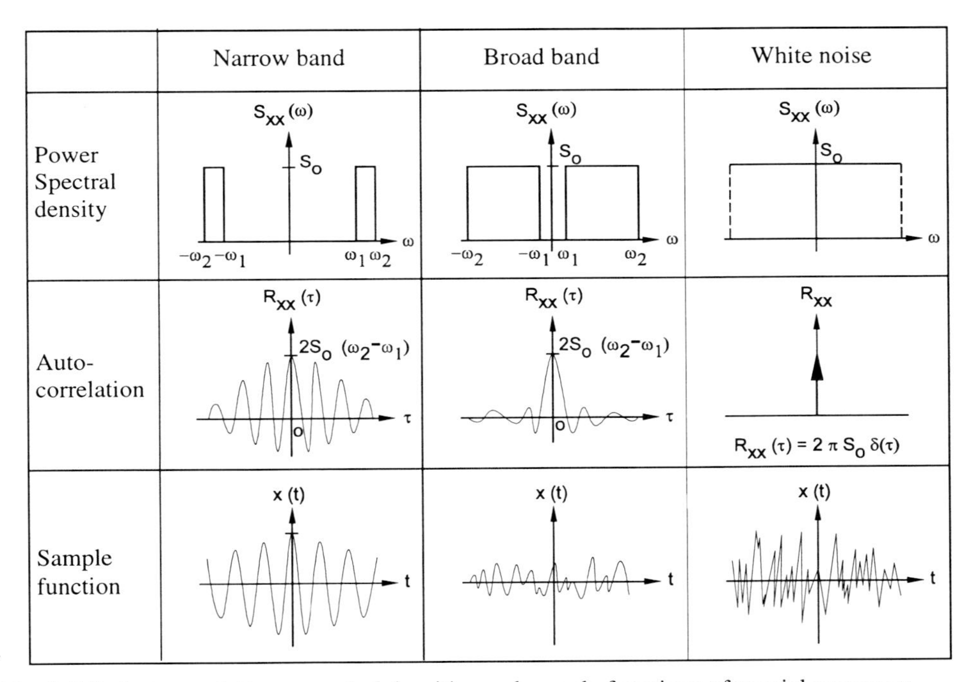

Fig. 2.7-5: Autocorrelations, spectral densities and sample functions of special processes

The limiting case of a broad-band stationary process is the so-called "white noise", in which all frequencies contribute equally to the (infinite) variance. The corresponding autocorrelation function is a DIRAC δ (delta) function, which means that the values of sample functions are totally uncorrelated. This process is described by a single parameter, namely the constant power spectral density S_0 , and can be used as a simple model for stationary stochastic excitations. Its non-stationary counterpart is often referred to as "shot noise". A simple method to produce non-stationary sample functions is to multiply stationary functions by some deterministic envelopes (amplitude modulation).

Fig. 2.7-5 shows autocorrelation functions, power spectral densities and sample functions of some typical processes.

The program AUTKOR can be used for computing the mean, variance, standard deviation and autocorrelation function of a time series. The input file is TIMSER.DAT, containing the (equidistantly sampled) time series in two columns, with the time values in the first and the ordinates in the second columnn (in a 2E14.7 format). The computed autocorrelation function is written in the file KORREL.DAT , with the correlation step in the first and the ordinates in the second column (also formatted as 2E14.7).

The program LININT serves for the linear interpolation of a (polygonal) function f(t). It reads from the input file FKT (in free format) N co-ordinate pairs [t, f(t)] and outputs in FKTINT the equidistantly interpolated function in a 2E14.7 format, with the time points in the first and the ordinates in the second column. The program names and the names of the input and output files are summarised below:

Program name	Input file	Output file
AUTKOR	TIMSER.DAT	KORREL.DAT
LININT	FKT	FKTINT

3 Single-degree-of-freedom systems

Single-degree-of-freedom (SDOF) systems are the simplest oscillators; in spite of their simplicity, they are well-suited for modelling many real-life cases. SDOF systems are discussed in some detail in this chapter, because they are also very suitable for introducing basic structural dynamics methods and concepts.

3.1 Free, undamped vibrations

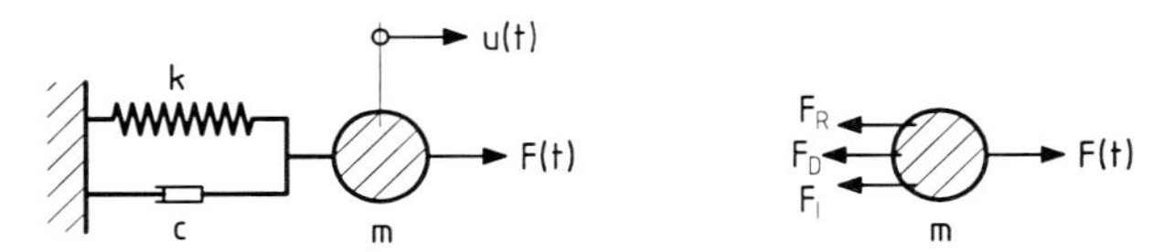

Fig. 3.1-1: Single-degree-of-freedom system with acting forces

Fig. 3.1-1 depicts the standard case of a viscously damped SDOF oscillator with an external load F(t). If both F(t) and F_D are equal to zero, we have the case of free, undamped vibrations, described by the following differential equation:

$$m\ddot{u} + ku = 0 \tag{3.1.1}$$

or

$$\ddot{u} + \frac{k}{m} u = 0 \tag{3.1.2}$$

This is a linear and homogeneous differential equation with constant coefficients. Its solutions must be functions with second derivatives having the same form as the function itself, in order for their sum to vanish. This suggests exponential or trigonometric functions, and the general solution can be written down as:

$$u = e^{\lambda t};\ \dot{u} = \lambda e^{\lambda t};\ \ddot{u} = \lambda^2 e^{\lambda t} \tag{3.1.3}$$

The so-called characteristic equation obtained by introducing Eqs. (3.1.3) into Eq. (3.1.2) is given by:

$$\lambda^2 + (k/m) = 0 \tag{3.1.4}$$

With $\omega_1{}^2 = \dfrac{k}{m}$ its solution is

$$\lambda_{1,2} = \pm\sqrt{\left(-\omega_1{}^2\right)};\quad \lambda_{1,2} = \pm i\,\omega_1 \tag{3.1.5}$$

leading to

$$u_1 = e^{\,i\,\omega_1 t};\ u_2 = e^{-i\,\omega_1 t} \tag{3.1.6}$$

ω_1 is the circular natural frequency of the SDOF system; the corresponding natural frequency f_1 and period T_1 are given by $f_1 = \frac{\omega_1}{2\pi}$, $T_1 = \frac{1}{f_1}$.

EULER's formula $e^{i\phi} = \cos\phi + i\sin\phi$ yields

$$\begin{aligned} u_1 &= \cos\omega_1 t + i\sin\omega_1 t \\ u_2 &= \cos\omega_1 t - i\sin\omega_1 t \end{aligned} \tag{3. 1. 7}$$

Linear combinations of u_1 and u_2 according to

$$\begin{aligned} \frac{1}{2}(u_1 + u_2) &= \cos\omega_1 t \\ \frac{1}{2i}(u_1 - u_2) &= \sin\omega_1 t \end{aligned} \tag{3. 1. 8}$$

yield real solutions of the original differential equation, and the general homogeneous solution can be written down as:

$$u(t) = C_1\cos\omega_1 t + C_2\sin\omega_1 t \tag{3. 1. 9}$$

The constants C_1 and C_2 are determined by considering the initial conditions for the dependent variable u and its time derivative at time $t = 0$. As an example, the case $u(0) = u_0$, $\dot{u}(0) = 0$ is considered. Starting with

$$\begin{aligned} u &= C_1\cos\omega_1 t + C_2\sin\omega_1 t \\ \dot{u} &= -C_1\omega_1\sin\omega_1 t + C_2\omega_1\cos\omega_1 t \end{aligned}$$

we obtain

$$\begin{aligned} u(0) &= C_1 \Rightarrow C_1 = u_0 \\ \dot{u}(0) &= C_2\omega_1 \Rightarrow C_2 = 0. \end{aligned}$$

The solution is

$$u(t) = u_0\cos\omega_1 t \tag{3. 1. 10}$$

as shown in Fig. 3.1-2 for $u_0 = 10$ units and $\omega_1 = \pi$ rad/s for the time interval $0 \le t \le 4s$. The natural period of the system is

$$T_1 = \frac{2\pi}{\omega_1} = \frac{2\pi}{\pi} = 2{,}0s$$

Another example is used to illustrate the practical applicability of the solution Eq. (3.1.9) for free undamped vibrations. Fig. 3.1-3 shows a single-story, single-bay steel frame subject to a transient load F(t) acting horizontally at roof level. The problem is to find an (approximate) solution for the maximum horizontal displacement max u and also for the time function u(t). The girder's bending stiffness is much higher than the bending stiffness of the columns $(EI_G \rightarrow \infty)$, and the girder's mass is given with 5t. The columns (considered as massless) are HEB 240 steel profiles with a bending stiffness EI_C of

$$EI_C = 2{,}1\cdot 10^8 \frac{kN}{m^2}\cdot 11260\cdot 10^{-8} m^4 = 23646\ kNm^2$$

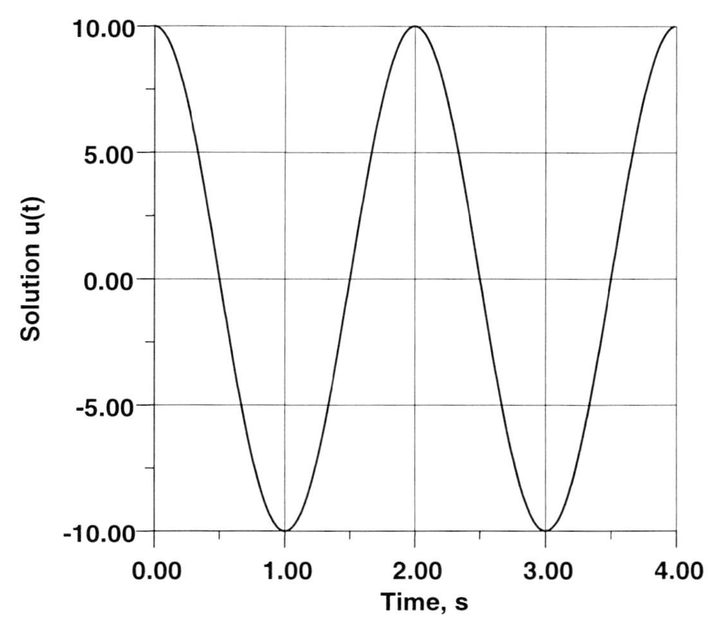

Fig. 3.1-2: Cosine function as solution for $u(0) = u_0 = 10, \quad \dot{u}(0) = 0$

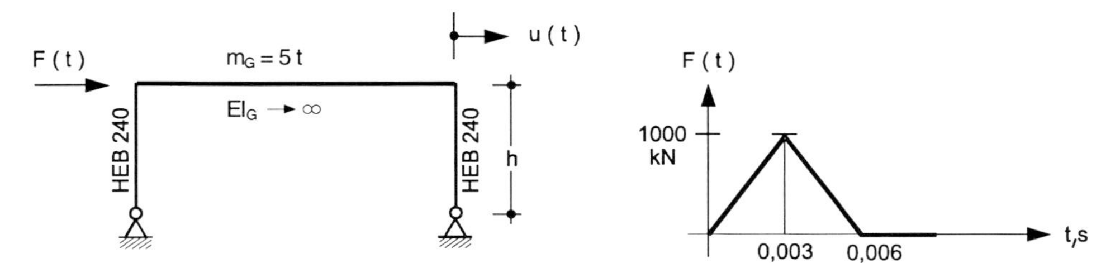

Fig. 3.1-3: Steel frame with transient loading

The height h of the frame is equal to 3.0m. A simple SDOF model of the frame is characterised by a mass of 5.0 t and a spring stiffness equal to

$$k = 2 \cdot \frac{3EI_S}{h^3} = 5254.67 \frac{kN}{m}$$

where the rigid girder has been taken into account. The circular natural frequency of the system is given by

$$\omega_1 = \sqrt{\frac{k}{m}} = \sqrt{\frac{5254.67}{5.0}} = 32.4 \frac{rad}{s}$$

with the corresponding values of the natural frequency and period equal to f_1=5.16 Hz and T_1=0.19s. The transient load imparts to the structure a momentum

$$I=\int_{t=0}^{t=0,006s} F(t)dt=\frac{1}{2}\cdot 1000\cdot 0.006=3.0\,kNs$$

which means that the mass of 5t obtains a velocity of

$$\dot{u}(0)=\frac{3.0\,kNs}{5\,t}=0.6\frac{m}{s}$$

at time t=0. In view of the short duration of the transient load, no appreciable error is involved in not considering damping when the maximum displacement is computed. Eq. (3.1.1) with initial conditions $u(0)=0,\ \dot{u}(0)=0.6\,m/s$ is used, and the coefficients C_1 and C_2 are computed as

$$C_1=0,\ \ C_2=\frac{\dot{u}(0)}{\omega_1}=\frac{0.6}{32.4}=0.0185\,m$$

The complete solution is given by

$$u(t)=\frac{\dot{u}(0)}{\omega_1}\sin\omega_1 t=0.0185\sin(32.4t) \tag{3. 1. 11}$$

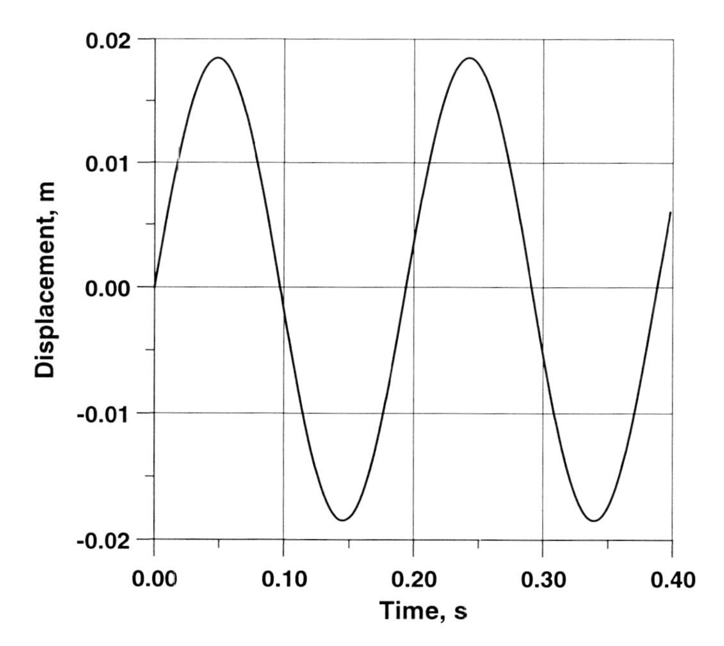

Fig. 3.1-4: Time history of the girder's horizontal displacement

This is a sine wave with a peak amplitude (corresponding to the maximum horizontal displacement of the girder) of 0.0185 m = 1.85 cm as shown in Fig. 3.1-4.

The peak value of the restoring force is given by

$$\max F_R = 0.0185\text{m} \cdot 5254.67 \frac{\text{kN}}{\text{m}} = 97.21 \text{ kN} \tag{3. 1. 12}$$

This is equal to the peak inertia force

$$F_I = m \cdot \max(\ddot{u}) = m \cdot \omega_1^2 \cdot \max(u) = 5 \cdot 32.4^2 \cdot 0.0185 = 97.21 \text{ kN} \tag{3. 1. 13}$$

and the maximum bending moment at the column head is

$$\max M = \frac{Ph}{2} = 145.82 \text{ kNm} \tag{3. 1. 14}$$

The stress values of the columns steel profile due to this bending moment are well below the yield stress, so that the implied linearity of the system is not violated.

3.2 Forced vibrations without damping

The pertinent differential equation is

$$\begin{aligned} m\ddot{u} + ku &= F(t) \\ \ddot{u} + \omega_1^2\, u &= \frac{F(t)}{m} = f(t) \end{aligned} \tag{3. 2. 1}$$

Its general solution is equal to the homogeneous part derived in the last section plus a particular solution u_P:

$$u = u_h + u_p = C_1 \cos \omega_1 t + C_2 \sin \omega_1 t + u_p \tag{3. 2. 2}$$

The particular solution can be determined, for instance, by variation of the constants. It can be shown to be of the form

$$u_p = \frac{1}{\omega_1} \int_0^t f(\tau) \sin \omega_1 (t - \tau)\, d\tau \tag{3. 2. 3}$$

This is the so-called DUHAMEL integral.

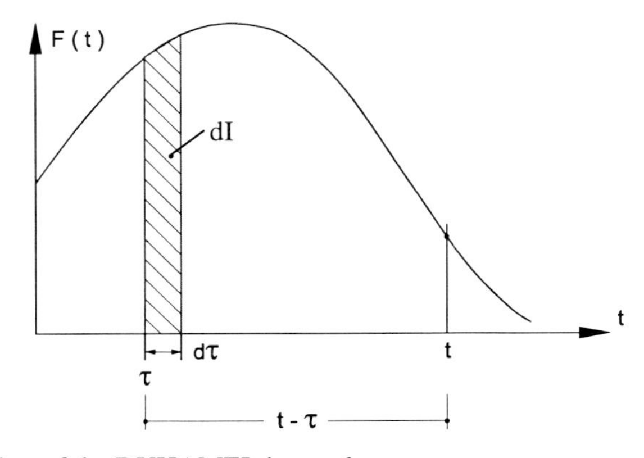

Fig. 3.2-1: Derivation of the DUHAMEL integral

A straightforward derivation of Eq. (3.2.3) can be carried out by considering the "system output" of the SDOF oscillator as sum of the single response time histories for unit impulse inputs. In Fig. 3.2-1 the cross-hatched area represents the differential impulse

$$dI = F(\tau)\,d\tau = m \cdot \dot{u}(\tau) \tag{3.2.4}$$

According to Eq. (3.1.11), the corresponding system response is given by

$$u(t) = \frac{\dot{u}(\tau)}{\omega_1} \sin \omega_1 (t-\tau) = \frac{1}{\omega_1} \frac{F(\tau)\,d\tau}{m} \sin \omega_1 (t-\tau) \tag{3.2.5}$$

Integrating over all differential impulses yields

$$u_p = \frac{1}{\omega_1} \int_0^t \frac{F(\tau)}{m} \sin \omega_1 (t-\tau)\, d\tau \tag{3.2.6}$$

This expression is equal to Eq. (3.2.3) because of $f(\tau) = F(\tau)/m$.

Attempting a closed-form solution of the DUHAMEL integral can be recommended only if the loading function is sufficiently simple. Normally, a numerical evaluation is in order, which can be carried out as follows. Denoting the integral in Eq. (3.2.3) by J

$$J = \int_0^t f(\tau) \sin \omega_1 (t-\tau)\, d\tau \tag{3.2.7}$$

and making use of the addition theorem

$$\sin(\omega_1 t - \omega_1 \tau) = \sin \omega_1 t \cos \omega_1 \tau - \cos \omega_1 t \sin \omega_1 \tau \tag{3.2.8}$$

a two-term sum is obtained:

$$J = \sin \omega_1 t \int_0^t f(\tau) \cos \omega_1 \tau\, d\tau - \cos \omega_1 t \int_0^t f(\tau) \sin \omega_1 \tau\, d\tau \tag{3.2.9}$$

The two definite integrals in Eq. (3.2.9) can be computed numerically by standard methods, e.g. SIMPSON- or trapezoidal rule integration.

The computer program DUHAMI can serve for this purpose. Its input file RHS contains the loading function $f(t) = \frac{F(t)}{m}$ in N rows, with the N data pairs [t, f(t)] in 2E14.7 format. A constant time step DT is assumed; if linear interpolation is necessary, it can be carried out by the program LININT. If the loading function contained in RHS is F(t) rather than f(t), the scaling with (1/m) can also be carried out automatically by entering the corresponding factor when prompted. Further data to be entered during the program run are the number N of loading function values, time step DT, period T and finally the damping D of the SDOF system (the latter being zero in the undamped case). The output file DUHOUT contains the displacement time history u(t) in 2E14.7 format with the time points in the first and the displacement values in the second column. The maximum absolute displacement is also output to the screen.

Program name	Input file	Output file
DUHAMI	RHS	DUHOUT

The steel frame of the last section is considered here as an example. It is subjected to the 0.006 s long impulse depicted in Fig. 3.1-3, and its initial conditions are $u(0) = 0$, $\dot{u}(0) = 0$. The input file RHS consists of 250 rows with the function values of f(t) given every 0.001s:

```
.000000E+00   .000000E+00
.100000E-02   .666667E+02
.200000E-02   .133333E+03
.300000E-02   .200000E+03
.400000E-02   .133333E+03
.500000E-02   .666666E+02
.600000E-02   .000000E+00
.700000E-02   .000000E+00
.800000E-02   .000000E+00
  ......        ......
```

The period T of the SDOF system to be input during run time is 0.1938 s, and its damping is zero; since RHS contains f(t) and not F(t), the scaling factor is equal to unity. The results are shown in Fig. 3.2-2. Here, the maximum displacement equals 1.849 cm, this being practically the same value as the one already computed in Section 3.1.

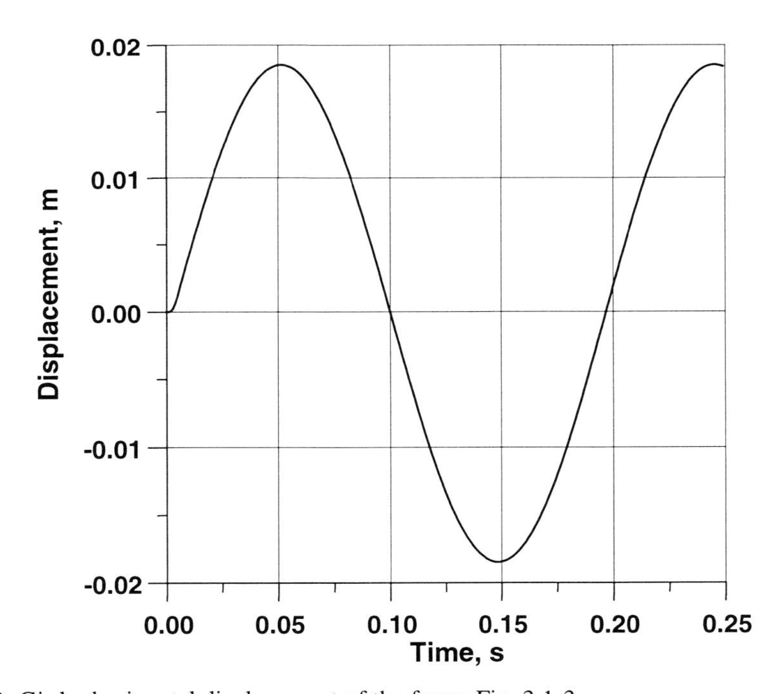

Fig. 3.2-2: Girder horizontal displacement of the frame Fig. 3.1-3

3.3 Damped free and forced vibrations

It is a well-known fact that vibration amplitudes decrease with time, if no external energy source is available, due to the gradual dissipation of the oscillator's energy by damping. Damping mechanisms can be very heterogeneous, such as, for instance, damping due to the interaction between the structure and its environment, damping due to friction at connections and joints, and of course also material damping. The simplest damping model is the so-called linear viscous damping, where the damping force F_D is assumed to be linearly proportional to the velocity $\dot{u}$ of the stationary harmonic vibration u(t):

$$F_D = c\dot{u} \quad (3.3.1)$$

In the case of free vibrations of a SDOF system, this leads to the differential equation

$$m\ddot{u} + c\dot{u} + ku = 0; \quad \ddot{u} + \frac{c}{m}\dot{u} + \frac{k}{m}u = 0 \quad (3.3.2)$$

In this expression, m is the mass (e.g. in t), c is the constant viscous damping coefficient (e.g. in kNs/m) and k is the spring constant (e.g. in kN/m).

Introducing the exponential solution

$$u = e^{\lambda t}; \dot{u} = \lambda e^{\lambda t}; \ddot{u} = \lambda^2 e^{\lambda t} \quad (3.3.3)$$

yields the characteristic equation

$$\lambda^2 + \frac{c}{m}\lambda + \omega_1{}^2 = 0; \quad \omega_1{}^2 = \frac{k}{m} \quad (3.3.4)$$

Its general solution is given by

$$u = c_1 e^{\lambda_1 t} + c_2 e^{\lambda_2 t} \quad (3.3.5)$$

where λ_1, λ_2 are the roots of the characteristic equation

$$\lambda_{1,2} = -\frac{c}{2m} \pm \sqrt{\left(\frac{c}{2m}\right)^2 - \omega_1{}^2} \quad (3.3.6)$$

The behaviour of the solution depends on whether the radical in Eq. (3.36) is less than, equal to or greater than zero, corresponding to the underdamped, critically damped and overdamped case, respectively. In the latter case, λ_1 and λ_2 are real and no vibration occurs. The critical damping is defined as the value of c, for which the quantity under the square-root sign equals zero:

$$\frac{c}{2m} = \omega_1 \rightarrow c_{krit} = 2m\,\omega_1 = 2\sqrt{km} \quad (3.3.7)$$

The dimensionless ratio D or ξ of the actual damping coefficient, c, to the critical damping, c_{krit}, called "damping ratio", is very often used for quantifying damping. The following expressions hold:

$$D = \xi = \frac{c}{c_{krit}} = \frac{c}{2m\omega_1}; \quad \frac{c}{m} = 2\xi\omega_1 \quad (3.3.8)$$

The differential equation (3.3.2) now takes the form:

$$\ddot{u} + 2\xi\omega_1\dot{u} + \omega_1{}^2 u = 0 \quad (3.3.9)$$

Its solution is given by

$$u(t) = e^{-\xi\omega_1 t}\left(C_1 \cos\sqrt{1-\xi^2}\,\omega_1 t + C_2 \sin\sqrt{1-\xi^2}\,\omega_1 t\right) \quad (3.3.10)$$

where C_1, C_2 are integration constants. For the general initial conditions $u(0)=u_0$, $\dot{u}(0)=\dot{u}_0$ this leads to

$$u(t) = e^{-\xi\omega_1 t}\left(u_0 \cos\sqrt{1-\xi^2}\,\omega_1 t + \frac{(\dot{u}_0 + \xi\omega_1 u_0)}{\omega_1\sqrt{1-\xi^2}} \sin\sqrt{1-\xi^2}\,\omega_1 t\right) \qquad (3.3.11)$$

This expression can be simplified by introducing the damped natural circular frequency ω_D and corresponding period T_D:

$$\omega_D = \omega_1\sqrt{1-\xi^2}; \quad T_D = 2\pi / \omega_D \qquad (3.3.12)$$

In the case of forced vibrations, Eq. (3.3.9) reads

$$\ddot{u} + 2\xi\omega_1\,\dot{u} + \omega_1^2\,u = f(t) \qquad (3.3.13)$$

The general solution of (3.3.13) is given by the sum of the homogeneous solution Eq. (3.3.11) and the particular integral (DUHAMEL integral)

$$u_p(t) = \frac{1}{\omega_D}\int_0^t f(\tau)\, e^{-\xi\omega_1(t-\tau)} \sin\omega_D(t-\tau)\, d\tau \qquad (3.3.14)$$

A numerical evaluation of this DUHAMEL integral can be conducted by the program DUHAMI as described in section 3.2.

Another widely used damping parameter, in addition to the critical damping ratio D or ξ according to Eq. (3.3.8), is the so-called logarithmic decrement Λ. It is defined as the natural logarithm of the ratio of the amplitudes of two successive positive (or negative) peaks:

$$\begin{aligned}\Lambda &= \ln\frac{u_i}{u_{i+1}} = \ln\frac{e^{-\xi\omega_1 t_i}\cos\omega_D t_i}{e^{-\xi\omega_1 t_{i+1}}\cos\omega_D t_{i+1}} = \ln e^{-\xi\omega_1(t_i - t_{i+1})} = \xi\omega_1(t_{i+1} - t_i)\\ &= \xi\omega_1\frac{2\pi}{\omega_D} = \xi\omega_1 T_D = \xi\omega_1\frac{2\pi}{\omega_1\sqrt{1-\xi^2}} = \xi\frac{2\pi}{\sqrt{1-\xi^2}}\end{aligned} \qquad (3.3.15)$$

For the lightly damped systems normally encountered in structural dynamics, it is sufficiently accurate to write

$$\xi = D \approx \frac{\Lambda}{2\pi} \qquad (3.3.16)$$

The logarithmic decrement can be experimentally determined from time-history measurements of free vibrations. Normally, two peaks u_1 and u_{n+1} occurring at times t_1 and t_{n+1} are considered, with n vibration cycles between them, in which case we obtain

$$\Lambda = \frac{1}{n}\ln\frac{u_1}{u_{n+1}} \qquad (3.3.17)$$

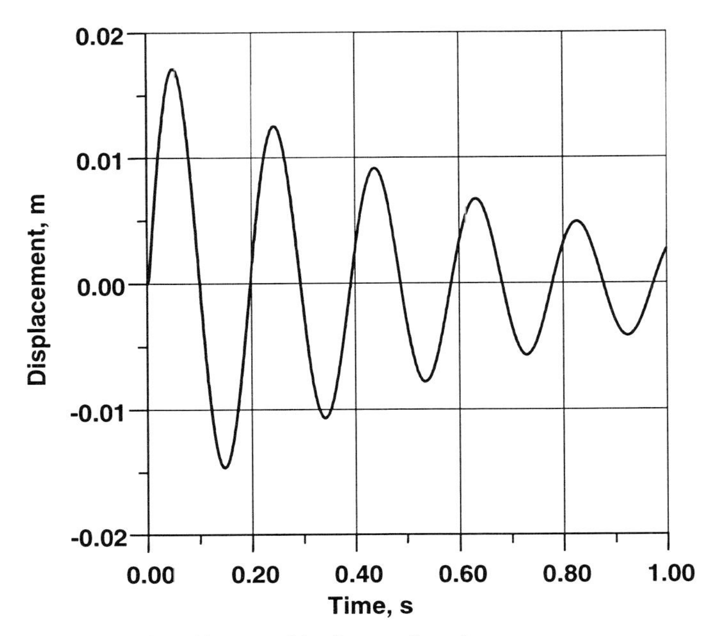

Fig. 3.3-1: Free vibration time history with viscous damping

As an example, Fig. 3.3-1 shows the displacement time history for the SDOF system of section 3.1, calculated with the program DUHAMI for a damping value of D = 5%. The positive peak amplitudes of the first 4 cycles are equal to 0.01714, 0.01251, 0.009137, 0.006671 and 0.004870 m, leading to a logarithmic decrement of

$$\Lambda = \ln \frac{0.01714}{0.01251} = \frac{1}{4} \ln \frac{0.01714}{0.00487} = 0.315; \quad D \approx \frac{\Lambda}{2\pi} = 0.05 \tag{3. 3. 18}$$

Now to some further remarks on the viscous damping model employed. Considering a SDOF system vibrating harmonically in its natural circular natural frequency ω_1, its response is given by $u = u_0 \sin \omega_1 t$. The work dissipated by viscous damping per cycle is given by

$$W_D = \int_0^{2\pi/\omega_1} c\dot{u}^2 \, dt = \pi \, c \, \omega_1 \, u_0^2 \tag{3. 3. 19}$$

Eq. (3.3.8) can also be written as

$$D = \xi = \frac{c}{2m\omega_1} = \frac{c}{2\dfrac{k}{\omega_1}} \tag{3. 3. 20}$$

Multiplying numerator and denominator of the first fraction by $\pi \, \omega_1 \, u_0^2$ leads to

$$D = \frac{\pi c \omega_1 u_0^2}{4\pi\left(\dfrac{1}{2} m \omega_1^2 u_0^2\right)} = \frac{\pi c \omega_1 u_0^2}{4\pi\left(\dfrac{1}{2} k u_0^2\right)} \tag{3. 3. 21}$$

The numerator of this fraction is the work dissipated per cycle and the term in brackets in the denominator describes the maximum elastically stored work, which is also equal to the potential energy. Eq. (3.3.21) can be used for defining an equivalent viscous damping D in the case of general non-viscous damping by computing the ratio of measured damping and elastically stored work amounts.

For a viscously damped SDOF system vibrating harmonically in its natural frequency, the velocity $\dot{u}$ is given by

$$\dot{u} = \omega_1 u_0 \cos \omega_1 t = \pm \omega_1 u_0 \sqrt{1 - \sin^2 \omega_1 t} = \pm \omega_1 \sqrt{u_0^{\,2} - u^2} \tag{3. 3. 22}$$

The corresponding damping force is equal to

$$F_D = \pm c\omega_1 \sqrt{u_0^{\,2} - u^2} \tag{3. 3. 23}$$

which leads to the expression

$$\left(\frac{F_D}{c\,\omega_1\,u_0}\right)^2 + \left(\frac{u}{u_0}\right)^2 = 1 \tag{3. 3. 24}$$

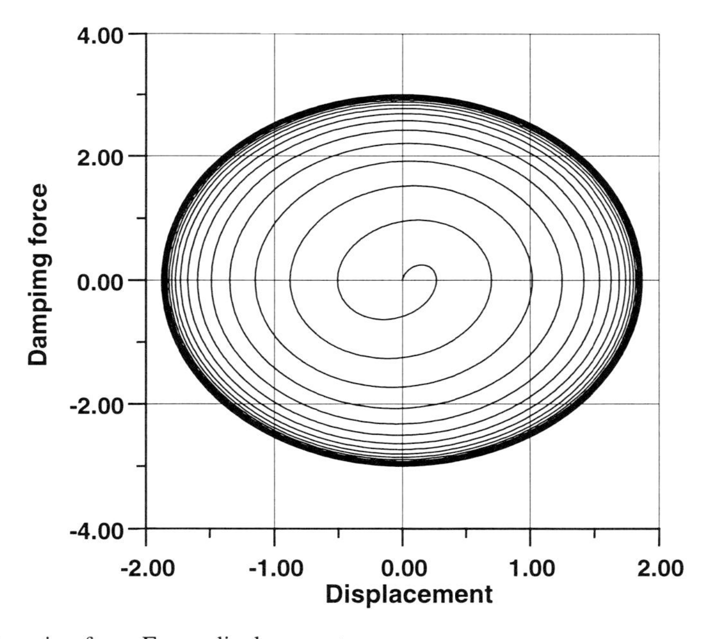

Fig. 3.3-2: Damping force F_D vs. displacement

This expression can be graphically depicted as an ellipse (Fig. 3.3-2), with half-axes equal to u_0 and $c\,\omega_1\,u_0$ and its area W_D equalling $\pi c\,\omega_1\,u_0^{\,2}$. The example shown in Fig. 3.3-2 is a SDOF system with D= 5%, m = 1, k = 16, c = 0.4, the natural circular frequency of which is equal to $\omega_1 = 4\ \text{rad/s}$. The differential equation of motion and the initial conditions are given by

$$\ddot{u}+2\cdot 0{,}05\cdot 4\cdot\dot{u}+4^2\cdot u=3{,}0\sin(4t);\ u(0)=\dot{u}(0)=0 \tag{3. 3. 25}$$

The peak amplitude of the response is equal to u_0 = 1,879 units (e.g. m), leading to a maximum damping force of

$$\max F_D = c\omega_1 u_0 = 0.4\cdot 4\cdot 1.879 = 3.00\ \text{kN} \tag{3. 3. 26}$$

The work dissipated by damping per vibration cycle is equal to

$$W_D = \pi\, c\, \omega_1\, u_0^{\,2} = 17.70\ \text{kNm} \tag{3. 3. 27}$$

If the restoring force, F_R = ku, is considered in addition to the damping force F_D, the hysteretic loop of Fig. 3.3-3 results.

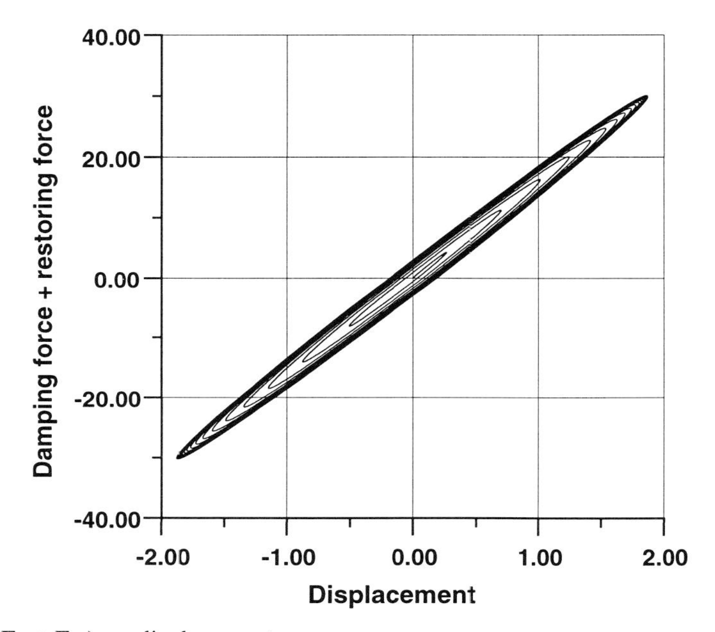

Fig. 3.3-3: (F_D + F_R) vs. displacement

According to Eq. (2.3.13), the "elastic" work per cycle corresponding to the restoring force is given by

$$W_E = \frac{1}{2}\,k\,u_0^{\,2} \tag{3. 3. 28}$$

For this example, W_E is equal to 28.23 kNm, and Eq. (3.3.21) leads to a critical damping ratio for the SDOF system of

$$D = \frac{1}{4\pi}\,\frac{17.70}{28.23} = 0.05 \tag{3. 3. 29}$$

as expected. The so-called loss factor d can be introduced as twice the damping ratio D, or the ratio of the work dissipated by damping to the "elastic" work multiplied by $\left(\frac{1}{2\pi}\right)$ for a SDOF system vibrating in its natural frequency:

$$d = \frac{W_D}{2\pi W_E} = \frac{\omega_1 c}{k} \tag{3. 3. 30}$$

Eq. (3.3.30) shows damping to be directly proportional to the excitation frequency for the viscous damping model, which does not agree with experimental evidence. The elliptic hysteretic loops of Figs. 3.3-2 and 3.3-3 are also typical for linear damping, with work dissipated by damping increasing with the square of the displacement amplitude. Since the "elastic" work Eq. (3.3.28) is also proportional to the square of the displacement amplitude, the loss factor d is independent of the vibration amplitude and the superposition law remains valid.

3.4 Direct Integration of the differential equation of motion

The standard differential equation of motion for the SDOF system in the form

$$m\ddot{u} + c\dot{u} + ku = F(t) \tag{3. 4. 1}$$

with the general initial conditions $u(0) = u_0$, $\dot{u}(0) = \dot{u}_0$ can also be solved by Direct Integration in the time domain. There exist many suitable algorithms for solving this classical initial-value problem; only the widely-used NEWMARK $\beta - \gamma$ operator [3.1] will be presented here.

Two issues are of central importance for the choice of an integration scheme, namely its stability and its accuracy. An unconditionally stable algorithm is present if the solution u(t) remains finite for arbitrary initial conditions and arbitrarily large $\Delta t / T$ ratios, Δt being the time step employed in the integration and T the period of the SDOF system, $T = 2\pi\sqrt{m/k}$. A conditionally stable algorithm (which is generally more accurate than an unconditionally stable one) guarantees that the solution remains finite, only if the ratio $\Delta t / T$ does not exceed a certain value. For SDOF systems with known T it is no problem to choose a suitable integration time step Δt; however, unconditionally stable algorithms are still preferred, especially if non-linearities are expected.

The accuracy of a Direct Integration algorithm depends on the loading function f(t), the system's properties and, most importantly, on the ratio of the time step Δt to the period T. The deviation of the computed from the true solution can be seen as an elongation of the period and a decay of the amplitude of the former, corresponding to a fictitious additional damping.

Integration algorithms are further divided into single-step and multi-step methods, which can also be implicit or explicit. Single-step methods, which are quite popular in structural dynamics, furnish the values of u, $\dot{u}$ and $\ddot{u}$ at time $t + \Delta t$ as functions of the same variables at time t alone, while multi-step methods also use values at times $t - \Delta t$, $t - 2\Delta t$ etc. Multi-step methods therefore require some additional initial computations (e.g. by a single-step method), while single-step methods are "self starting".

Explicit algorithms furnish the solution at time $t + \Delta t$ directly, while in implicit methods the unknowns appear on both sides of algebraic equations and must be determined by solving a linear equation system (or just one equation for a SDOF system). This shortcoming of implicit algorithms is offset by their better stability properties.

The NEWMARK $\beta - \gamma -$ algorithm is an implicit, single-step scheme with two parameters β and γ which determine its stability and accuracy properties. Considering the time points t_1 and t_2 with $t_2 = t_1 + \Delta t$, the dynamic equilibrium of the SDOF system at time t_2 is given by

$$m\ddot{u}_2 + c\dot{u}_2 + ku_2 = F(t_2) = F_2 \tag{3.4.2}$$

Introducing increments of the displacement, velocity, acceleration and external force ($u, \dot{u}, \ddot{u}$ and F) according to $\Delta u = u_2 - u_1$, $\Delta\dot{u} = \dot{u}_2 - \dot{u}_1$ etc. leads to the incremental version of Eq. (3.4.2):

$$m\,\Delta\ddot{u} + c\,\Delta\dot{u} + k\,\Delta u = \Delta F \tag{3.4.3}$$

The increments $\Delta\ddot{u}$, $\Delta\dot{u}$ can be given as functions of the displacement increment Δu and the known values of velocity and acceleration at time t_1 :

$$\begin{aligned} \Delta\dot{u} &= \frac{\gamma}{\beta\Delta t}\Delta u - \frac{\gamma}{\beta}\dot{u}_1 - \Delta t\,(\frac{\gamma}{2\beta} - 1)\,\ddot{u}_1 \\ \Delta\ddot{u} &= \frac{1}{\beta(\Delta t)^2}\Delta u - \frac{1}{\beta\Delta t}\dot{u}_1 - \frac{1}{2\beta}\ddot{u}_1 \end{aligned} \tag{3.4.4}$$

Values of $\beta = 1/4$ and $\gamma = 1/2$ for the two parameters yield an unconditionally stable scheme which assumes a constant acceleration $\ddot{u}$ between t_1 and t_2. For $\beta = 1/6$ and $\gamma = 1/2$ the integrator is only conditionally stable and the acceleration may vary linearly between t_1 and t_2. Introducing Eqs. (3.4.4) into Eq. (3.4.3) yields the displacement increment Δu as

$$\Delta u = \frac{f^*}{k^*} \tag{3.4.5}$$

with

$$k^* = m\frac{1}{\beta\Delta t^2} + c\frac{\gamma}{\beta\Delta t} + k \tag{3.4.6}$$

$$f^* = \Delta F + m(\frac{\dot{u}_1}{\beta\Delta t} + \frac{\ddot{u}_1}{2\beta}) + c\left(\frac{\gamma\dot{u}_1}{\beta} + \ddot{u}_1\,\Delta t\,(\frac{\gamma}{2\beta} - 1)\right) \tag{3.4.7}$$

These expressions have been programmed in the program LEINM. Its input file RHS is identical to the input file of program DUHAMI (time points and values f(t) = F(t)/m of the loading function in two columns in 2E14.7 format) and the remaining data are entered interactively. If the loading function in the file RHS is F(t) rather than f(t), it is scaled by the factor (1/m) to be given separately. The output file THNEW contains four columns with the time points and the computed values for the displacement, velocity and acceleration, the latter in units of g. For this conversion it is necessary to enter the value of g in the unit system employed (e.g. 9.81 if m and s are used). Additionally, the (absolute) maximum displacement, velocity and acceleration values together with the corresponding times are output to the monitor.

Program name	Input file	Output file
LEINM	RHS	THNEW

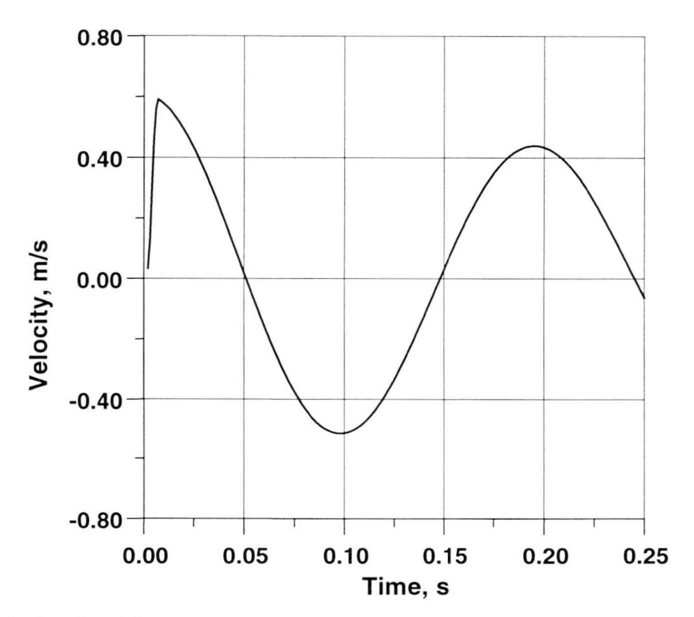

Fig. 3.4-1: Velocity time history

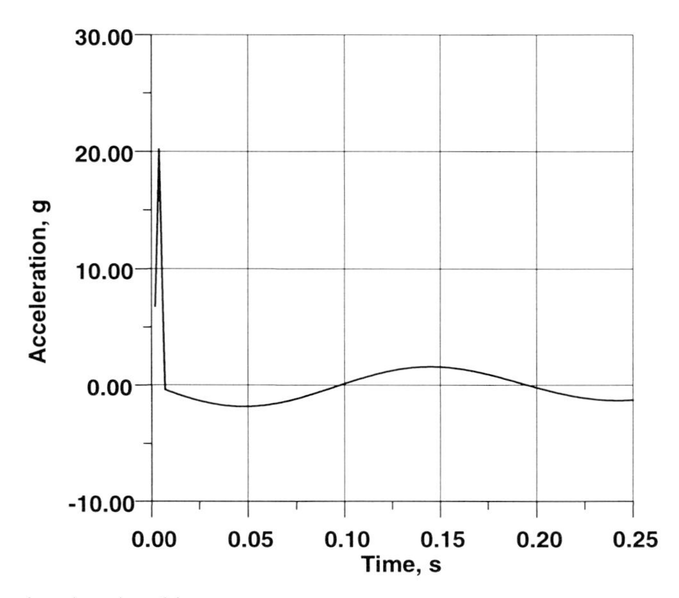

Fig. 3.4-2: Acceleration time history

As an example, Fig. 3.4-1 shows the velocity time history (in m/s), Fig. 3.4-2 the acceleration time history (in g) of the girder of the single-story frame depicted in Fig. 3.1-3 subject to the triangular loading function given. A damping D = 5% critical was assumed.

3.5 Frequency domain methods

Ordinary differential equations as, for instance, the differential equation of motion for a SDOF system

$$m\ddot{u} + c\dot{u} + ku = F(t) \tag{3.5.1}$$

or, written differently

$$\ddot{u} + 2\xi\omega_1\dot{u} + \omega_1^2 u = \frac{F}{m} = f(t) \tag{3.5.2}$$

can be transformed into algebraic equations by utilising integral transform methods such as the FOURIER or the LAPLACE transform. Having solved the equations in the frequency domain, the solutions must be returned to the time domain via the pertinent inverse transforms. As an example for a transformation, the DUHAMEL integral yielding the particular solution of Eq. (3.5.2) can be viewed as transforming the excitation F(t) or f(t) in the system's displacement response u(t):

$$u_p(t) = \frac{1}{\omega_D}\int_0^t f(\tau)\, e^{-\xi\omega_1(t-\tau)} \sin\omega_D(t-\tau)\, d\tau \tag{3.5.3}$$

The upper limit t of this integral may be put equal to the duration t_D of the transient excitation f(t), $t = t_D$. For $t \geq t_D$ f(t) vanishes so that no additional contributions to the value of the integral appear, making it possible to formally extend the upper limit to infinity. The DUHAMEL integral then effectively "folds" the excitation function f(t) with the so-called impulse response function h(t), with

$$h(t) = \frac{e^{-\xi\omega_1 t}}{\omega_D}\sin\omega_D t \tag{3.5.4}$$

This can be formally expressed as

$$u_p(t) = f(t) * h(t) = \int_{-\infty}^{\infty} f(\tau)\, h(t-\tau)\, d\tau \tag{3.5.5}$$

The impulse response function h(t) is defined as the displacement response u(t) of a SDOF system, Eq. (3.5.2), to a unit impulse (DIRAC impulse). It can be derived by considering the homogeneous solution of Eq. (3.5.2)

$$u_h(t) = e^{-\xi\omega_1 t}\left(C_1 \cos\sqrt{1-\xi^2}\,\omega_1 t + C_2 \sin\sqrt{1-\xi^2}\,\omega_1 t\right) \tag{3.3.6}$$

For the general set of initial conditions $u(0) = u_0$, $\dot{u}(0) = \dot{u}_0$ this leads to:

$$u_h(t) = e^{-\xi\omega_1 t}\left(u_0 \cos\sqrt{1-\xi^2}\,\omega_1 t + \frac{(\dot{u}_0 + \xi\omega_1 u_0)}{\omega_1\sqrt{1-\xi^2}} \sin\sqrt{1-\xi^2}\,\omega_1 t\right) \tag{3.3.7}$$

A unit impulse excitation at time t=0 corresponds to initial conditions $u_0 = 0$, $\dot{u}_0 = 1$ and yields the expression

$$u(t) = h(t) = \frac{e^{-\xi\omega_1 t}}{\omega_D} \sin \omega_D t \tag{3.5.8}$$

Fig. 3.5-1 shows a typical SDOF impulse reaction function, in this case with 5% critical damping and a natural circular frequency of 30 rad/s.

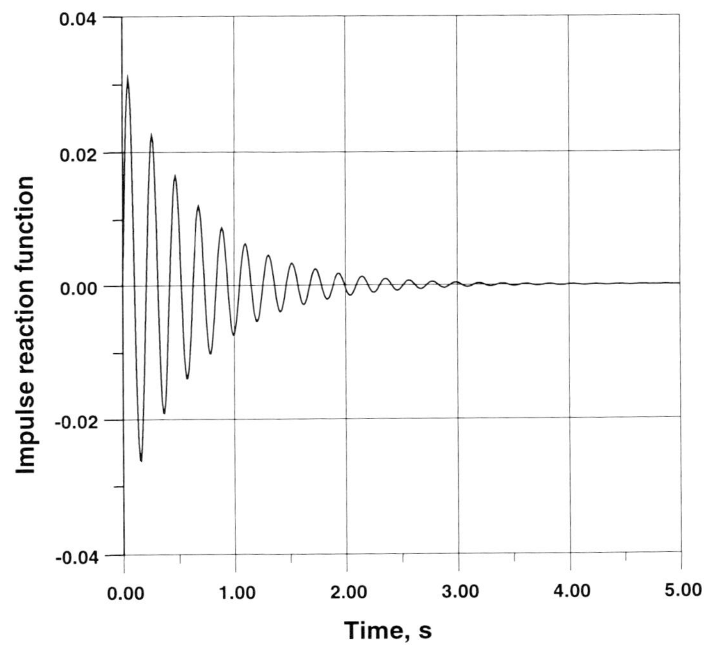

Fig. 3.5-1: Impulse reaction function with $\omega_1 = 30\,\text{rad/s}$, $\xi = 0.05$

Introducing the FOURIER transform of the displacement response u(t) as U(f), of the excitation function f(t) as F(f) and of the impulse reaction function h(t) as H(f) according to the general expression

$$X(f) = \int_{-\infty}^{\infty} x(t)\, e^{-i2\pi ft} dt \tag{3.5.9}$$

leads to the FOURIER transform U(f) in the form

$$U(f) = \int_{-\infty}^{\infty} u(t)\, e^{-i2\pi ft} dt = \int_{-\infty}^{\infty} \left(\int_{-\infty}^{\infty} f(\tau)\, h(t-\tau)\, d\tau \right) e^{-i2\pi ft} dt \tag{3.5.10}$$

Reversing the order of integration and introducing $\lambda = t - \tau$ yields:

$$U(f)= \int_{-\infty}^{\infty} f(\tau)\left(\int_{-\infty}^{\infty} h(t-\tau)e^{-i2\pi ft}dt \right)d\tau = \int_{-\infty}^{\infty} f(\tau)\left(\int_{-\infty}^{\infty} h(\lambda)e^{-i2\pi f(\lambda+\tau)}d\lambda \right)d\tau \qquad (3.5.11)$$

which can be also written as

$$U(f)= \left(\int_{-\infty}^{\infty} f(\tau)e^{-i2\pi f\tau}\, d\tau \right)\left(\int_{-\infty}^{\infty} h(\lambda)e^{-i2\pi f\lambda} d\lambda \right) \qquad (3.5.12)$$

or, in short

$$U(f)=F(f)\cdot H(f) \qquad (3.5.13)$$

This has shown that a "folding" of the functions f(t) and h(t) in the time domain is equivalent to a multiplication of their FOURIER transforms F(f) and H(f) in the frequency domain. The function H(f), or $H(\omega)$, which has been derived as the FOURIER transform of the impulse reaction function is called the transfer function of the system. It can be visualised as the output/input ratio for a certain response quantity if the input is stationary and harmonic. For the general harmonic unit excitation

$$x(t) = 1\cdot e^{i\omega t} \qquad (3.5.14)$$

the displacement response of the SDOF oscillator is thus defined to be

$$u(t)=H(\omega)\cdot e^{i\omega t} \qquad (3.5.15)$$

Considering Eq. (3.5.2) and $f(t)=1\cdot e^{i\omega t}$ this leads to

$$\begin{aligned} u(t)&=H(\omega)e^{i\omega t}\\ \dot{u}(t)&=i\omega H(\omega)e^{i\omega t}\\ \ddot{u}(t)&=-\omega^2 H(\omega)e^{i\omega t} \end{aligned} \qquad (3.5.16)$$

Introducing Eqs. (3.5.16) into Eq. (3.5.2) yields an expression for the transfer function:

$$H(\omega) = \frac{1}{\omega_1{}^2 - \omega^2 + i\, 2\xi\omega_1\omega} \qquad (3.5.17)$$

Fig. 3.5-2 shows the real and Fig. 3.5-3 the imaginary part of $H(\omega)$ for values of $\omega_1 = 30\frac{\text{rad}}{\text{s}}$ and $\xi = 0{,}05$. Fig. 3.5-4 shows the absolute value $|H(\omega)|$ vs. circular frequency with a peak for $\omega = 30\frac{\text{rad}}{\text{s}}$ equal to $|H(\omega_1)| = \frac{1}{2\xi\omega\omega_1} \approx 0.0111$.

The general approach for the solution of the differential equation of motion in the frequency domain is as follows: Having determined the complex FOURIER transform of the loading function f(t), $F(\omega)$, the latter is multiplied by the transfer function Eq. (3.5.17), yielding the FOURIER transform $U(\omega)$ of the response u(t); the latter must be transformed back to the time domain.

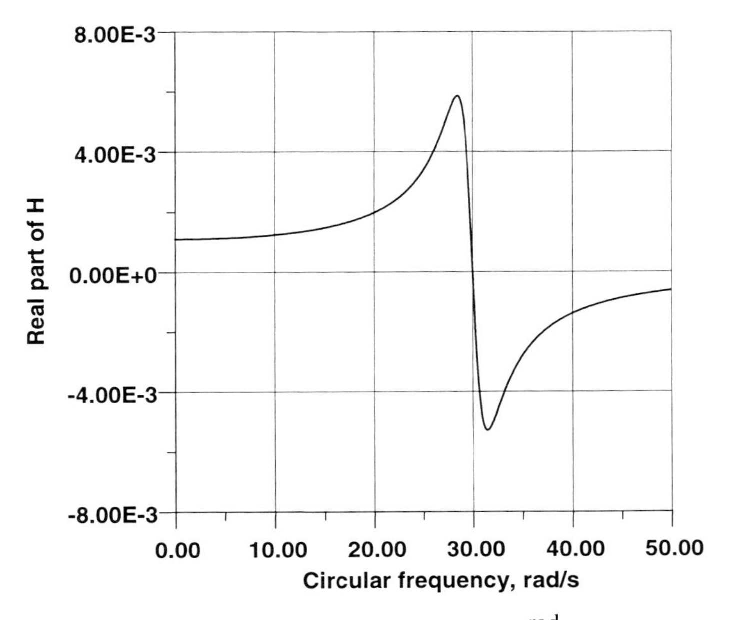

Fig. 3.5-2: Real part of the SDOF transfer function for $\omega_1 = 30\frac{\text{rad}}{\text{s}}$, $\xi = 0.05$

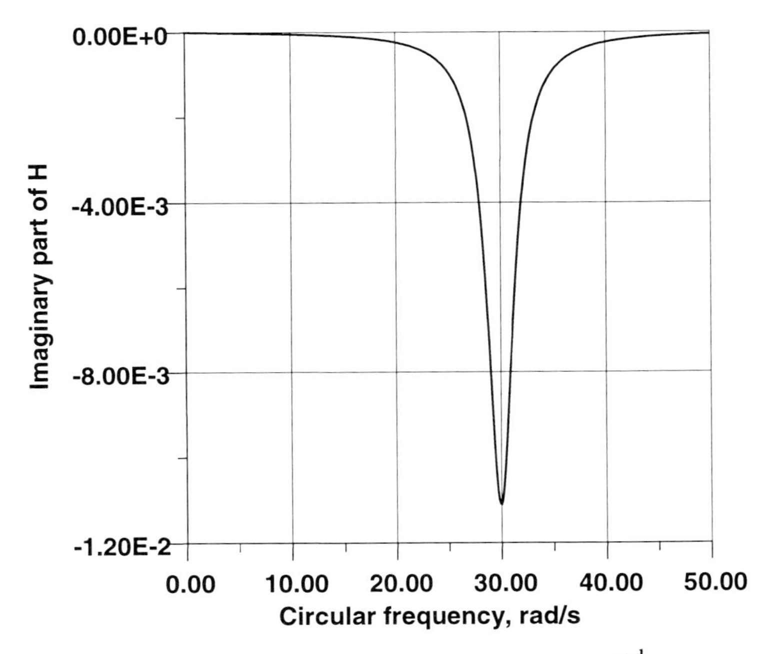

Fig. 3.5-3: Imaginary part of the SDOF transfer function for $\omega_1 = 30\frac{\text{rad}}{\text{s}}$, $\xi = 0.05$

The transformation of an "input signal" f(t) into an "output signal" u(t) can be generally considered as a filtering process, which modifies the frequency content of the original signal according to the transfer function characterising the system. Fig. 3.5-5 schematically depicts this filtering process; the input signal shown to the left is put through a low-pass filter which almost eliminates the high-frequency components, and, alternatively, through a high-pass filter which eliminates the low-frequency components.

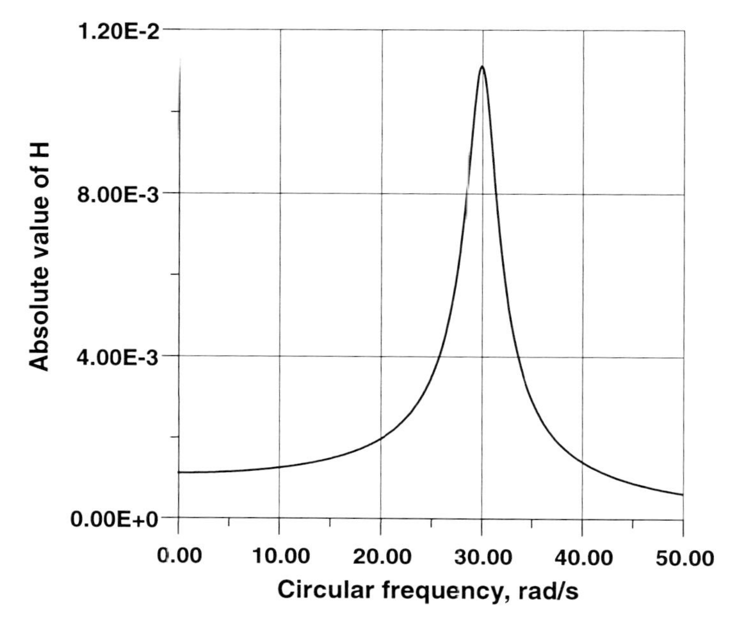

Fig. 3.5-4: Absolute value of the SDOF transfer function for $\omega_1 = 30\frac{\text{rad}}{\text{s}}$, $\xi = 0.05$

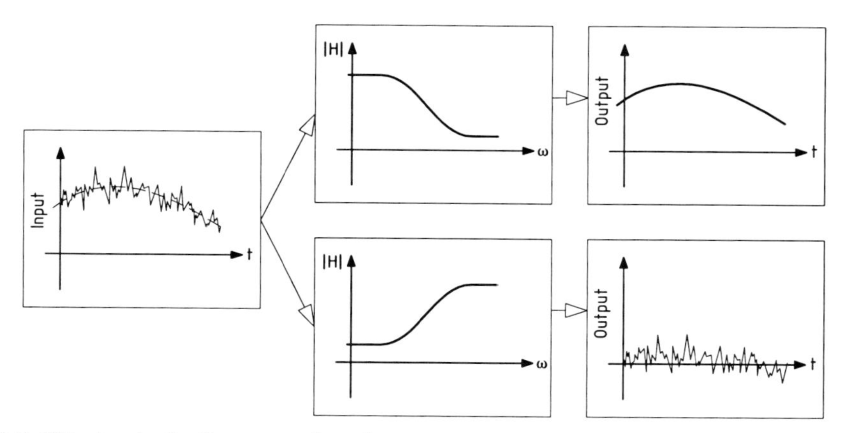

Fig. 3.5-5: Filtering in the frequency domain

Three simple filters (high-pass, low-pass and band-pass) are considered as examples:

- High-pass filter as a linear first order system with a transfer function given by

$$H(\omega)=\frac{\omega^2+i\omega\omega_H}{\omega_H{}^2+\omega^2} \quad (3.5.18)$$

This filter is defined by the single parameter ω_H (its so-called corner frequency).

- Low-pass filter as a SDOF oscillator (second order system); its transfer function is given by

$$H(\omega)=\frac{1+\frac{\omega^2}{\omega_0^2}(4\xi_0^2-1)-i2\xi_0\frac{\omega^3}{\omega_0^3}}{\left[1-\left(\frac{\omega^2}{\omega_0^2}\right)^2\right]^2+4\xi_0^2\left(\frac{\omega^2}{\omega_0^2}\right)^2} \quad (3.5.19)$$

This is the well-known KANAI-TAJIMI filter, characterised by the two parameters ω_0 and ξ_0. It is often used in earthquake engineering applications, where the parameters ω_0 and ξ_0 may be considered as the natural circular frequency and the critical damping of the ground.

- A band-pass filter can be constructed as a product of the transfer functions of the high-pass filter given by Eq. (3.5.18) and a first order low-pass filter with the transfer function

$$H(\omega)=\frac{\omega_T^2-i\omega\omega_T}{\omega_T^2+\omega^2} \quad (3.5.20)$$

The band-pass filter thus obtained is characterised by the corner frequencies ω_T and ω_H and can be utilised for selectively amplifying frequency components in the range $\omega_H<\omega<\omega_T$ considered to be of special importance to a site or a structure.

As an example, the signal shown in Fig. 3.5-6 is considered, the equation of which is given by

$$f(t)=1\cdot\sin\frac{2\pi t}{1.024}+0.5\cdot\sin\frac{2\pi t}{0.1024}+0.25\cdot\sin\frac{2\pi t}{0.01024} \quad (3.5.21)$$

A low-pass filtering by means of a KANAI-TAJIMI filter with $\omega_0=10\frac{\text{rad}}{\text{s}}$ and $\xi_0=0.30$ yields the output shown in Fig. 3.5-7, which obviously contains only the low-frequency components. A high-pass filtering of the signal Eq. (3.5.21) by means of the filter given by Eq. (3.5.18) with $\omega_H=600\frac{\text{rad}}{\text{s}}$ yields the signal shown in Fig. 3.5-8, which is almost free of low-frequency components.

These computations have been carried out by the program FILTER, which contains the three filters presented in this section. Its input file, TIMSER.DAT, contains the input signal as time series sampled with a constant time step in a 2E14.7 format, with the time points in the first and the ordinates in the second column. The resulting time series is written in the output file FILT.DAT in the same 2E14.7 format with time points in the first and ordinates in the second column. The number N of points of the time series may not exceed 4096; if N is not already a power of 2, zeros are added until the next power of 2. All further data, such as the number N of the points of the time series, the time step, the type and the parameters of the filter, are entered interactively.

Program name	Input file	Output file
FILTER	TIMSER.DAT	FILT.DAT

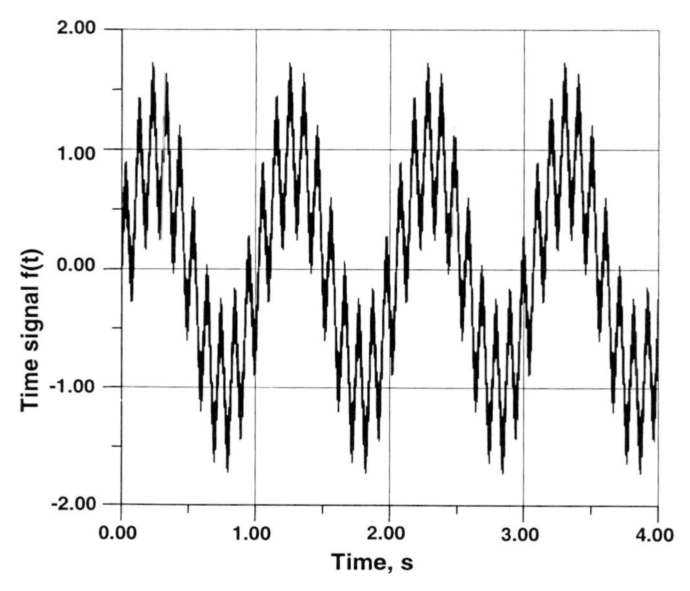

Fig. 3.5-6: Stationary harmonic signal

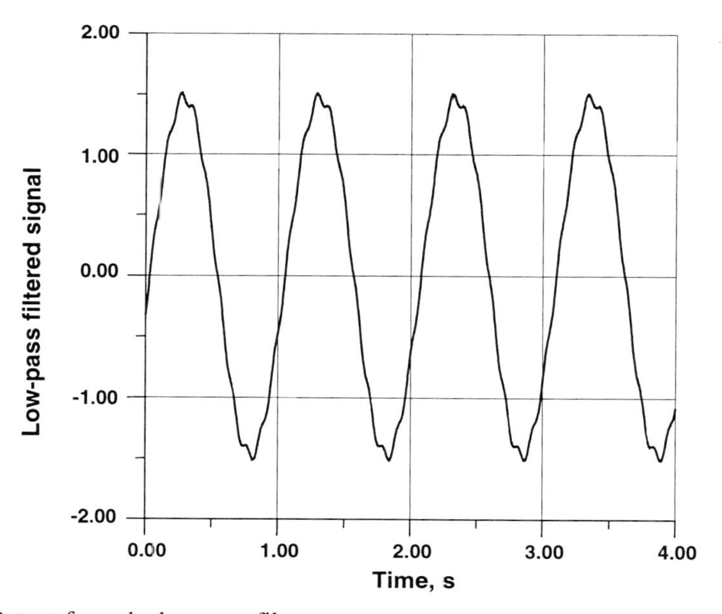

Fig. 3.5-7: Output from the low-pass filter

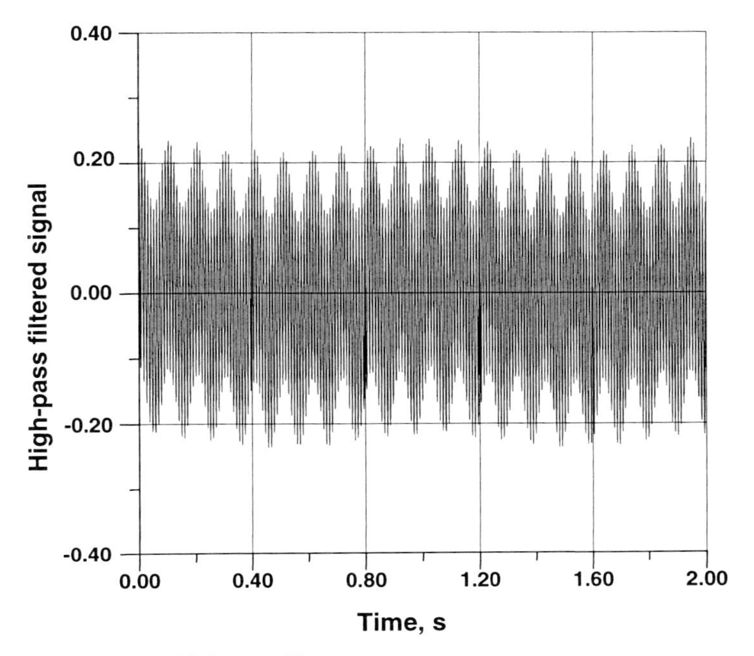

Fig. 3.5-8: Output from the high-pass filter

3.6 Harmonic excitation, vibration isolation for harmonic loads

A SDOF system subject to a harmonic excitation with amplitude F_o is considered. The differential equation of motion is given by

$$\ddot{u} + 2\xi\omega_1\dot{u} + \omega_1^2 u = \frac{F_o}{m}\sin\Omega t \tag{3.6.1}$$

Its general solution is equal to the sum of the homogeneous solution

$$u_h(t) = e^{-\xi\omega_1 t}(C_1\cos\omega_D t + C_2\sin\omega_D t) \tag{3.6.2}$$

which depends on the initial conditions $u(0), \dot{u}(0)$, and a particular solution which can be assumed as a combination of a sine and a cosine wave with circular frequencies equal to the excitation frequency Ω:

$$u_p(t) = C_3\sin\Omega t + C_4\cos\Omega t \tag{3.6.3}$$

By introducing the particular solution into the differential equation and comparing the coefficients of the trigonometric functions the following values for C_3 and C_4 are obtained, with $\beta = \dfrac{\Omega}{\omega_1}$ as frequency ratio and k as spring stiffness of the SDOF oscillator ($k = m\,\omega_1^2$):

$$C_3 = \frac{F_0}{k}\frac{1-\beta^2}{(1-\beta^2)^2 + (2\xi\beta)^2} \tag{3.6.4}$$

$$C_4 = \frac{F_0}{k} \frac{-2\xi\beta}{(1-\beta^2)^2 + (2\xi\beta)^2} \tag{3.6.5}$$

This leads to the particular solution

$$u_p(t) = \frac{F_0}{k} \frac{(1-\beta^2)\sin\Omega t - 2\xi\beta\cos\Omega t}{(1-\beta^2)^2 + (2\xi\beta)^2} \tag{3.6.6}$$

Since the homogeneous part of the solution which depends on the initial conditions will be damped out sooner or later, it is only the part given by Eq. (3.6.6) that must be considered. Eq. (3.6.3) can be also expressed as follows by introducing the resultant amplitude u_R and the phase lag φ:

$$u_p(t) = u_R \sin(\Omega t - \varphi) \tag{3.6.7}$$

Here we have

$$u_R = \sqrt{C_3{}^2 + C_4{}^2} = \frac{F_0}{k}[(1-\beta^2)^2 + (2\xi\beta)^2]^{-0,5} \tag{3.6.8}$$

$$\varphi = \arctan\frac{-C_4}{C_3} = \arctan\frac{2\xi\beta}{1-\beta^2} \tag{3.6.9}$$

Concerning the sign of the phase angle in Eqs. (3.6.7) and (3.6.9) it must be noted that $\arctan(-\varphi) = -\arctan(\varphi)$.

Next, the so-called dynamic magnification factor V is introduced as ratio of the maximum dynamic displacement u_R to its static value $\frac{F_0}{k}$. It is given by

$$V = \frac{1}{\sqrt{(1-\beta^2)^2 + (2\xi\beta)^2}} \tag{3.6.10}$$

and we obtain

$$\max\ u_p(t) = u_R = \frac{F_0}{k} V \tag{3.6.11}$$

In order to determine the maximum value of the magnification factor V as a function of the frequency ratio β, the derivative of Eq. (3.6.10) with respect to β is computed and set equal to zero:

$$\frac{dV}{d\beta} = \frac{2\beta - 2\beta^3 - 4\beta\xi^2}{\left[(1-\beta^2)^2 + 4\xi^2\beta^2\right]^{\frac{3}{2}}} = 0 \tag{3.6.12}$$

The corresponding solutions are

$$\beta_1 = 0,\ \beta_2 = -\sqrt{1-2\xi^2},\ \beta_3 = \sqrt{1-2\xi^2} \tag{3.6.13}$$

from which β_3 is the only one of interest. For this frequency ratio, the peak value of the magnification factor is given by

$$\max V = \frac{1}{2} \frac{1}{\xi\sqrt{1-\xi^2}} \tag{3.6.14}$$

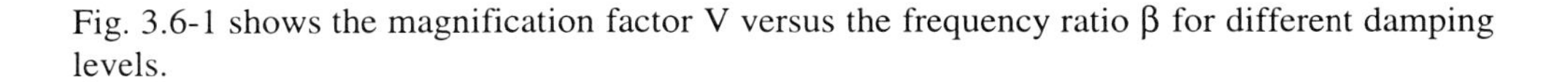

Fig. 3.6-1 shows the magnification factor V versus the frequency ratio β for different damping levels.

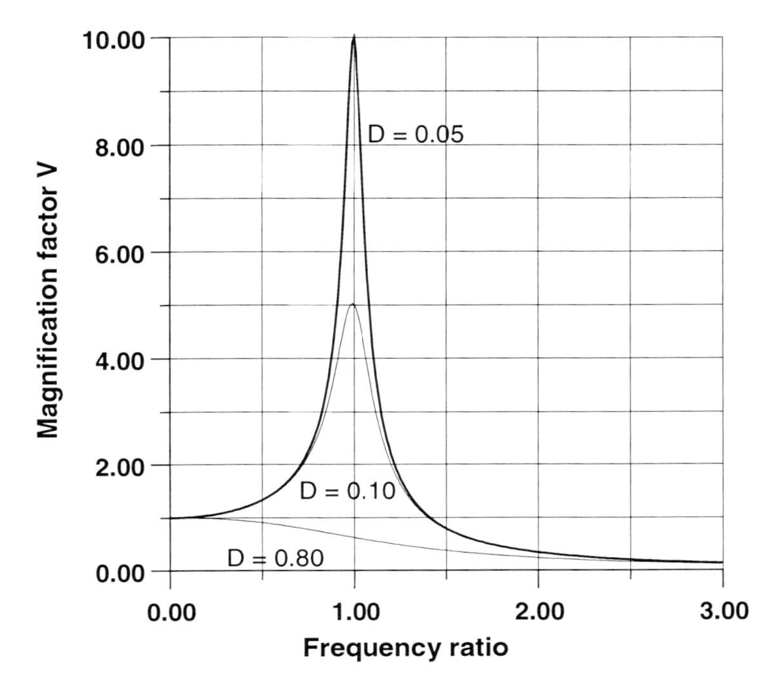

Fig. 3.6-1: Magnification factor V for damping values D = 0.05, 0.10 and 0.8 critical

In practice, damping values are sufficiently small for the following expression to be valid:

$$\max V = \frac{1}{2\xi} \text{ for } \beta = 1 \qquad (3.6.15)$$

If an experimentally determined curve depicting the peak amplitudes u_R versus the frequency ratio is available, it can serve for extracting the corresponding damping value D or ξ. This so-called half-power method furnishes the damping value as $\frac{1}{2}(\beta_2 - \beta_1)$, that is, half the distance between the points β_1, β_2 with ordinates equal to $\frac{\max u_R}{\sqrt{2}}$, max u_R being the ordinate for $\beta = 1$. Fig. 3.6-2 shows an example of a SDOF system with 10% critical damping; the distance $\beta_2 - \beta_1 = 0.20$ can be seen to separate the points with ordinates equal to $5/\sqrt{2} = 3.54$ units.

Fig. 3.6-3 schematically shows a machine with a harmonic load $F = F_0 \sin \Omega t$ supported by spring elements, with additional viscous damping present, as depicted by the dashpot. The total load acting on the foundation is equal to the sum of the restoring and damping forces

$$F_R = k\,u(t), \qquad F_D = c\,\dot{u}(t) \qquad (3.6.16)$$

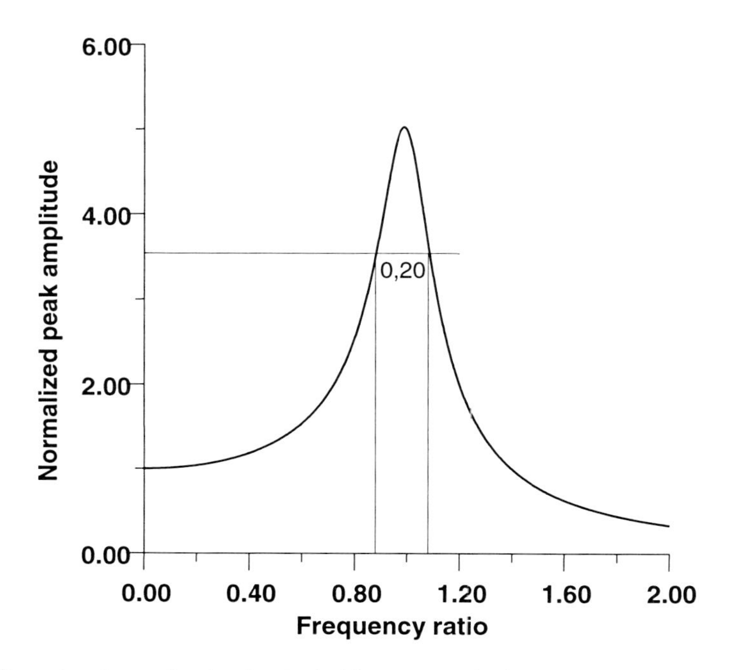

Fig. 3.6-2: Damping determination by the half-power method

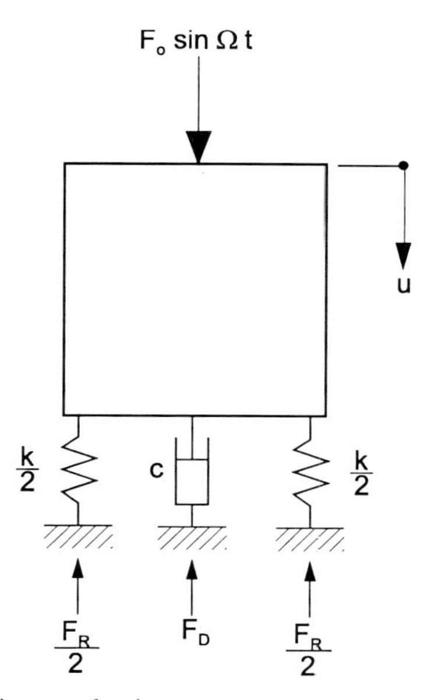

Fig. 3.6-3: Scheme of a machine on elastic supports

With Eq. (3.6.7) for u(t) this leads to

$$F_R = kV\frac{F_o}{k}\sin(\Omega t - \varphi) \tag{3. 6. 17}$$

$$F_D = cV\frac{F_o}{k}\Omega\cos(\Omega t - \varphi) = 2\xi\beta V F_o \cos(\Omega t - \varphi) \tag{3. 6. 18}$$

There exists a phase lag of 90° between the restoring force and the damping force, so that their resultant is given by

$$\sqrt{F_R^2 + F_D^2} = VF_o\sqrt{1+(2\xi\beta)^2} \tag{3. 6. 19}$$

This is the maximum load to act on the foundation. Its ratio to the peak amplitude F_o of the harmonic excitation is the so-called transmissibility V_F. We get

$$V_F = V\sqrt{1+(2\xi\beta)^2} \tag{3. 6. 20}$$

with the magnification factor V after Eq. (3.6.10). Fig. 3.6-4 shows transmissibility curves versus the frequency ratio for some damping levels.

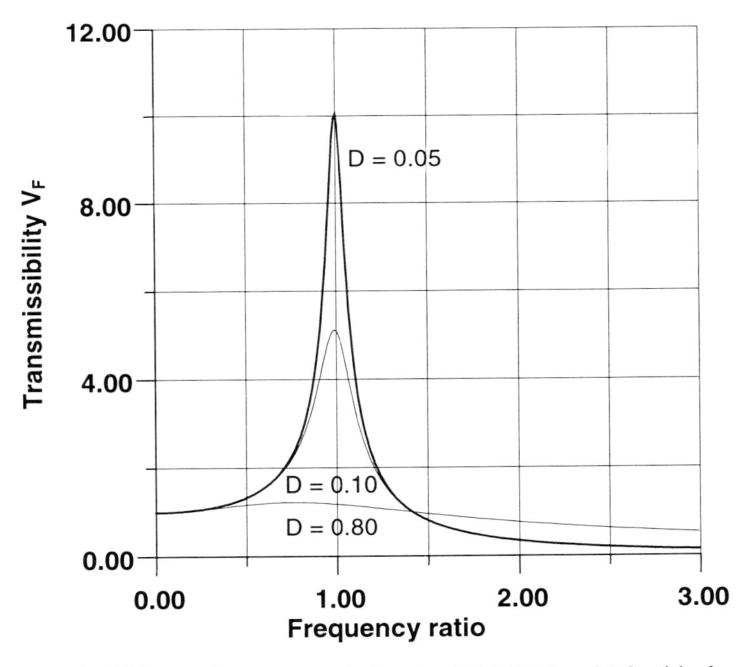

Fig. 3.6-4: Transmissibility vs. frequency ratio for D = 0.05, 0.10 and 0.8 critical.

Please note that for $\beta = \sqrt{2}$, all curves have the same ordinate $V_F = 1$. For $\beta < 1$ the excitation frequency is less than the natural frequency of the system, for $\beta > 1$ it is larger. No vibration isolation is possible as long as $\beta < 1$. For $\beta > \sqrt{2}$ vibration isolation becomes possible ($V_F < 1$), however it is not very effective for $\beta \approx \sqrt{2}$. In order to quantify vibration isolation, an efficiency factor J in percent is introduced according to

$$J = 100(1 - V_F) \tag{3.6.21}$$

As an example, for zero damping and a frequency ratio of 1.6, the efficiency factor is equal to 36%, while for $\beta = 4$ it reaches 93%. Frequency ratios about 3 or 4 are sufficient in practice, because for larger values the law of diminishing results applies.

Fig. 3.6-4 illustrates another interesting aspect, namely that damping actually counteracts vibration isolation measures for frequency ratios larger than $\sqrt{2}$. Therefore, vibration isolation systems with springs as shown in Fig. 3.6-3 should have as little damping as possible. For zero damping ($\xi = 0$) and $\beta > \sqrt{2}$ the efficiency factor J is equal to

$$J = 100\frac{\beta^2 - 2}{\beta^2 - 1} \tag{3.6.22}$$

The following expression also holds:

$$\beta^2 = \frac{\frac{J}{100} - 2}{\frac{J}{100} - 1} \tag{3.6.23}$$

As an example, the spring constant for an isolating support of a machine is determined. An efficiency factor of 90% is required for a 24 kN heavy block supported on 6 bearings carrying equal loads (4 kN each). The machine operates at 1200 rpm, corresponding to a circular frequency Ω of

$$\Omega = 2\pi f = 2\pi\frac{1200\,U}{60\,s} = 125.66\frac{rad}{s} \tag{3.6.24}$$

For the efficiency factor J to be equal to 90%, the frequency ratio after Eq. (3.6.23) is equal to

$$\beta = \sqrt{\frac{0.90 - 2}{0.90 - 1}} = 3.317 \tag{3.6.25}$$

and the circular natural frequency of the system is given by

$$\omega_1 = \frac{\Omega}{\beta} = \frac{125.66}{3.317} = 37.89\frac{rad}{s} \tag{3.6.26}$$

With a mass of 0.4 t per support this leads to a spring constant for each bearing equal to

$$k = m \cdot \omega_1^2 = 0.4 \cdot 37.89^2 = 574.2\frac{kN}{m} \tag{3.6.27}$$

3.7 SDOF system with material non-linearity

In this section, a SDOF system is considered, in which the restoring force $F_R(u)$ is a non-linear function of the displacement u. Its differential equation of motion is given by

$$m\ddot{u} + c\dot{u} + F_R(u) = F(t) \tag{3.7.1}$$

Direct Integration methods in the time domain are best for solving Eq. (3.7.1). Frequency domain methods (which depend on the superposition of system response components due to harmonic excitations) as well as the time-domain DUHAMEL integral approach (which depends on the superposition of system response components due to differential impulses) are not applicable because the superposition principle is not valid for non-linear system behaviour. For time domain direct integration, on the other hand, it is perfectly straightforward to update the system configuration from time step to time step and to compute the currently valid restoring force corresponding to the actual displacement.

Non-linear curves expressing the relationship between restoring force and displacement can be quite complicated in real life. However, for many situations of practical importance, simple multi-linear (e.g. bilinear) laws are quite useful and sufficiently accurate if calibrated accordingly. The models depicted in Figs. 3.7-1 and 3.7-2 are quite popular. The elastic-perfectly plastic model of Fig. 3.7-1 is characterised by the stiffness k (as slope of the F_R-u-curve) becoming zero after a displacement limit value u_{el} or $-u_{el}$ has been reached. The corresponding restoring force is constant, in contrast to the bilinear model of Fig. 3.7-2, where the stiffness beyond the elastic limit decreases to pk, with $p < 1$. Both the elastic-perfectly plastic and the more general bilinear model do not exhibit stiffness degradation effects, in the sense that the stiffness values k or pk are independent of the number of deformation cycles. A simple model including degradation effects is the origin-oriented UMEMURA law shown in Fig. 3.7-3, where each unloading branch points to the origin.

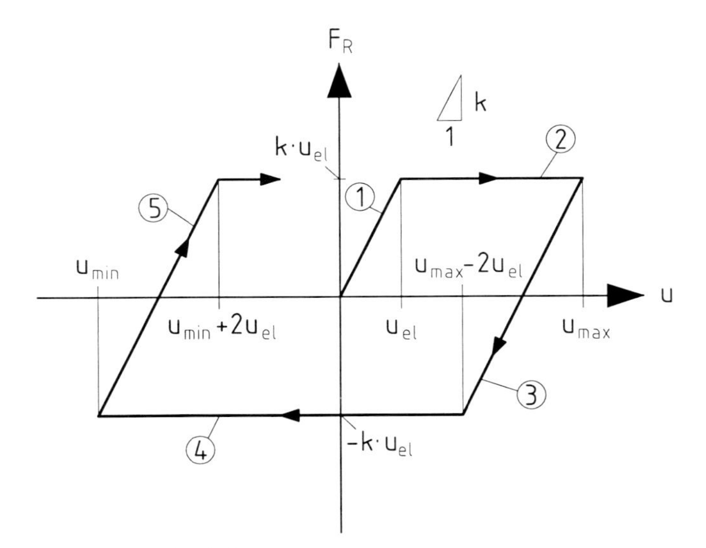

Fig. 3.7-1: Elastic-perfectly plastic law

The procedure for determining the correct actual value of the restoring force is best demonstrated for the simple elastic-perfectly plastic model with the regions 1 up to 5 as shown in Fig. 3.7-1:

- Region 1, elastic: $-u_{el} \leq u \leq u_{el}$, Restoring force $F_R = k\ u$.
- Region 2, perfectly plastic: $u > u_{el}$, $\dot{u} > 0$ (Increasing displacement), $F_R = k\ u_{el}$.
  As soon as the displacement stops increasing,
  $\dot{u} < 0$, Region 3 is entered; simultaneously, the value of
  u_{max} is defined.

- Region 3, elastic: $u_{max} - 2\ u_{el} \leq u \leq u_{max}$, $F_R = k \cdot (u_{el} - u_{max} + u)$.
- Region 4, perfectly plastic: $u < u_{max} - 2\,u_{el}$, $\dot{u} < 0$, $F_R = -k \cdot u_{el}$.

  As soon as $\dot{u} > 0$, Region 5 is entered; simultaneously, the value of u_{min} is defined.
- Region 5, elastic: $u_{min} \leq u \leq u_{min} + 2\,u_{el}$, $F_R = k \cdot (u - u_{el} - u_{min})$.

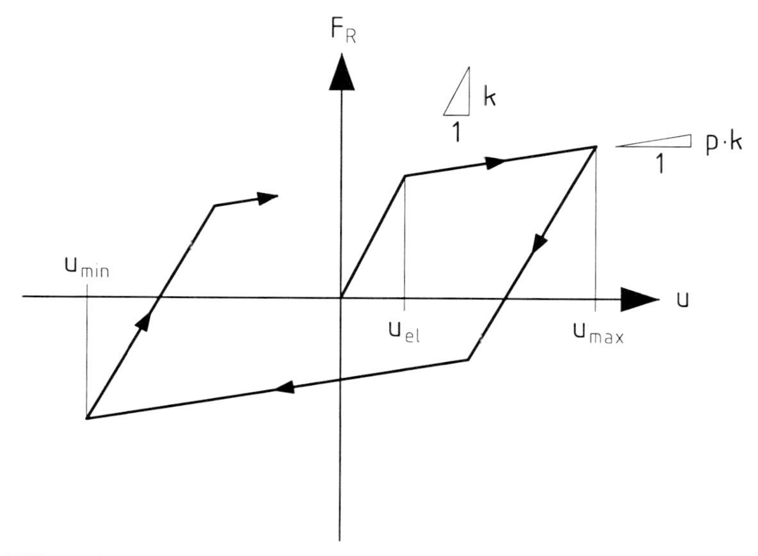

Fig. 3.7-2: Bilinear law

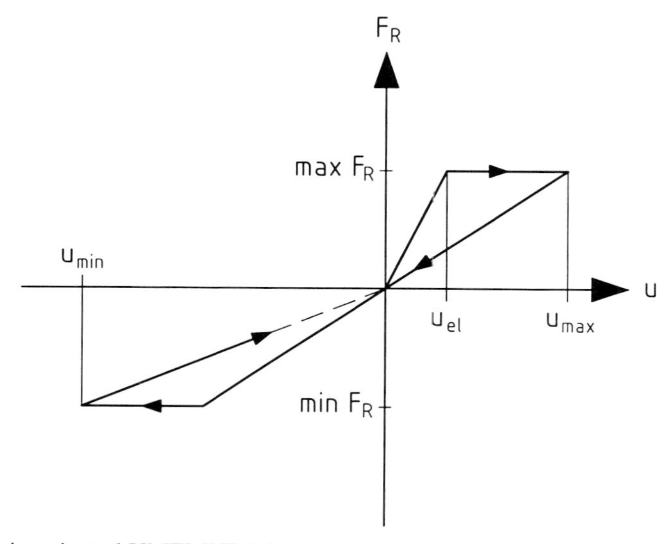

Fig. 3.7-3: Origin-oriented UMEMURA law

The computation of the non-linear system response u(t) by Direct Integration is carried out as in the linear case, with the difference that special steps must be taken if the state of the system changes during a time step. Let us suppose that a change in the system has occurred between t_1 and t_2 (with $\Delta t = t_2 - t_1$), e.g. because u has exceeded u_{el}. The equilibrium equation between the inertia, damping and restoring forces and the external loading at time t_2 is given by

$$F_I + F_D + F_R = F(t_2) \tag{3. 7. 2}$$

It is no longer satisfied, because the actual restoring force $F_R(t_2)$ is now different from the value it would have, if no system change had taken place. This leads to an unbalanced force R according to

$$R = F(t_2) - \left[m\ddot{u}(t_2) + c\dot{u}(t_2) + F_R(t_2)\right] \tag{3. 7. 3}$$

There exist various alternatives for taking the unbalanced force into account:

1. R is simply ignored and the computed displacement $u(t_2)$ is not modified; only the stiffness of the system is updated before continuing the computation for the next time step.

2. The computed solution $u(t_2)$ is not modified; however, R is considered in the external loading of the next time step.

3. The solution $u(t_2)$ is corrected iteratively, until the absolute value of R does not exceed a certain tolerance limit.

In the first two cases there is always the possibility that the computed and the correct solution diverge to a certain degree. A simple method for checking the accuracy of the computed solution is to repeat the calculation with a smaller time step. Such checks are quite important for non-linear systems, since even Direct Integration algorithms, which are unconditionally stable for linear systems, may produce unacceptable results in the presence of non-linearities.

The third alternative suggested above is generally the best, especially in the case of multi-linear hysteretic laws. In these cases, the iteration process normally ends after a single cycle, since there is no further change in stiffness. On the other hand, if the hysteretic law contains curved branches, more than one iteration cycle is usually necessary until the absolute value of the unbalanced force R drops below the required threshold.

In order to arrive at the pertinent expressions for carrying out these computations, we derive the incremental form of Eq. (3.7.1). Assuming that the spring and damping coefficient do not change during the time step in question, the force equilibrium equations at the time instants t_1 and $t_2 = t_1 + \Delta t$ are:

$$m\ddot{u}_1 + c\dot{u}_1 + ku_1 = F_1 \tag{3. 7. 4}$$

and

$$m(\ddot{u}_1 + \Delta\ddot{u}) + c(\dot{u}_1 + \Delta\dot{u}) + k(u_1 + \Delta u) = F_1 + \Delta F \tag{3. 7. 5}$$

Subtracting Eq. (3.7.4) from Eq. (3.7.5) leads to the incremental form of the differential equation of motion for a SDOF system:

$$m\,\Delta\ddot{u} + c\,\Delta\dot{u} + k\,\Delta u = \Delta F \tag{3. 7. 6}$$

This equation will be solved by the implicit NEWMARK integration scheme (with constant or linear acceleration during a time step). The single steps of the algorithm are as follows:

1. Choice of the parameters β, γ :

   $\beta = \frac{1}{4}, \ \gamma = \frac{1}{2}$ for the constant acceleration method or $\beta = \frac{1}{6}, \ \gamma = \frac{1}{2}$ for the linear acceleration method

2. Computing the constants c_1 to c_6:

$$c_1 = \frac{1}{\beta(\Delta t)^2}; \quad c_2 = \frac{1}{\beta \Delta t}; \quad c_3 = \frac{1}{2\beta}; \quad c_4 = \frac{\gamma}{\beta \Delta t}; \quad c_5 = \frac{\gamma}{\beta}; \quad c_6 = \Delta t\left(\frac{c_5}{2} - 1\right) \tag{3.7.7}$$

3. Evaluating the generalised stiffness k* and the loading F*:

$$k^* = k_{updated} + c_1 \cdot m + c_4 \cdot c \tag{3.7.8}$$

$$F^* = \Delta F + m(c_2 \dot{u}_1 + c_3 \ddot{u}_1) + c(c_5 \dot{u}_1 + c_6 \ddot{u}_1) \tag{3.7.9}$$

4. Computing the displacement increment $\Delta u = \frac{F^*}{k^*}$.

5. Determining the displacement, velocity and acceleration values at time t_2:

$$\begin{aligned} \ddot{u}_2 &= \ddot{u}_1 + c_1 \Delta u - c_2 \dot{u}_1 - c_3 \ddot{u}_1 \\ \dot{u}_2 &= \dot{u}_1 + c_4 \Delta u - c_5 \dot{u}_1 - c_6 \ddot{u}_1 \\ u_2 &= u_1 + \Delta u \end{aligned} \tag{3.7.10}$$

6. Determining the current restoring force $F_R(u)$ at time t_2 from the hysteretic law.
7. Computation of the unbalanced force R according to Eq. (3.7.3). If $|R|$ exceeds a given threshold, an iteration cycle is entered (Step 8), else the computation proceeds with step 3 by determining the next loading value F*.

8. A new generalised stiffness k^*_{new} is computed, based on the current system stiffness $k_{updated}$ at time t_2 :

$$k^*_{new} = k_{updated} + c_1 m + c_4 c \tag{3.7.11}$$

   This leads to a correction of the displacement increment:

$$\delta u = \frac{R}{k^*_{new}} \tag{3.7.12}$$

   The updated values for u_2, $\dot{u}_2$, $\ddot{u}_2$ are:

$$\begin{aligned} \ddot{u}_2 &= \ddot{u}_{2,old} + c_1 \cdot \delta u \\ \dot{u}_2 &= \dot{u}_{2,old} + c_4 \cdot \delta u \\ u_2 &= u_{2,old} + \delta u \end{aligned} \tag{3.7.13}$$

With these updated values a new unbalanced force R according to Eq. (3.7.3) is computed. If the convergence criterion is met, the computation proceeds with the next time step.

The computer program NLM evaluates time responses of SDOF systems with non-linear spring laws of elastoplastic, bilinear or UMEMURA type. The input file RHS contains the ordinates of the loading function $\frac{F(t)}{m}$ at constant time intervals (2E14.7 format with the time points in the first and the ordinates in the second column). All further data are input interactively following corresponding prompts. They refer to the type of the spring law employed, the circular natural frequency ω_1 and the damping of the SDOF system, the maximum elastic spring deformation u_{el}, the factor p for the spring stiffness reduction, and finally the number of loading function ordinates and the time step. Results are output in the file THNLM and consist of time histories of the displacement, velocity and acceleration values as well as the restoring force $F_R(t)$.

Program name	Input file	Output file
NLM	RHS	THNLM

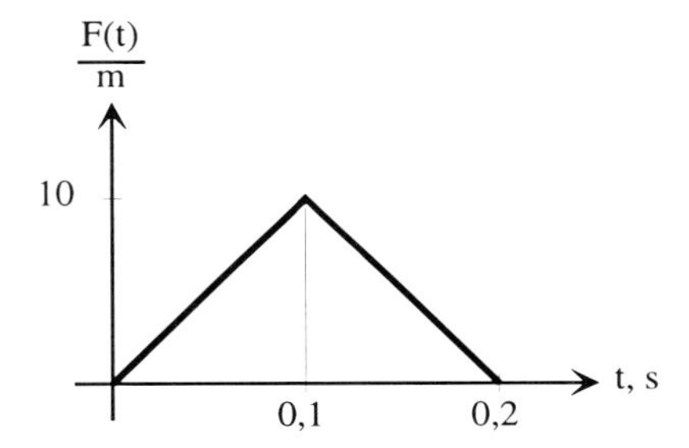

Fig. 3.7-4: Triangular impulse

As an example, a SDOF system with $m = 1\,t$, natural circular frequency $\omega_1 = 20\frac{rad}{s}$ and damping $\xi = 5\%$ is subjected to the triangular force impulse shown in Fig. 3.7-4 with $\max F = 10\ kN$. The maximum elastic spring deformation is assumed as u_{el} = 1.5 cm. The calculation of the system response for spring laws of the bilinear (p = 0.01) and UMEMURA type furnish the results shown in Figs. 3.7-5 and 3.7-6 for the time histories of the displacement and the restoring force. The period elongation inherent to the UMEMURA hysteretic law due to the decrease in stiffness is evident; Fig. 3.7-5 illustrates the residual permanent displacement typical of the bilinear law.

With the stiffness of the SDOF system equal to $k = m \cdot \omega_1^2 = 1 \cdot 400 = 400\frac{kN}{m}$, the static displacement for the peak load equals $u_{stat} = \frac{\max F}{k} = \frac{10}{400} = 0.025\,m$. For a linear system the peak dynamic displacement is computed as 0.0328 m, which corresponds to a restoring force of $400 \cdot 0.0328 = 13.13\ kN$. Limiting the elastic elongation to 0.015 m causes a reduction of the peak restoring force to $0.015 \cdot 400 = 6.0\ kN$ for the elastoplastic and the UMEMURA law, while the bilinear model with p = 1% leads to a small additional increase of the restoring force, as can be seen in Fig. 3.7-6. The reduction of the restoring force to about half of its linear value leads, however, to increased displacements. The peak displacements are 0.0485 m for the elastoplastic and the UMEMURA law and 0.0481m for the bilinear law, that is about double the values for the linear system. Fig. 3.7-7 shows the restoring force vs. displacement for the bilinear and the UMEMURA hysteretic law.

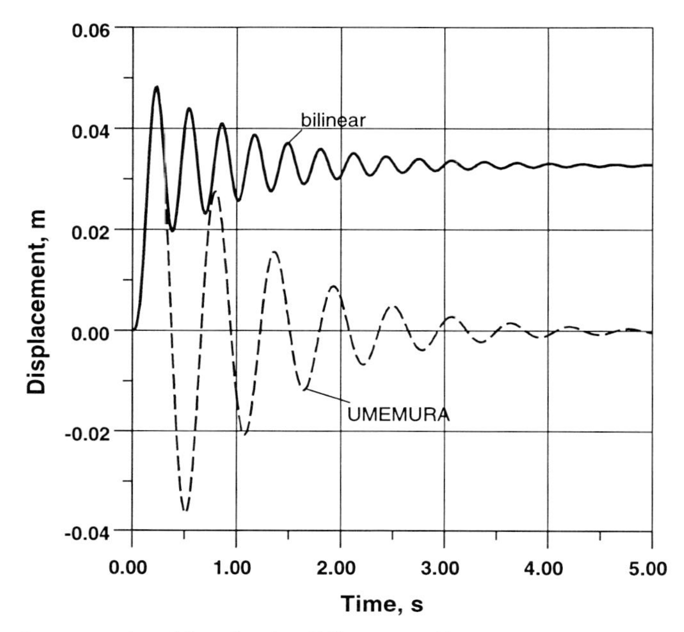

Fig. 3.7.5: Displacement time histories for different nonlinearity laws

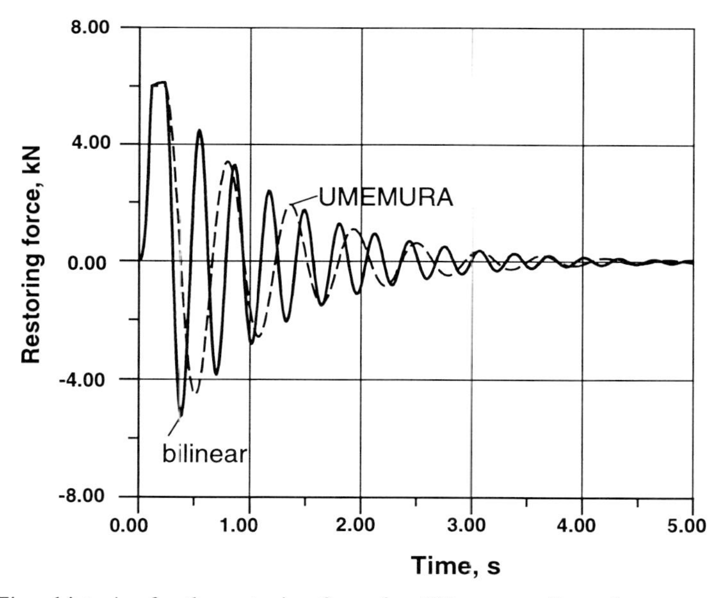

Fig. 3.7-6: Time histories for the restoring force for different nonlinear laws

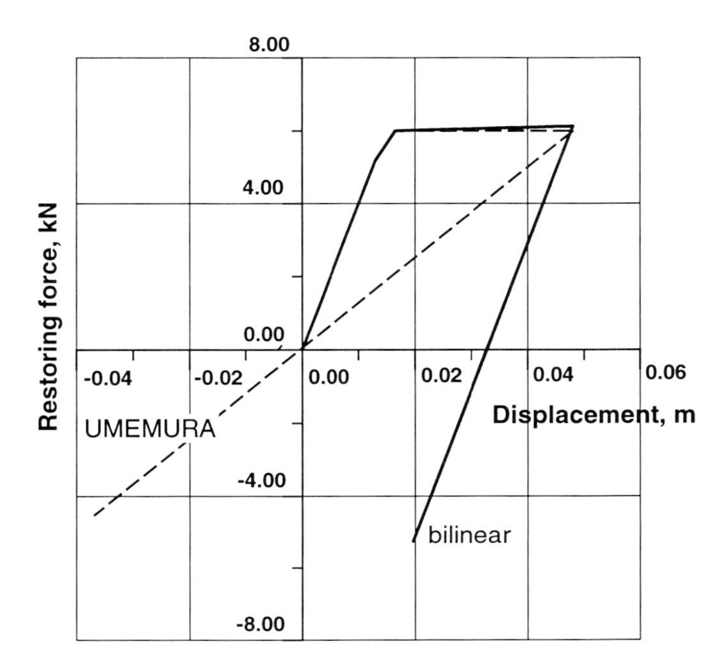

Fig. 3.7-7: Restoring force vs. displacement

4 Systems with several degrees of freedom

4.1 General

Some multi-degree-of-freedom (MDOF) systems have already been introduced in section 2.2 in connection with the derivation of their differential equations of motion with the help of d'ALEMBERT's principle. These systems consisted of discrete masses, springs and dashpots and the resulting differential equation system was of the form

$$\underline{F}_I + \underline{F}_D + \underline{F}_R = \underline{P} \tag{4. 1. 1}$$

In this equation $\underline{P}$ is the loading vector, $\underline{F}_I$ the vector of inertial forces, $\underline{F}_D$ the vector of damping forces and $\underline{F}_R$ the vector of restoring forces. Eq. (4.1.1) is completely analogous to the equilibrium condition of the SDOF system with vectors replacing the corresponding scalar quantities. The actual differential equation system expressing the equilibrium between inertia, damping and restoring forces and the external loading is given by

$$\underline{M}\,\ddot{\underline{V}} + \underline{C}\,\dot{\underline{V}} + \underline{K}\,\underline{V} = \underline{P} \tag{4. 1. 2}$$

The (n,1) column vector $\underline{V}$ contains the displacements in the n degrees of freedom of the MDOF system while $\dot{\underline{V}}, \ddot{\underline{V}}$ denote the corresponding velocities and accelerations. $\underline{K}$ is the stiffness matrix of the system, $\underline{M}$ its mass matrix and $\underline{C}$ the damping matrix, all matrices being square (n,n) matrices. For systems consisting of discrete springs, masses and (viscous) dashpots, as considered in Section 2.2, the derivation of these matrices is, at least in principle, a straightforward task. However, it is not immediately obvious how to arrive at a meaningful discretisation for real-life structures, such as beams, frames, plates or general 3-dimensional systems. Real-life structures possess distributed mass, stiffness and damping properties and an infinite number of degrees of freedom, which, however, can be expressed by a finite number of suitably chosen generalised co-ordinates. This can be carried out by a formal approach (e.g. by discretising the continuum through finite element formulations) or on a semi-empirical basis by introducing discrete point masses, stiffness matrices derived from analytical or experimental results and meaningful damping models. In the interest of reducing the work load associated with dynamic analyses is important to minimise the number n of the system's degrees of freedom. This can be done with the aid of condensation and substructuring techniques as described in Section 4.3.

In the following, emphasis is placed on MDOF systems with discrete point masses (so-called „lumped-mass" models) because of their versatility and wide practical applicability. For these models the mass matrix is a diagonal matrix and if the additional assumption of "modal damping" is introduced, the solution of Eq. (4.1.2) becomes quite straightforward. The next section deals with the general formulation of the equations of motions of viscoelastic continua and their displacement-based discretisation.

4.2 Fundamental equations and discretisation for viscoelastic continua

We consider the general case of a continuum with $\underline{p}(x^i,t)$ as external forces, $\underline{u}(x^i,t)$ as external displacements, $\underline{\sigma}(x^i,t)$ as internal forces (CAUCHY stresses), $\underline{\varepsilon}(x^i,t)$ as internal deformations (GREEN-LAGRANGE strains) and $\underline{f}(\rho, \underline{\ddot{u}})$ as d'ALEMBERT's inertia forces, with ρ as density. The dynamic equilibrium is given by [4.1]

$$-(\underline{p}^0 + \underline{f}) = \underline{D}_e \cdot \underline{\sigma} \tag{4. 2. 1}$$

where $\underline{p}^0$ are given loads and $\underline{D}_e$ the equilibrium operator. The kinematic equations are expressed as

$$\underline{\varepsilon} = \underline{D}_k \cdot \underline{u} \tag{4. 2. 2}$$

with the kinematic operator $\underline{D}_k$. Finally, the viscoelastic material law is given by

$$\underline{\sigma} = \underline{E}(\varepsilon, \dot{\varepsilon}, t) + \underline{D}(\varepsilon, \dot{\varepsilon}, t) \cdot \underline{\dot{\varepsilon}} \tag{4. 2. 3}$$

with the "elastic" matrix $\underline{E}$ and the "viscous" matrix $\underline{D}$. Introducing the boundary force variables $\underline{t}(x^i, t)$ and the boundary displacement variables $\underline{r}(x^i, t)$ leads to the following formulation of the boundary conditions:

$$\underline{t} = \underline{R}_t \cdot \underline{\sigma} \tag{4. 2. 4}$$

$$\underline{r} = \underline{R}_r \cdot \underline{u} \tag{4. 2. 5}$$

$\underline{R}_t$ and $\underline{R}_r$ are the corresponding operators along the pertinent boundaries C_t and C_r of the domain F. The principle of virtual displacements can be expressed in the following form for the viscoelastic material law Eq. (4.2.3):

$$\int_F \rho^* \, \underline{\ddot{u}}^T \cdot \delta\underline{u} \, dF + \int_F \underline{\dot{\varepsilon}}^T \cdot \underline{D} \cdot \underline{\dot{\varepsilon}} \, dF + \int_F \underline{\varepsilon}^T \cdot \underline{E} \cdot \underline{\varepsilon} \, dF - \int_F \underline{p}^{0T} \cdot \delta\underline{u} \, dF - \int_{C_t} \underline{t}^{0T} \cdot \delta\underline{r} \, dC = 0 \tag{4. 2. 6}$$

Here, ρ^* is the density function in the domain F.

Numerical analyses are usually carried out using finite element methods, preferably in their displacement (strain) formulation. To this end, the domain is divided into finite elements and the displacement field $\underline{u}^p$ of the pth element is approximated by

$$\underline{u}^p = \underline{\varphi}^p \, \underline{\hat{u}}^p \tag{4. 2. 7}$$

The matrix $\underline{\varphi}^p$ contains suitable functions, usually low-order polynomials, and the vector $\underline{\hat{u}}^p$ the corresponding coefficients. Introducing the co-ordinates of the element nodes in Eq. (4.2.7) yields the element nodal displacements $\underline{v}^p$ as element degrees of freedom:

$$\underline{v}^p = \underline{\hat{\varphi}}^p \, \underline{\hat{u}}^p, \quad \underline{\hat{u}}^p = \left(\underline{\hat{\varphi}}^p\right)^{-1} \underline{v}^p \tag{4. 2. 8}$$

Inserting Eq. (4.2.8) into Eq. (4.2.7) leads to

$$\underline{u}^p = \underline{\varphi}^p \left(\underline{\hat{\varphi}}^p\right)^{-1} \underline{v}^p = \underline{\Omega}^p \, \underline{v}^p \tag{4. 2. 9}$$

where the matrix $\underline{\Omega}^p$ contains the shape functions of the pth element. Introducing Eq. (4.2.9) into Eq. (4.2.6) finally yields the discretised principle of virtual displacements in the form

$$\delta \underline{v}^{pT} \cdot \left(\underline{m}^p \, \underline{\ddot{v}}^p + \underline{c}^p \, \underline{\dot{v}}^p + \underline{k}^p \, \underline{v}^p - \underline{p}^p\right) = 0 \tag{4. 2. 10}$$

with the abbreviations

$$\underline{m}^p = \int_{F^p} \rho * \underline{\Omega}^{pT} \underline{\Omega}^p \, dF^p \tag{4. 2. 11}$$

$$\underline{c}^p = \int_{F^p} \underline{\Omega}^{pT} \underline{D}_k{}^T \underline{D} \, \underline{D}_k \; \underline{\Omega}^p \, dF^p \tag{4. 2. 12}$$

$$\underline{k}^p = \int_{F^p} \underline{\Omega}^{pT} \underline{D}_k{}^T \underline{E} \, \underline{D}_k \; \underline{\Omega}^p \, dF^p \tag{4. 2. 13}$$

$$\underline{p}^p = \int_{F^p} \underline{\Omega}^{pT} \underline{p}^0 \; dF^p + \int_{C_t{}^p} \underline{\Omega}^{pT} \underline{R}_r{}^T \underline{t}^0 \; dC_t{}^p \tag{4. 2. 14}$$

These are, in succession, the „consistent" element mass matrix, the viscous element damping matrix, the element stiffness matrix and finally the element load vector. It should be noted that the damping matrix is not usually determined through Eq. (4.2.12) but by assuming plausible damping values for the whole structure, as explained later.

As an example, Fig. 4.2-1 depicts a plane straight beam element with constant mass and stiffness distribution.

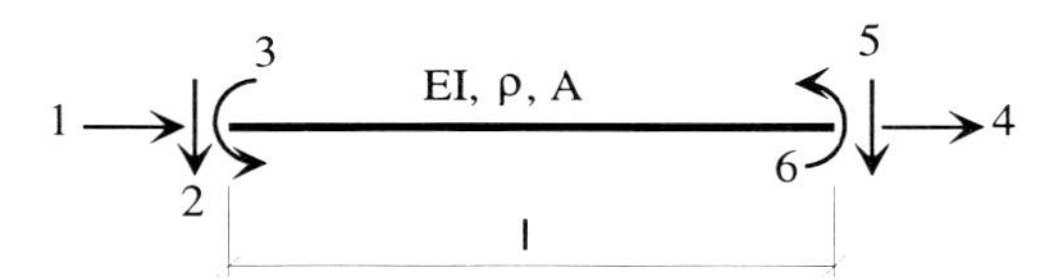

Fig. 4.2-1: Plane beam element with its degrees of freedom

Its consistent mass and stiffness matrices in the six degrees of freedom of Fig. 4.2-1 (neglecting shear effects) are given by:

$$\underline{m} = \frac{\rho A \ell}{420} \begin{bmatrix} 140 & & & & & \\ 0 & 156 & & \text{symm.} & & \\ 0 & -22\ell & 4\ell^2 & & & \\ 70 & 0 & 0 & 140 & & \\ 0 & 54 & -13\ell & 0 & 156 & \\ 0 & 13\ell & -3\ell^2 & 0 & 22\ell & 4\ell^2 \end{bmatrix} \tag{4. 2. 15}$$

$$\underline{k} = \frac{EI}{\ell^3} \begin{bmatrix} \frac{A\ell^2}{I} & & & & & \\ 0 & 12 & & \text{symm.} & & \\ 0 & -6\ell & 4\ell^2 & & & \\ -\frac{A\ell^2}{I} & 0 & 0 & \frac{A\ell^2}{I} & & \\ 0 & -12 & 6\ell & 0 & 12 & \\ 0 & -6\ell & 2\ell^2 & 0 & 6\ell & 4\ell^2 \end{bmatrix} \tag{4. 2. 16}$$

From Eq. (4.2.10) the differential equation of motion for a single finite element is given by

$$\underline{m}^p \underline{\ddot{v}}^p + \underline{c}^p \underline{\dot{v}}^p + \underline{k}^p \underline{v}^p = \underline{p}^p \tag{4. 2. 17}$$

The system degrees of freedom are contained in the vector $\underline{V}$, with the kinematic transformation matrix $\underline{a}^p$ combining element and system degrees of freedom:

$$\underline{v}^p = \underline{a}^p \underline{V} \tag{4. 2. 18}$$

This leads finally to the equation of motion of the whole system

$$\underline{M}\underline{\ddot{V}} + \underline{C}\underline{\dot{V}} + \underline{K}\underline{V} = \underline{P} \tag{4. 2. 19}$$

where the matrices $\underline{M}$, $\underline{C}$ and $\underline{K}$ and the loading vector $\underline{P}$ are given by

$$\underline{M} = \sum_{p=1}^{n} \underline{a}^{pT} \underline{m}^p \underline{a}^p \tag{4. 2. 20}$$

$$\underline{C} = \sum_{p=1}^{n} \underline{a}^{pT} \underline{c}^p \underline{a}^p \tag{4. 2. 21}$$

$$\underline{K} = \sum_{p=1}^{n} \underline{a}^{pT} \underline{k}^p \underline{a}^p \tag{4. 2. 22}$$

$$\underline{P} = \sum_{p=1}^{n} \underline{a}^{pT} \underline{p}^p \tag{4. 2. 23}$$

Computing these matrices does not actually involve the setting up of (sparse) kinematic transformation matrices $\underline{A}$ and carrying out the multiplication shown above; it is much simpler to directly add the contributions of each new element from their element matrices (expressed in global co-ordinates) to the corresponding matrices (e.g. $\underline{K}$) for the whole structure. To illustrate, the system shown in Fig. 4.2-2 is considered, consisting of two beam elements. Both element and system degrees of freedom are given, without yet considering the support conditions.

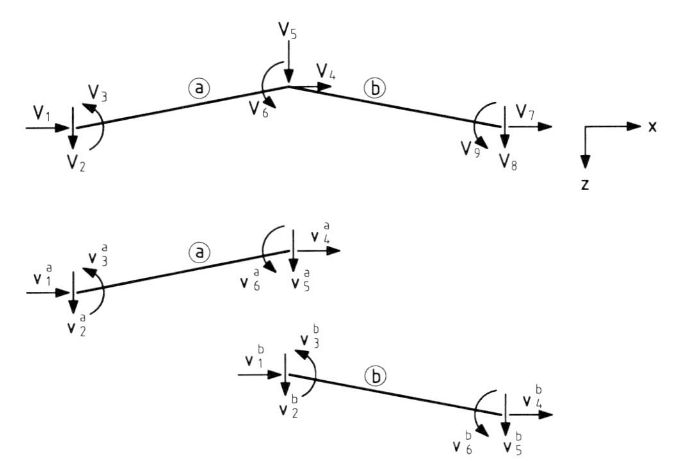

Fig. 4.2-2: Element and system degrees of freedom in the global co-ordinate system

The relationship $\underline{v} = \underline{a}\,\underline{V}$ is given by:

$$\begin{bmatrix} v_1^a \\ v_2^a \\ v_3^a \\ v_4^a \\ v_5^a \\ v_6^a \\ v_1^b \\ v_2^b \\ v_3^b \\ v_4^b \\ v_5^b \\ v_6^b \end{bmatrix} = \begin{bmatrix} 1 & 0 & 0 & 0 & 0 & 0 & 0 & 0 & 0 \\ 0 & 1 & 0 & 0 & 0 & 0 & 0 & 0 & 0 \\ 0 & 0 & 1 & 0 & 0 & 0 & 0 & 0 & 0 \\ 0 & 0 & 0 & 1 & 0 & 0 & 0 & 0 & 0 \\ 0 & 0 & 0 & 0 & 1 & 0 & 0 & 0 & 0 \\ 0 & 0 & 0 & 0 & 0 & 1 & 0 & 0 & 0 \\ 0 & 0 & 0 & 1 & 0 & 0 & 0 & 0 & 0 \\ 0 & 0 & 0 & 0 & 1 & 0 & 0 & 0 & 0 \\ 0 & 0 & 0 & 0 & 0 & 1 & 0 & 0 & 0 \\ 0 & 0 & 0 & 0 & 0 & 0 & 1 & 0 & 0 \\ 0 & 0 & 0 & 0 & 0 & 0 & 0 & 1 & 0 \\ 0 & 0 & 0 & 0 & 0 & 0 & 0 & 0 & 1 \end{bmatrix} \begin{bmatrix} V_1 \\ V_2 \\ V_3 \\ V_4 \\ V_5 \\ V_6 \\ V_7 \\ V_8 \\ V_9 \end{bmatrix} \qquad (4.2.24)$$

It can be seen that the coefficient at the intersection of row i and column j is equal to 1 if the element degree of freedom i is identical with the system degree of freedom j, else it is zero. In other words, $\underline{a}$ expresses a "Yes/No" relationship between element and system degrees of freedom which can be given much more succinctly by a so-called "incidence matrix". The matrix products of Eq. (4.2.22) do not have to be actually carried out and the transformation matrix $\underline{a}$ does not even have to be set up explicitly, if the information needed for the correct adding of the single contributions of the element matrices to the system matrices is given by the incidence matrix. This also permits a simple consideration of the boundary and support conditions since only active kinematic degrees of freedom are introduced. The overall system stiffness matrix $\underline{K}$ of a stable structure is positive definite while it is singular if rigid-body movements are possible (that is, before support conditions have been taken into account).

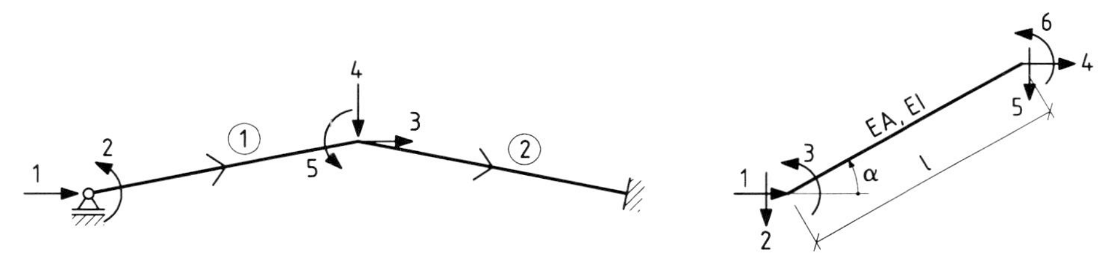

Fig. 4.2-3: Two-element system

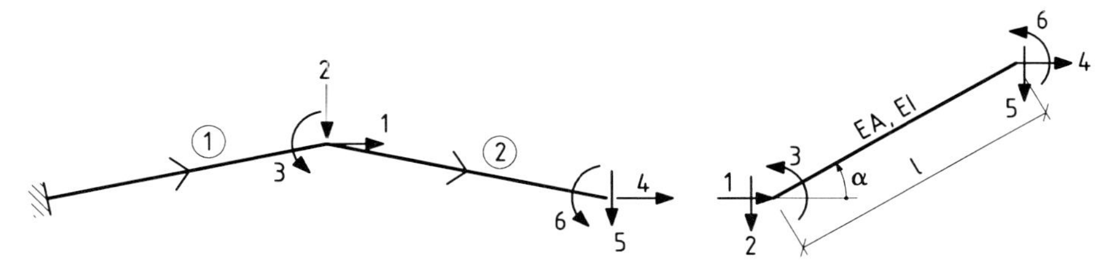

Fig. 4.2-4: Another two-element system

The setting up of an incidence matrix is illustrated by means of the structure depicted in Fig. 4.2-3 which corresponds to the system of Fig. 4.2-2 with added supports. The number of rows of the incidence matrix is generally equal to the number of (beam) elements while its number of columns equals the total number of the element degrees of freedom (e.g. six for a plane beam element). The integer coefficient at the intersection of row i and column j of the incidence matrix is the number of the system degree of freedom corresponding to the jth element degree of freedom of the ith element. For the system of Fig. 4.2-3 we obtain the incidence matrix:

$$\begin{pmatrix} 1 & 0 & 2 & 3 & 4 & 5 \\ 3 & 4 & 5 & 0 & 0 & 0 \end{pmatrix} \tag{4. 2. 25}$$

In the case of the system of Fig. 4.2-4 the incidence matrix is:

$$\begin{pmatrix} 0 & 0 & 0 & 1 & 2 & 3 \\ 1 & 2 & 3 & 4 & 5 & 6 \end{pmatrix} \tag{4. 2. 26}$$

The ith row of an incidence matrix is also called the incidence vector of the ith element. This incidence vector may be written down in an extra column and row to the left and over the corresponding element stiffness matrix, as shown in Eq. (4.2.27) for element 1 of the system depicted in Fig. 4.2-4:

$$\begin{array}{c} \\ 0 \\ 0 \\ 0 \\ 1 \\ 2 \\ 3 \end{array} \begin{array}{c} \begin{matrix} 0 & \;\;0 & \;\;0 & \;\;1 & \;\;2 & \;\;3 \end{matrix} \\ \begin{bmatrix} k_{11} & k_{12} & k_{13} & k_{14} & k_{15} & k_{16} \\ k_{21} & k_{22} & k_{23} & k_{24} & k_{25} & k_{26} \\ k_{31} & k_{32} & k_{33} & k_{34} & k_{35} & k_{36} \\ k_{41} & k_{42} & k_{43} & k_{44} & k_{45} & k_{46} \\ k_{51} & k_{52} & k_{53} & k_{54} & k_{55} & k_{56} \\ k_{61} & k_{62} & k_{63} & k_{64} & k_{65} & k_{66} \end{bmatrix} \end{array} \tag{4. 2. 27}$$

The coefficients k_{ij} of the element stiffness matrix must be added to the coefficients already present in the system stiffness matrix $\underline{K}$ at the positions given by the row and column numbers of the incidence vector. As an example, the coefficient k_{45} in Eq. (4.2.27) must be added at the intersection of the first row and the second column of $\underline{K}$, while the coefficient k_{65} is added at position (3,2). The total contribution of the element No. 1 of example Fig. 4.2-4 to the stiffness matrix $\underline{K}$ is given by:

$$\underline{K}^1 = \begin{bmatrix} k_{44} & k_{45} & k_{46} & 0 & 0 & 0 \\ k_{54} & k_{55} & k_{56} & 0 & 0 & 0 \\ k_{64} & k_{65} & k_{66} & 0 & 0 & 0 \\ 0 & 0 & 0 & 0 & 0 & 0 \\ 0 & 0 & 0 & 0 & 0 & 0 \\ 0 & 0 & 0 & 0 & 0 & 0 \end{bmatrix} \tag{4. 2. 28}$$

As an additional example we consider the contribution of element no. 1 of the structure shown in Fig. 4.2-3 to its (5,5) stiffness matrix $\underline{K}$. The result is:

$$\underline{K}^1 = \begin{bmatrix} k_{11} & k_{13} & k_{14} & k_{15} & k_{16} \\ k_{31} & k_{33} & k_{34} & k_{35} & k_{36} \\ k_{41} & k_{43} & k_{44} & k_{45} & k_{46} \\ k_{51} & k_{53} & k_{54} & k_{55} & k_{56} \\ k_{61} & k_{63} & k_{64} & k_{65} & k_{66} \end{bmatrix} \tag{4. 2. 29}$$

Based on the information contained in the incidence matrix, an automatic addition of the coefficients of the single element matrices to the system matrix may be carried out by a program subroutine. More information on that point is given in Section 4.4, where the important case of a lumped-mass multi-degree-of-freedom system is discussed.

The system matrices $\underline{M}$, $\underline{K}$ and $\underline{C}$ are typically sparse, and this property should be put to use in the course of the electronic computation in order to save CPU time and storage space. The coefficients of a symmetric band matrix may be stored, for example, as shown in Fig. 4.2-5 showing a possible storage sequence of the coefficients of a symmetric (10,10) matrix. A "half-bandwidth" of 3 has been assumed, this value being half the actual bandwidth without considering the coefficient on the matrix diagonal. The so-defined "half-bandwidth" is equal to the maximum difference of the of the numbers of the system degrees of freedom connected by the element in question which are present in the pertinent incidence vector (or row of the incidence matrix). To illustrate, the "half-bandwidth" for the structure shown in Fig. 4.2-3 is equal to 4 (=5-1 from element 1), for the structure shown in Fig. 4.2-4 it equals 5 (= 6-1 from element 2).

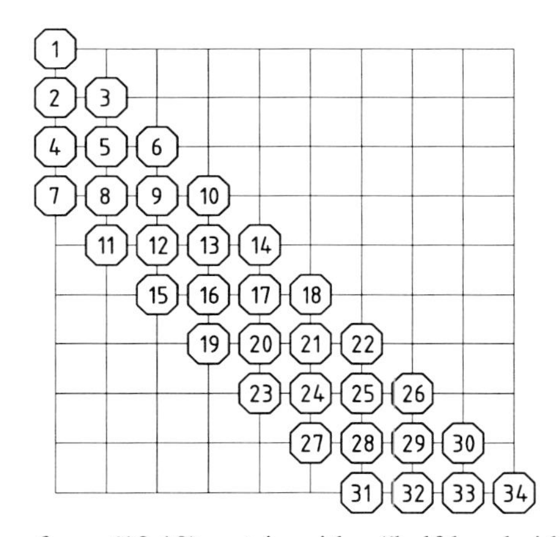

Fig. 4.2-5: Storage scheme for a (10,10) matrix with a "half-bandwidth" of 3

The computation of displacements and internal forces of an arbitrary plane frame with or without spring supports subject to (static) loading at its nodes only can be carried out by the program RAHMEN :

Program name	Input files	Output file
RAHMEN	ERAHM INZFED FEDMAT	ARAHM

The input file ERAHM contains information on the geometrical and mechanical properties of the structure, followed by its incidence matrix and the load vector. The program also permits the coupling of degrees of freedom through generalised spring matrices $\underline{k}_{Fed}$, which are typically of the form

$$\underline{k}_{Fed} = \begin{pmatrix} k_{11} & \cdot & \cdot & \cdot & k_{1n} \\ \cdot & & & & \cdot \\ \cdot & & & & \cdot \\ \cdot & & & & \cdot \\ k_{n1} & \cdot & \cdot & \cdot & k_{nn} \end{pmatrix} \tag{4. 2. 30}$$

These matrices are supplied by the input file FEDMAT and the corresponding incidence vector (giving the numbers of the system degrees of freedom coupled by this stiffness matrix) is read from the input file INZFED. This also permits consideration of (rotational or longitudinal) spring supports, by coupling the pertinent system degrees of freedom to the rigid earth disc (with 0 as corresponding degree of freedom). More than one (NFED) spring matrix of arbitrary dimensions may be considered; the matrices themselves and their incidence vectors are entered in the same (arbitrary) sequence in the input files FEDMAT and INZFED.

The output file ARAHM contains the displacements and internal forces of all beam end sections, referred to the global (x,z) co-ordinate system, in the following order: Horizontal component (displacement or force), vertical component (displacement or force), and finally the rotation or bending moment values for end 1 and end 2 of the beam element. As an example, the structure shown in Fig. 4.2-3 is considered, subject to a vertical load P = 10 kN corresponding to the system degree of freedom no. 4. Both beam elements have a bending stiffness of EI = 20 000 kNm^2, an axial stiffness EA=100 000 kN, lengths of 2.50 m and they are inclined by 12° with reference to the horizontal axis. We have two beam elements (NELEM=2) and 5 active kinematic degrees of freedom (NDOF=5); no spring matrices are considered as yet (NFED=0). These values (NDOF, NELEM, NFED) must be input interactively during the session at the computer terminal.

The input file ERAHM contains first NELEM rows with the values of EI, ℓ, EA and α for every beam element. These values are the bending stiffness (e.g. in kNm^2), the element length (e.g. in m), the axial stiffness (e.g. in kN) and the angle (in degrees, positive counter-clockwise) between the global horizontal x-axis and the beam axis. These values are entered format-free as follows:

```
20000., 2.5, 100000.,      12.0
20000., 2.5, 100000.,     -12.0
```

Next, the incidence matrix is entered in ERAHM. It consists of NELEM rows with 6 integer values in each, expressing the numbers of the system degrees of freedom present at the end sections of the beam element in question (see Eq. 4.2-25):

```
1,0,2, 3,4,5
3,4,5, 0,0,0
```

Finally, the load vector must be input in ERAHM, consisting of NDOF values which correspond to the NDOF active kinematic system degrees of freedom:

```
0.0, 0.0, 0.0, 10.0, 0.0
```

The output file ARAHM is as follows:

```
ELEMENT NO.          1
-.1812E-03   .0000E+00  -.3833E-03  -.8064E-04   .5512E-03   .9459E-04
-.7109E-15  -.3127E+01  -.3162E-14   .7109E-15   .3127E+01   .7647E+01
ELEMENT NO.          2
-.8064E-04   .5512E-03   .9459E-04   .0000E+00   .0000E+00   .0000E+00
-.3993E-15   .6873E+01  -.7647E+01   .3993E-15  -.6873E+01  -.9160E+01
```

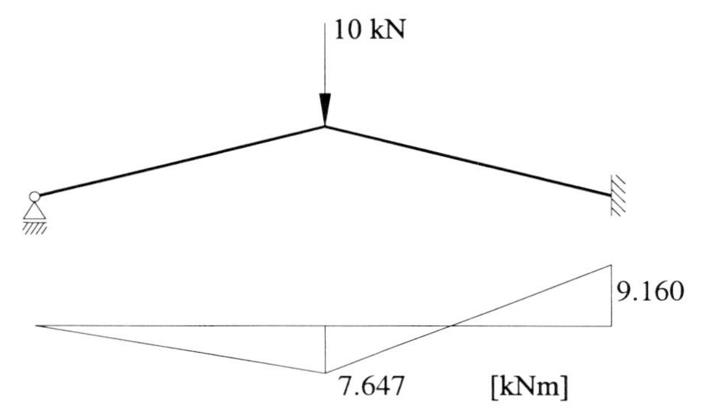

Fig. 4.2-6: Bending moment diagram

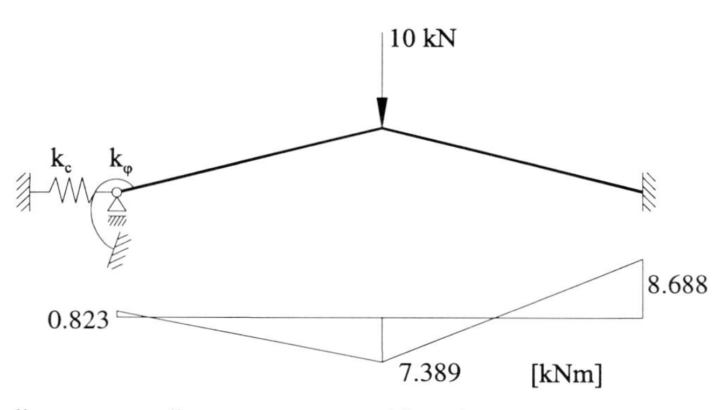

Fig. 4.2-7: Bending moment diagram, structure with spring supports

The bending moment diagram is depicted in Fig. 4.2-6. If additional springs are introduced at the left support as shown in Fig. 4.2-7 with constants equal to k_c = 1000 kN/m and k_φ = 2500 kNm/rad, the resulting internal forces and displacements are given by

```
ELEMENT NO.          1
  -.1586E-03   .0000E+00  -.3291E-03  -.7076E-04   .5168E-03   .8122E-04
   .1586E+00  -.3392E+01   .8229E+00  -.1586E+00   .3392E+01   .7389E+01
  ELEMENT NO.          2
  -.7076E-04   .5168E-03   .8122E-04   .0000E+00   .0000E+00   .0000E+00
   .1586E+00   .6608E+01  -.7389E+01  -.1586E+00  -.6608E+01  -.8688E+01
```

and the corresponding bending moment diagram is as shown in Fig. 4.2-7. The input file FEDMAT in this case is given by

```
1000.0, -1000.0, -1000.0, 1000.0
2500.0, -2500.0, -2500.0, 2500.0
```

with the spring matrices

$$\underline{k}_{long} = \begin{bmatrix} k_c & -k_c \\ -k_c & k_c \end{bmatrix}, \ \underline{k}_{rot} = \begin{bmatrix} k_\varphi & -k_\varphi \\ -k_\varphi & k_\varphi \end{bmatrix} \qquad (4.\,2.\,31)$$

and, in the input file INZFED, the two incidence vectors

```
0,1
0,2
```

The program VOLLST may be used in order to compute the stiffness matrix and the consistent mass matrix of an arbitrary plane frame. Its input and output files are given below:

Program name	Input files	Output files
VOLLST	ERAHM EMAS INZFED FEDMAT	KVOLL MVOLL

The program input consists of the already discussed files ERAHM, INZFED and FEDMAT (as for the program RAHMEN) and an additional input file, EMAS, which contains the (constant) mass per unit length value for all NELEM beam elements (preferably in t/m). The output files KVOLL and MVOLL contain the quadratic (NDOF, NDOF) stiffness and mass matrix respectively.

4.3 Substructuring techniques, static condensation

In structural dynamics it is very important that in the differential equation system

$$\underline{M}\,\ddot{\underline{V}} + \underline{C}\,\dot{\underline{V}} + \underline{K}\,\underline{V} = \underline{P} \qquad (4.\,3.\,1)$$

the size of the matrices, that is the number of components of the vectors $\underline{V}$, $\dot{\underline{V}}$ and $\ddot{\underline{V}}$, is as small as possible, in order to minimise the overall effort. One way to achieve this is to consider only the degrees of freedom associated with large, or at any rate substantial inertia forces, which are denoted as „master“ degrees of freedom. This does not mean that the displacements, velocities and accelerations in the remaining „slave“ degrees of freedom are neglected but simply that they are given as functions of the corresponding values of the master degrees of freedom. The concept of expressing a group of degrees of freedom as functions of other degrees of freedom is also quite useful when different parts of a system (substructures) are being investigated separately before being integrated in the overall structure. In this case the degrees of freedom connecting the different substructures must be considered as master degrees of freedom; the substructure itself can be regarded as a „macroelement“ in these master degrees of freedom which express the contribution of the particular substructure to the whole assemblage. Formally, the first step lies in distinguishing between master and slave degrees of freedom, with the latter, $\underline{V}_{\varphi}$, to be expressed as functions of the former, $\underline{V}_u$. The corresponding partitioning of the differential equation system leads to the following equation:

$$\begin{pmatrix} \underline{M}_{uu} & \underline{M}_{u\varphi} \\ \underline{M}_{\varphi u} & \underline{M}_{\varphi\varphi} \end{pmatrix} \cdot \begin{pmatrix} \ddot{\underline{V}}_u \\ \ddot{\underline{V}}_{\varphi} \end{pmatrix} + \begin{pmatrix} \underline{K}_{uu} & \underline{K}_{u\varphi} \\ \underline{K}_{\varphi u} & \underline{K}_{\varphi\varphi} \end{pmatrix} \cdot \begin{pmatrix} \underline{V}_u \\ \underline{V}_{\varphi} \end{pmatrix} = \begin{pmatrix} \underline{P}_u \\ \underline{P}_{\varphi} \end{pmatrix} \tag{4. 3. 2}$$

So far, damping has not been considered; it will be introduced later as „proportional damping“.

Expressing the slave degrees of freedom by the master degrees of freedom corresponds to a linear transformation of the form

$$\begin{aligned} \underline{V}_{\varphi} &= \underline{a}\,\underline{V}_u , \\ \dot{\underline{V}}_{\varphi} &= \underline{a}\,\dot{\underline{V}}_u , \\ \ddot{\underline{V}}_{\varphi} &= \underline{a}\,\ddot{\underline{V}}_u \end{aligned} \tag{4. 3. 3}$$

This leads to

$$\underline{V} = \begin{bmatrix} \underline{V}_u \\ \underline{V}_{\varphi} \end{bmatrix} = \begin{bmatrix} \underline{V}_u \\ \underline{a}\,\underline{V}_u \end{bmatrix} = \begin{bmatrix} \underline{I} \\ \underline{a} \end{bmatrix} \underline{V}_u = \underline{A}\,\underline{V}_u \tag{4. 3. 4}$$

with similar expressions for the velocity and acceleration vectors $\dot{\underline{V}}, \ddot{\underline{V}}$. The original system of differential equations (still without damping) is reduced to

$$\begin{aligned} &\underline{M}\,\underline{A}\,\ddot{\underline{V}}_u + \underline{K}\,\underline{A}\,\underline{V}_u = \underline{P} \\ &\underline{A}^T\underline{M}\,\underline{A}\,\ddot{\underline{V}}_u + \underline{A}^T\,\underline{K}\,\underline{A}\,\underline{V}_u = \underline{A}^T\underline{P} \end{aligned} \tag{4. 3. 5}$$

or

$$\tilde{\underline{M}}\,\ddot{\underline{V}}_u + \tilde{\underline{K}}\,\underline{V}_u = \tilde{\underline{P}} \tag{4. 3. 6}$$

with

$$\begin{aligned} \tilde{\underline{M}} &= \underline{A}^T\underline{M}\,\underline{A} \\ \tilde{\underline{K}} &= \underline{A}^T\,\underline{K}\,\underline{A} \\ \tilde{\underline{P}} &= \underline{A}^T\underline{P} \end{aligned} \tag{4. 3. 7}$$

The matrix $\tilde{\underline{K}}$ is the condensed or reduced stiffness matrix in the master degrees of freedom and $\tilde{\underline{P}}$ is the corresponding reduced load vector. Once the vectors $\underline{V}_u, \dot{\underline{V}}_u, \ddot{\underline{V}}_u$ have been determined, the remaining variables in the slave degrees of freedom can be readily computed through Eq. (4.3.3).

In the case of the well-known „static condensation" after GUYAN [4.2] and IRONS [4.3], static equations are used for eliminating the slave degrees of freedom, that is, expressing them as functions of a subset of master degrees of freedom. Writing down the second row of Eq. (4.3.2) and noting that $\underline{K}^T{}_{u\varphi} = \underline{K}_{\varphi u}$ leads to

$$\underline{P}_\varphi = \underline{M}_{\varphi u}\, \ddot{\underline{V}}_u + \underline{M}_{\varphi\varphi}\, \ddot{\underline{V}}_\varphi + \underline{K}^T_{u\varphi}\, \underline{V}_u + \underline{K}_{\varphi\varphi}\, \underline{V}_\varphi \tag{4.3.8}$$

Having assumed that the inertia forces in the slave degrees of freedom are small, it is permissible to retain only the „static" part in this expression:

$$\begin{aligned} \underline{P}_\varphi &= \underline{K}^T_{u\varphi}\, \underline{V}_u + \underline{K}_{\varphi\varphi}\, \underline{V}_\varphi \\ \underline{V}_\varphi &= \underline{K}^{-1}_{\varphi\varphi}\,(\underline{P}_\varphi - \underline{K}^T_{u\varphi}\, \underline{V}_u) \end{aligned} \tag{4.3.9}$$

A further restriction being that no external forces are allowed to act in the direction of slave degrees of freedom, $\underline{P}_\varphi = 0$, (which simply means that all degrees of freedom with associated load components must be considered as master degrees of freedom), we obtain

$$\underline{V}_\varphi = -\underline{K}^{-1}_{\varphi\varphi}\, \underline{K}^T_{u\varphi}\, \underline{V}_u \tag{4.3.10}$$

This further leads to

$$\underline{V} = \begin{bmatrix} \underline{V}_u \\ \underline{V}_\varphi \end{bmatrix} = \begin{bmatrix} \underline{I} \\ -\underline{K}^{-1}_{\varphi\varphi}\, \underline{K}^T_{u\varphi} \end{bmatrix} \underline{V}_u = \underline{A}\, \underline{V}_u \tag{4.3.11}$$

The reduced (or condensed) mass and stiffness matrices are given by

$$\tilde{\underline{M}} = \underline{M}_{uu} - \underline{K}_{u\varphi}\, \underline{K}^{-1}_{\varphi\varphi}\, \underline{M}^T_{u\varphi} - \underline{M}_{u\varphi}\, \underline{K}^{-1}_{\varphi\varphi}\, \underline{K}^T_{u\varphi} + \underline{K}_{u\varphi}\, \underline{K}^{-1}_{\varphi\varphi}\, \underline{M}_{\varphi\varphi}\, \underline{K}^{-1}_{\varphi\varphi}\, \underline{K}^T_{u\varphi} \tag{4.3.12}$$

$$\tilde{\underline{K}} = \underline{K}_{uu} - \underline{K}_{u\varphi}\, \underline{K}^{-1}_{\varphi\varphi}\, \underline{K}^T_{u\varphi} \tag{4.3.13}$$

while the loading vector is simply

$$\tilde{\underline{P}} = \underline{P}_u \tag{4.3.14}$$

A very elegant method for obtaining the condensed stiffness matrix for $\underline{P}_\varphi = 0$ is the following: Unit loads corresponding to the n master degrees of freedom are applied sequentially and the resulting deformations saved as columns of the (n,n) flexibility matrix of the structure. Its inverse can be readily computed and is equal to the condensed stiffness matrix $\tilde{\underline{K}}$.

The numerical determination of the condensed stiffness matrix and the corresponding transformation matrix A (see Eq. 4.3.11) for arbitrary plane frames is carried out by the program KONDEN:

Program name	Input files	Output files
KONDEN	EKOND INZFED FEDMAT	KMATR AMAT

The input file EKOND is almost the same as the input file ERAHM of the program RAHMEN, the only difference being that the numbers of the NDU master degrees of freedom are input instead of the loading vector. The output file KMATR contains the condensed (ndu,ndu) stiffness matrix while the corresponding transformation matrix A is output in the file AMAT.

The structure depicted in Fig. 4.3-1 is analysed as an example. The upper horizontal beam (divided into five elements of equal length) is regarded as substructure A and the problem is to determine its condensed stiffness matrix in terms of the four vertical degrees of freedom acting at the nodes. While it is no problem to determine this matrix directly for this simple example, a step-by-step solution will be carried out in order to demonstrate the application of the substructuring concept. Here the supporting structure shown in Fig. 4.3-2 is regarded as substructure B.

The first step consists in discretising the system by introducing elements, nodes and, most important, the present active kinematic degrees of freedom. The chosen discretisation for substructure B is shown in Fig. 4.3-3; the arrows serve to distinguish end 1 from end 2 of the single beam elements. Please note that an additional rotational degree of freedom must be introduced in order to simulate a hinge.

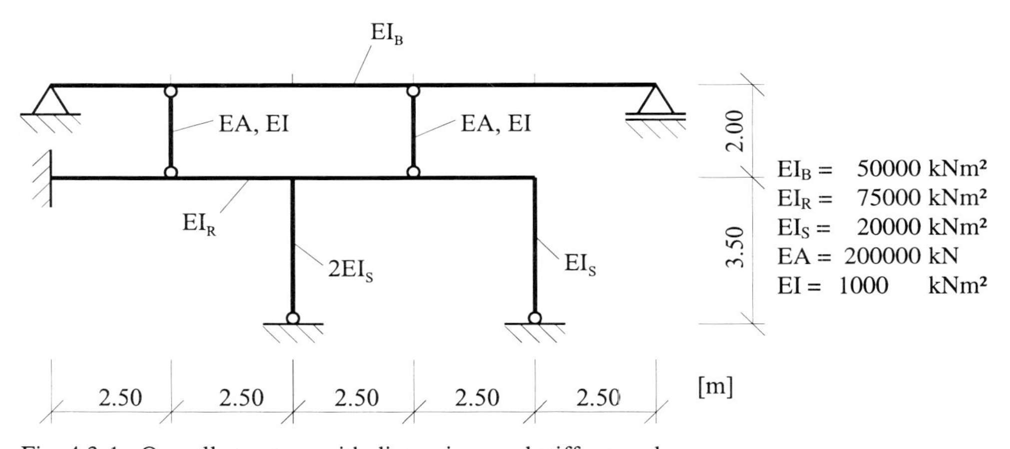

Fig. 4.3-1: Overall structure with dimensions and stiffness values

The degrees of freedom with numbers 1 and 3 are coupling the substructures A and B and are regarded as master degrees of freedom for substructure B (NDU = 2). The substructure consists of 8 beam elements (NELEM = 8) and has 14 active kinematic degrees of freedom (NDOF = 14) while no additional springs are present (NFED = 0). These values must be input interactively while running the program.

In the input file EKOND the first NELEM rows contain the parameters EI, ℓ, EA, α for each beam element. For simplicity's sake, the axial stiffness EA is input as zero for the beam elements 3 to 8, where no corresponding active kinematic degrees of freedom have been introduced so that their axial stiffness is, in effect, infinite.

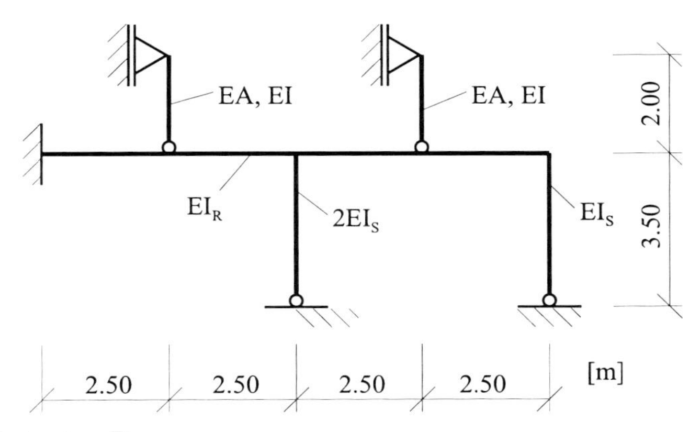

Fig. 4.3-2: Substructure B

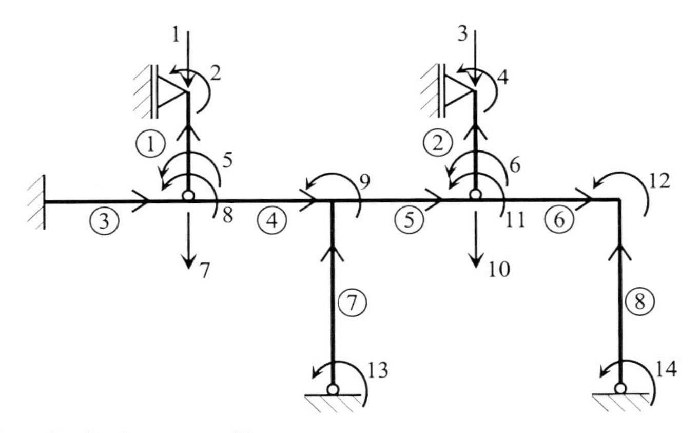

Fig. 4.3-3: Discretised substructure B

```
1000., 2.,200000., 90.0
1000., 2.,200000., 90.0
75000.,2.5, 0., 0.,
75000.,2.5, 0., 0.,
75000.,2.5, 0., 0.,
75000.,2.5, 0., 0.,
40000.,3.5, 0., 90.
20000.,3.5, 0., 90.
```

Next, the incidence vectors for all NELEM elements are input and finally the numbers of the master degrees of freedom (here 1 and 3) are entered:

```
0,7,5,  0,1,2
0,10,6, 0,3,4
0,0,0,  0,7,8
0,7,8,  0,0,9
0,0,9,   0,10,11
0,10,11, 0,0,12
0,0,13,  0,0,9
0,0,14,  0,0,12
1,3
```

The resulting (2,2) stiffness matrix for substructure B in the degrees of freedom 1 and 2 is output in file KMATR as:

$$K_B = \begin{bmatrix} 47798.93 & 6264.063 \\ 6264.063 & 35270.81 \end{bmatrix} \tag{4. 3. 15}$$

This matrix represents substructure B and can be treated as a spring matrix coupling the active kinematic degrees of freedom 2 and 6 of substructure A (see Fig. 4.3-4). The input file EKOND for substructure A is given by

```
50000., 2.5, 0., 0.,
50000., 2.5, 0., 0.,
50000., 2.5, 0., 0.,
50000., 2.5, 0., 0.,
50000., 2.5, 0., 0.,
0,0,1, 0,2,3,
0,2,3, 0,4,5,
0,4,5, 0,6,7,
0,6,7, 0,8,9,
0,8,9, 0,0,10,
2,4,6,8,
```

where the numbers 2, 4, 6 and 8 in the last row denote the master degrees of freedom.

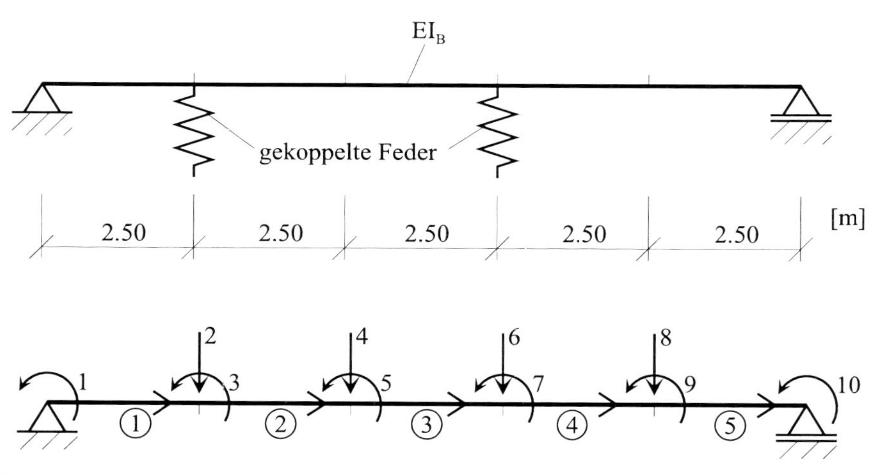

Fig. 4.3-4: Discretised substructure A

The total number of active kinematic degrees of freedom NDOF is 10, the number of master degrees of freedom NDU equals 4, the number of elements NELEM is 5 and finally the number NFED of spring matrices to be considered is 1. The input file INZFED contains the numbers 2 and 6 of the degrees of freedom to be coupled by the (2,2) spring matrix Eq. (4.3.15) which must be copied into the input file FEDMAT. The resulting condensed stiffness matrix of the whole system is given by:

$$K = \begin{bmatrix} 79400.81 & -30407.66 & 19492.71 & -3307.18 \\ -30407.66 & 44830.62 & -33714.83 & 13228.71 \\ 19492.71 & -33714.83 & 80101.42 & -30407.66 \\ -3307.18 & 13228.71 & -30407.66 & 31601.91 \end{bmatrix} \qquad (4.3.16)$$

A direct computation of this matrix through a discretisation of the whole system would of course lead to the same result.

4.4 Lumped-mass multi-degree-of-freedom systems

Lumped-mass idealisations of multi-degree-of-freedom systems are very popular in practice because of their simplicity. Their central feature is that the mass matrix $\underline{M}$ is a diagonal matrix, with each coefficient on the diagonal expressing the mass associated with the corresponding active kinematic degree of freedom. These values are usually determined "by hand" and the degrees of freedom present are simply the master degrees of freedom associated with medium or large inertia forces. The first step in setting up the lumped-mass idealisation consists in identifying these n "master" degrees of freedom to be retained in the equation and eliminating the remaining "slave" degrees of freedom, e.g. by static condensation as explained in Section 4.3. For plane frame structures, the determination of the pertinent (n,n) condensed stiffness matrix can be carried out with the program KONDEN, the use of which is demonstrated with the following example.

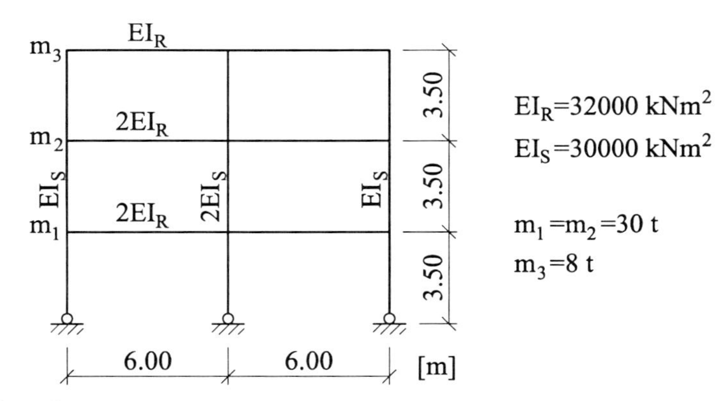

Fig. 4.4-1: Plane frame structure

The first step is, as always, the discretisation of the structure by introducing elements, nodes and active kinematic degrees of freedom. It is recommended to draw arrowheads on the single beam elements as in Fig. 4.4-2 in order to be able to clearly distinguish end section 1 from end section 2.

Fig. 4.4-1 depicts a simple frame, the condensed stiffness matrix of which is to be determined. The horizontal displacements of the single floors are clearly the active kinematic degrees of freedom associated the important inertia forces (degrees no. 4, 8 and 12), and they are chosen to be the master degrees of freedom (NDU=3). There are 15 beam elements present (NELEM=15) and a total of 15 active kinematic degrees of freedom (NDOF=15) have been introduced. The axial deformations of both girders and columns have been neglected by omitting the corresponding degrees of freedom, which amounts to infinite axial stiffness EA of all members. The values of NDU, NELEM and NDOF are input interactively at the terminal during the program execution. The first data block in the input file EKOND consists of NELEM rows, each containing the bending stiffness, length, axial stiffness and angle of each member:

```
30000.0, 3.5, 0.0, 90.0
30000.0, 3.5, 0.0, 90.0
30000.0, 3.5, 0.0, 90.0
30000.0, 3.5, 0.0, 90.0
30000.0, 3.5, 0.0, 90.0
30000.0, 3.5, 0.0, 90.0
60000.0, 3.5, 0.0, 90.0
60000.0, 3.5, 0.0, 90.0
60000.0, 3.5, 0.0, 90.0
64000.0, 6.0, 0.0, 0.0
64000.0, 6.0, 0.0, 0.0
64000.0, 6.0, 0.0, 0.0
64000.0, 6.0, 0.0, 0.0
32000.0, 6.0, 0.0, 0.0
32000.0, 6.0, 0.0, 0.0
```

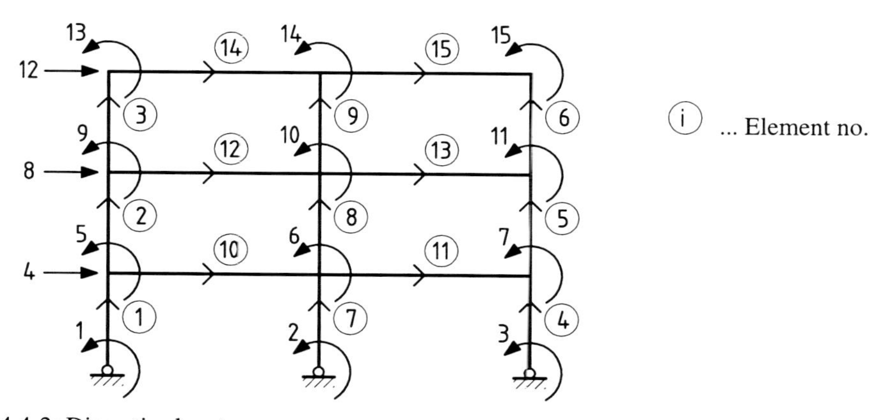

Fig. 4.4-2: Discretised system

Here, axial stiffness values are not needed since no relevant degrees of freedom are present, so the corresponding numerical values have been arbitrarily put equal to zero. The second data block consists of the incidence matrix (NELEM rows with six degrees-of-freedom numbers in each one):

```
0,0,1, 4,0,5,
4,0,5, 8,0,9,
8,0,9, 12,0,13,
0,0,3, 4,0,7,
4,0,7, 8,0,11,
8,0,11,12,0,15,
0,0,2, 4,0,6,
4,0,6, 8,0,10,
```

```
8,0,10,12,0,14,
4,0,5, 4,0,6,
4,0,6, 4,0,7,
8,0,9, 8,0,10,
8,0,10,8,0,11,
12,0,13,12,0,14,
12,0,14,12,0,15,
```

Finally, the NDU numbers of the master degrees of freedom are entered, in this case also

```
4,8,12
```

The resulting (3,3) condensed stiffness matrix, written in the file KMATR, is given by:

```
 .3429090E+05  -.2933414E+05   .4739090E+04
-.2933414E+05   .4707862E+05  -.2116977E+05
 .4739090E+04  -.2116977E+05   .1678259E+05
```

The output file AMAT contains the (15,3) transformation matrix $\underline{A}$ which allows the subsequent computation of the displacements, velocities and accelerations in all NDOF degrees of freedom once the corresponding values in the NDU master degrees of freedom are known (Eq. 4.3.4). It is given by:

```
 -.45068    .05828   -.00599
 -.45068    .05828   -.00599
 -.45068    .05828   -.00599
 1.00000    .00000    .00000
  .04421   -.11655    .01197
  .04421   -.11655    .01197
  .04421   -.11655    .01197
  .00000   1.00000    .00000
  .10876   -.01406   -.08662
  .10876   -.01406   -.08662
  .10876   -.01406   -.08662
  .00000    .00000   1.00000
 -.02813    .22531   -.19927
 -.02813    .22531   -.19927
 -.02813    .22531   -.19927
```

The masses associated with the three master degrees of freedom (their numbers 1, 2 and 3 of the reduced system corresponding to the numbers 4, 8 and 12 of the original structure) are equal to 30, 30 and 8t respectively (see Fig. 4.4-1). These values are entered in the file MDIAG which contains the diagonal of the mass matrix:

```
30.0, 30.0, 8.0
```

The matrices $\underline{M}$ and $\underline{K}$ of Eq. (4.3.1) have now been determined; the computation of the damping matrix $\underline{C}$ will be discussed in Sections 4.5 and 4.7.

It is also possible to determine the full stiffness matrix (in all NDOF active kinematic degrees of freedom) and the corresponding consistent mass matrix by using the program VOLLST.

The solution of the differential equation system (4.3.1) can be carried out by the same methods that have already been used in dealing with single-degree-of-freedom systems, such as Direct Integration in the time domain or integral transform methods in the frequency domain. Another very popular method is the so-called "modal analysis" which leads to uncoupled equations of motion that can be dealt with separately. It is considered in the next section.

4.5 Modal analysis for lumped-mass systems

The starting point of the discussion is the differential equation system (4.3.1) which is repeated as Eq. (4.5.1) for convenience. While the mass matrix $\underline{M}$ is a diagonal matrix due to the lumped-mass idealisation, the condensed stiffness matrix is not, leading to a "stiffness coupling" of the single degrees of freedom. In order to arrive at independent (uncoupled) differential equations of motion, new generalised co-ordinates $\underline{\eta}$ must be introduced to replace the physical co-ordinates $\underline{V}$. These generalised co-ordinates may be visualised as amplitudes of mutually orthogonal system deflection shapes, the latter being usually chosen to be the "free vibration shapes" or eigenvectors of the undamped system as determined by solving the eigenvalue problem. The single steps are illustrated in the following.

Starting with the differential equation system

$$\underline{M}\,\ddot{\underline{V}} + \underline{C}\,\dot{\underline{V}} + \underline{K}\,\underline{V} = \underline{P}(t) \tag{4.5.1}$$

with the initial values

$$\begin{aligned} \underline{V}(0) &= \underline{V}_0 \\ \dot{\underline{V}}(0) &= \dot{\underline{V}}_0 \end{aligned} \tag{4.5.2}$$

we introduce "modal Co-ordinates" $\underline{\eta}$ by means of the linear transformation

$$\begin{aligned} \underline{V} &= \underline{\Phi}\,\underline{\eta} \\ \dot{\underline{V}} &= \underline{\Phi}\,\dot{\underline{\eta}} \\ \ddot{\underline{V}} &= \underline{\Phi}\,\ddot{\underline{\eta}} \end{aligned} \tag{4.5.3}$$

The (n,r) matrix $\underline{\Phi}$ is termed "modal matrix". Its coefficients are time-independent and its r columns, with r typically much less than the number n of rows of $\underline{V}$ are the eigenvectors of the structure. Introducing Eq. (4.5.3) into Eq. (4.5.1) leads to

$$\underline{M}\,\underline{\Phi}\,\ddot{\underline{\eta}} + \underline{C}\,\underline{\Phi}\,\dot{\underline{\eta}} + \underline{K}\,\underline{\Phi}\,\underline{\eta} = \underline{P}(t) \tag{4.5.4}$$

and further

$$\underline{\Phi}^T\underline{M}\,\underline{\Phi}\,\ddot{\underline{\eta}} + \underline{\Phi}^T\underline{C}\,\underline{\Phi}\,\dot{\underline{\eta}} + \underline{\Phi}^T\underline{K}\,\underline{\Phi}\,\underline{\eta} = \underline{\Phi}^T\underline{P}(t) \tag{4.5.5}$$

In order to obtain uncoupled differential equations, we now demand that the transformation inherent in Eq. (4.5.5) must lead to diagonal matrices. For simplicity, we also demand that the transformation of the diagonal mass matrix by the modal matrix as in eq. (4.5.5) lead to a unit matrix, thus obtaining unit "modal masses" for the r „modal contributions":

$$\underline{\Phi}^T\,\underline{M}\,\underline{\Phi} = \underline{I} \tag{4.5.6}$$

$$\underline{\Phi}^T\,\underline{K}\,\underline{\Phi} = \underline{\omega}^2 = \mathrm{diag}\left[\omega_i^2\right] \tag{4.5.7}$$

The diagonalisation of the damping matrix $\underline{C}$ is deferred until later. Eqs. (4.5.6) and (4.5.7) can be combined by pre-multiplying Eq. (4.5.7) with the unit matrix which is subsequently replaced on the right-hand side by the left-hand side of Eq. (4.5.6):

$$\underline{\Phi}^T \underline{K} \underline{\Phi} = \underline{\Phi}^T \underline{M} \underline{\Phi} \underline{\omega}^2 \tag{4.5.8}$$

This leads to the general eigenvalue problem

$$\underline{K} \underline{\Phi} = \underline{M} \underline{\Phi} \underline{\omega}^2 \tag{4.5.9}$$

the solution of which (modal matrix $\underline{\Phi}$) can be used to diagonalise the stiffness matrix $\underline{K}$. We further note that

- $\underline{\omega}^2$ is a diagonal matrix with the r eigenvalues ω_i^2 (squares of the natural circular frequencies) as coefficients, and
- $\underline{\Phi}$ the (n,r) modal matrix, the columns of which are the r eigenvectors or mode shapes.

Solution algorithms for the eigenvalue problem are discussed in the next section. The easiest way to include damping is to assume that the damping matrix $\underline{C}$ can also be brought to diagonal form by transforming it with the modal matrix $\underline{\Phi}$ according to

$$\underline{\tilde{C}} = \underline{\Phi}^T \underline{C} \underline{\Phi} = \mathrm{diag}\left[\tilde{c}_{ii}\right] \tag{4.5.10}$$

As with the single-degree-of-freedom system, the diagonal term $\tilde{c}_{ii}$ is set equal to

$$\tilde{c}_{ii} = 2 D_i \omega_i \tag{4.5.11}$$

Here, D_i is the damping ratio and ω_i the circular natural frequency. This leads to the uncoupled system of differential equations

$$\ddot{\underline{\eta}} + \underline{\tilde{C}} \dot{\underline{\eta}} + \underline{\omega}^2 \underline{\eta} = \underline{\Phi}^T \underline{P} \tag{4.5.12}$$

It consists of r 2nd order differential equations of the general form

$$\ddot{\eta}_i + 2D_i \omega_i \dot{\eta}_i + \omega_i^2 \eta_i = \underline{\Phi}_i^T \underline{P}, \quad i = 1, 2, ..r \tag{4.5.13}$$

Each equation can be solved by itself by means of the methods discussed earlier. However, we still need the correct initial conditions for the velocity and displacement values of the modal co-ordinates. To this effect, Eq. (4.5.6) is formally rewritten for a square modal matrix as

$$\underline{\Phi}^{-1} = \underline{\Phi}^T \underline{M} \tag{4.5.14}$$

which, together with the definitions of the modal co-ordinates of Eq. (4.5.3), yields

$$\underline{\eta}(0) = \underline{\eta}_0 = \underline{\Phi}^T \underline{M} \underline{V}_0 \tag{4.5.15}$$

$$\dot{\underline{\eta}}(0) = \dot{\underline{\eta}}_0 = \underline{\Phi}^T \underline{M} \dot{\underline{V}}_0 \tag{4.5.16}$$

The special advantage of modal analysis lies in the fact that quite accurate solutions can be arrived at with the help of only a few modal contributions. The relative importance of the ith modal contribution can be judged by the magnitude of the "generalised load" $\underline{\Phi}_i^T \underline{P}$ and also by the closeness of the frequencies contained in the loading to the natural frequency ω_i of the mode. On the negative side, the solution of the eigenvalue problem is quite time-consuming for large systems. Due to the underlying superposition principle, modal analysis is strictly only applicable to linear systems, although some degree of non-linearity may be accommodated by modifying the mode shapes accordingly. Once the time histories $\eta_i(t)$, $i = 1,2,\ldots r$ of the modal co-ordinates have been determined (together with their time derivatives), the displacements, velocities and accelerations in the original degrees of freedom can be computed through Eq. (4.5.3), with the number r of modal contributions generally significantly lower than the number n of the master degrees of freedom of the system.

If the damping matrix $\underline{C}$ in Eq. (4.5.1) cannot be diagonalised by the mode shapes of the undamped system, an alternative is to introduce

$$\underline{V}(t) = \hat{\underline{\Phi}} e^{\hat{\lambda} t} \tag{4.5.17}$$

with complex mode shapes $\hat{\underline{\Phi}}$ and eigenvalues $\hat{\lambda}$. Introducing Eq. (4.5.17) into the homogeneous differential equation system yields the quadratic eigenvalue problem

$$(\hat{\lambda}^2 \underline{M} + \hat{\lambda} \underline{C} + \underline{K}) \hat{\underline{\Phi}} = 0 \tag{4.5.18}$$

This can be transformed into a linear eigenvalue problem with twice as many unknowns by rewriting Eq. (4.5.1) as a 1st order system with a new variable $\tilde{\underline{V}}^T = \begin{bmatrix} \underline{V} & \dot{\underline{V}} \end{bmatrix}$:

$$\begin{bmatrix} \underline{C} & \underline{M} \\ \underline{M} & 0 \end{bmatrix} \frac{d}{dt} \begin{bmatrix} \underline{V} \\ \dot{\underline{V}} \end{bmatrix} + \begin{bmatrix} \underline{K} & 0 \\ 0 & -\underline{M} \end{bmatrix} \begin{bmatrix} \underline{V} \\ \dot{\underline{V}} \end{bmatrix} = 0 \tag{4.5.19}$$

Some details on solution strategies for the eigenvalue problem are given in the following section. The numerical computation is done e.g. with the program JACOBI; its input files are KMATR, containing the condensed (n,n) system stiffness matrix, and MDIAG, containing the n masses associated with these master degrees of freedom. The output files AUSJAC, OMEG and PHI contain all n natural periods and mode shapes of the system.

The (3,3) condensed stiffness matrix for the frame depicted in Fig. 4.4-1 has already been computed in the last section. With masses equal to

```
30.0, 30.0, 8.0
```

tons, to be read from the file MDIAG, the program JACOBI yields the following results in the output file AUSJAC:

```
Eigenvector No.              1 Period=  7.55128E-001
1     .1000492E+00
2     .1332279E+00
3     .1445747E+00
Eigenvector No.              2 Period=  1.81723E-001
1     .1388166E+00
2    -.4212581E-01
3    -.2146686E+00
Eigenvector No.              3 Period=  1.05522E-001
1     .6366659E-01
2    -.1175121E+00
3     .2408641E+00
```

The circular natural frequencies of the three modes are equal to $\omega_1 = \frac{2\pi}{0.755} = 8.32 \frac{\text{rad}}{\text{s}}$, $\omega_2 = 34.58 \frac{\text{rad}}{\text{s}}$ and $\omega_3 = 59.55 \frac{\text{rad}}{\text{s}}$; the corresponding mode shapes are shown in Fig. 4.5-1.

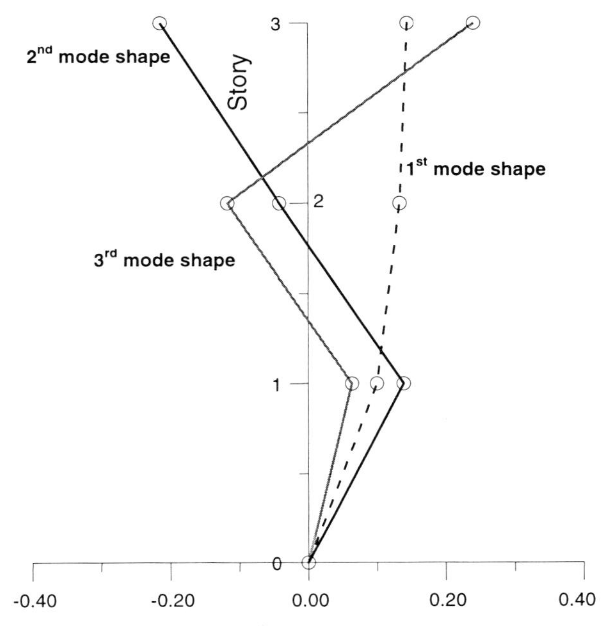

Fig. 4.5-1: Mode shapes of the three-story frame

Next, the response of this structure to a loading

$$\underline{P}(t)=\begin{bmatrix}10.0\\10.0\\5.0\end{bmatrix} f(t) \tag{4. 5. 20}$$

is investigated, with the time function f(t) as depicted in Fig. 4.5-2. The participation factors $\underline{\Phi}_i^T\,\underline{P}$ for the three modal contributions are given by:

$$\underline{\Phi}_1^T\underline{P}=\begin{pmatrix}0.10 & 0.322 & 0.446\end{pmatrix}\begin{pmatrix}10.0\\10.0\\5.0\end{pmatrix}=3.045 \tag{4. 5. 21}$$

$$\underline{\Phi}_2^T\underline{P}=\begin{pmatrix}0.139 & -0.042 & -0.215\end{pmatrix}\begin{pmatrix}10.0\\10.0\\5.0\end{pmatrix}=-0.105 \tag{4. 5. 22}$$

$$\underline{\Phi}_3^T\underline{P}=\begin{pmatrix}0.0637 & -0.118 & 0.241\end{pmatrix}\begin{pmatrix}10.0\\10.0\\5.0\end{pmatrix}=0.666 \tag{4. 5. 23}$$

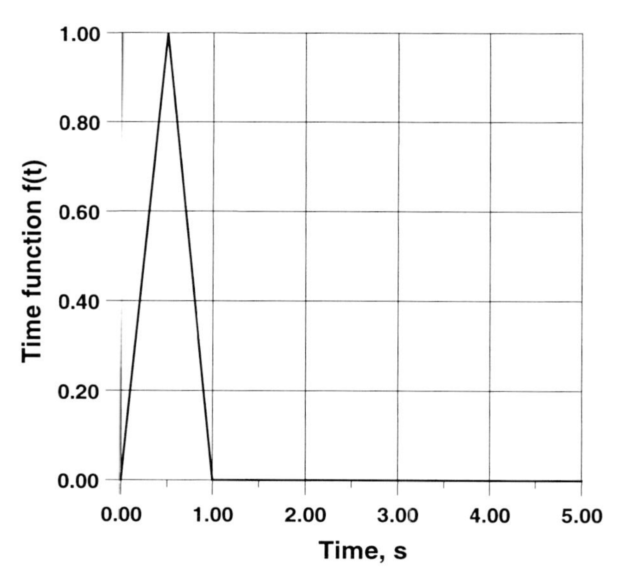

Fig. 4.5-2: Triangular loading function f(t)

Assuming a damping value of 2% for all three modes, we obtain the following differential equations in the modal co-ordinates η_1, η_2, η_3:

$$\ddot{\eta}_1 + 2 \cdot 0.02 \cdot 8.32 \dot{\eta}_1 + 8.32^2 \eta_1 = 3.045 \cdot f(t) \tag{4. 5. 24}$$

$$\ddot{\eta}_2 + 2 \cdot 0.02 \cdot 34.58 \dot{\eta}_2 + 34.48^2 \eta_2 = -0.105 \cdot f(t) \tag{4. 5. 25}$$

$$\ddot{\eta}_3 + 2 \cdot 0.02 \cdot 59.55 \dot{\eta}_3 + 59.55^2 \eta_3 = 0.666 \cdot f(t) \tag{4. 5. 26}$$

They can be solved for instance with the program LEINM and the resulting time histories are shown in Fig. 4.5-3, where the ordinates for the second and third modal co-ordinate have been scaled by the factor 100 for better visibility.

The modal analysis of a frame structure can also be carried out directly by means of the program MODAL. It requires information on the mode shapes and natural frequencies of the structure as computed by JACOBI and written by it in the files OMEG and PHI, furthermore the initial value vectors $\underline{V}_0$ and $\dot{\underline{V}}_0$ (n values in the input files V0 and VP0, respectively) and finally the system loading $\underline{P}(t)$ (to be read from the input file LASTV). LASTV contains n times NT values of the n time histories $P_1(t), \ldots, P_n(t)$ at NT equally spaced time points, in the form of NT data blocks containing the n loading components at each time step. If all components of the loading vector share the same time function, LASTV can be set up with the program INTERP. INTERP requires the piecewise linear time function to be interpolated with a given time step (described in the input file FKT) and also the amplitudes of all n components of $\underline{P}$ (input file AMPL); the output file is LASTV. If LASTV is to be used directly as input for the program MODAL, the output option without printing the time values must be chosen in INTERP.

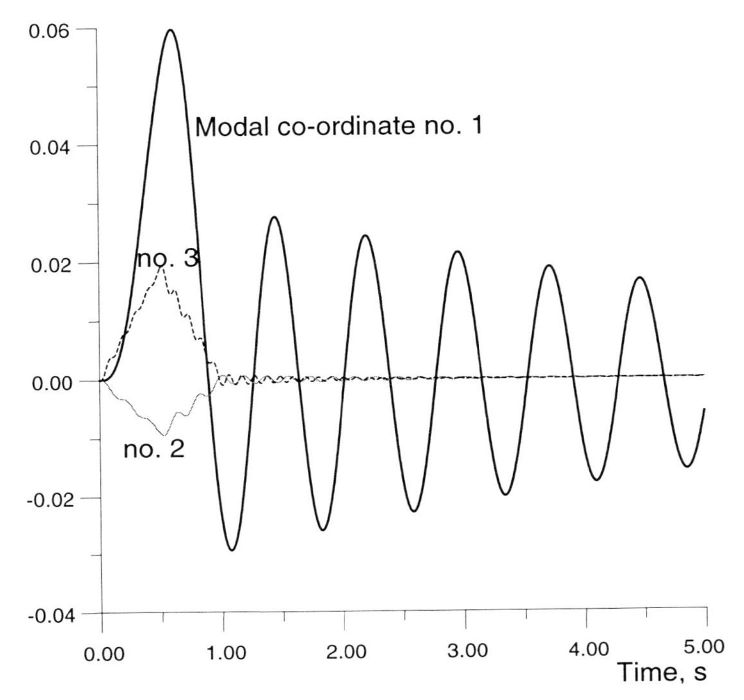

Fig. 4.5-3: Time histories of the modal co-ordinates, No. 2 and 3 scaled by 100

The program MODAL creates three output files, THIS.MOD, THISDG and THISDU. The first contains time histories of displacements in the master degrees of freedom with the corresponding time points in the first column and the displacements in the remaining NDU columns, THISDG contains the displacements in all (NDU + NDPHI) degrees of freedom without time points and, finally, THISDU presents the NDU displacement time histories in the master degrees of freedom without the corresponding time points. Fig. 4.5-4 shows the displacement time history for the roof of the three-story frame as computed by the program MODAL; the contribution of the fundamental mode alone yields sufficiently accurate results for this type of structure and loading.

In order to compute the internal forces in the structure, the deformations $\underline{V}$ in all active kinematic degrees of freedom must be determined next, from the known deformations $\underline{V}_u$ in the master degrees of freedom. This is done by means of Eq. (4.3.4):

$$\underline{V} = \underline{A}\ \underline{V}_u \tag{4. 5. 27}$$

The deformations of each beam element are now identified and the element internal forces $\underline{s}$ are computed by multiplying the beam stiffness matrix $\underline{k}$ with the element deformation vector $\underline{v}$:

$$\underline{s} = \underline{k} \cdot \underline{v} \tag{4. 5. 28}$$

This procedure has been coded in the program INTFOR which yields time histories of internal forces and deformations in the plane frame as well as their maximum values. Its input files are THISDU (as output file of the program MODAL), EKOND (as input file of the program KONDEN) and AMAT (as output file of the same program). Results are maximum and minimum values of internal forces complete with the time points at which they occurred and the values of all other internal forces at that time (in the file MAXMIN), the complete distribution of internal forces and displacements in the structure at a given time (in the file FORSTA), and, optionally, the time history of an arbitrary internal force or displacement component (in the file THHVM). The time step number for which the distribution of internal forces and displacements in the structure is to be output in the file FORSTA is entered interactively during the program run. Usually, the time step chosen is the one when the absolute maximum displacement is reached; this time point and the corresponding displacement value are output to the monitor by the program MODAL.

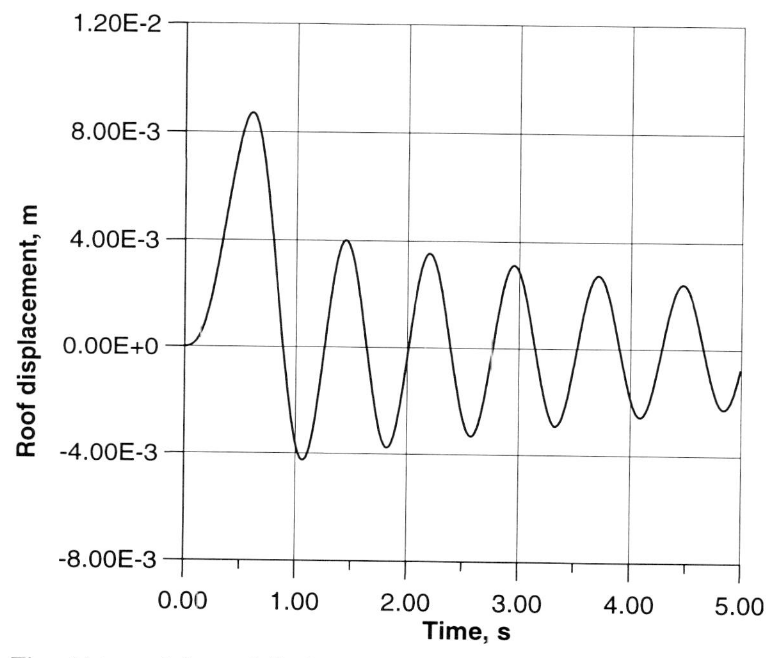

Fig. 4.5-4: Time history of the roof displacement

For the three-story frame being considered, the horizontal displacement of the roof attains its maximum value at t=0,61 s, which corresponds to the 61st time step for the time step $\Delta t = 0.01\,\mathrm{s}$ being employed. The output file FORSTA of the program INTFOR presents for each beam element (for the element numbering see Fig. 4.4-2) the internal forces and deformations expressed in the global co-ordinate system. The deformations at the end sections are given by the values of $u_1, w_1, \varphi_1, u_2, w_2, \varphi_2$ with the horizontal displacement u positive from left to right, the vertical displacement w positive downwards and the rotation φ positive counter-clockwise, while the internal forces (given in the next row) consist of the horizontal and vertical force components and the bending moments ($H_1, V_1, M_1, H_2, V_2, M_2$, also expressed in the global co-ordinate system). The output file FORSTA for our example is as follows:

```
Element No. 1
  .0000E+00   .0000E+00  -.2291E-02   .5998E-02   .0000E+00  -.5595E-03
 -.8480E+01  -.5192E-15  -.5365E-05   .8480E+01   .5192E-15   .2968E+02

Element No. 2
  .5998E-02   .0000E+00  -.5595E-03   .7971E-02   .0000E+00  -.2146E-03
 -.5190E+01  -.3178E-15   .6126E+01   .5190E+01   .3178E-15   .1204E+02

Element No. 3
  .7971E-02   .0000E+00  -.2146E-03   .8715E-02   .0000E+00  -.1095E-03
 -.1486E+01  -.9101E-16   .1700E+01   .1486E+01   .9101E-16   .3503E+01

Element No. 4
  .0000E+00   .0000E+00  -.2291E-02   .5998E-02   .0000E+00  -.5595E-03
 -.8480E+01  -.5192E-15  -.5365E-05   .8480E+01   .5192E-15   .2968E+02

Element No. 5
  .5998E-02   .0000E+00  -.5595E-03   .7971E-02   .0000E+00  -.2146E-03
 -.5190E+01  -.3178E-15   .6126E+01   .5190E+01   .3178E-15   .1204E+02

Element No. 6
  .7971E-02   .0000E+00  -.2146E-03   .8715E-02   .0000E+00  -.1095E-03
 -.1486E+01  -.9101E-16   .1700E+01   .1486E+01   .9101E-16   .3503E+01

Element No. 7
  .0000E+00   .0000E+00  -.2291E-02   .5998E-02   .0000E+00  -.5595E-03
 -.1696E+02  -.1038E-14  -.1073E-04   .1696E+02   .1038E-14   .5936E+02

Element No. 8
  .5998E-02   .0000E+00  -.5595E-03   .7971E-02   .0000E+00  -.2146E-03
 -.1038E+02  -.6355E-15   .1225E+02   .1038E+02   .6355E-15   .2408E+02

Element No. 9
  .7971E-02   .0000E+00  -.2146E-03   .8715E-02   .0000E+00  -.1095E-03
 -.2973E+01  -.1820E-15   .3399E+01   .2973E+01   .1820E-15   .7006E+01

Element No.10
  .5998E-02   .0000E+00  -.5595E-03   .5998E-02   .0000E+00  -.5595E-03
  .0000E+00   .1194E+02  -.3581E+02   .0000E+00  -.1194E+02  -.3581E+02

Element No.11
  .5998E-02   .0000E+00  -.5595E-03   .5998E-02   .0000E+00  -.5595E-03
  .0000E+00   .1194E+02  -.3581E+02   .0000E+00  -.1194E+02  -.3581E+02

Element No.12
  .7971E-02   .0000E+00  -.2146E-03   .7971E-02   .0000E+00  -.2146E-03
  .0000E+00   .4579E+01  -.1374E+02   .0000E+00  -.4579E+01  -.1374E+02

Element No.13
  .7971E-02   .0000E+00  -.2146E-03   .7971E-02   .0000E+00  -.2146E-03
  .0000E+00   .4579E+01  -.1374E+02   .0000E+00  -.4579E+01  -.1374E+02

Element No.14
  .8715E-02   .0000E+00  -.1095E-03   .8715E-02   .0000E+00  -.1095E-03
  .0000E+00   .1168E+01  -.3503E+01   .0000E+00  -.1168E+01  -.3503E+01

Element No.15
  .8715E-02   .0000E+00  -.1095E-03   .8715E-02   .0000E+00  -.1095E-03
  .0000E+00   .1168E+01  -.3503E+01   .0000E+00  -.1168E+01  -.3503E+01
```

Fig. 4.5-5 depicts the bending moment distribution in the structure at time t=0.61 s. As a comparison, Fig. 4.5-6 shows the bending moment diagram of the static case, with horizontal loads equal to 10.0, 10.0 and 5.0 kN acting at the girders, as computed with the program RAHMEN introduced in Section 4.2. The input file ERAHM of the latter is practically the same as EKOND, with the last row (containing the numbers of the master degrees of freedom) being replaced by the load vector which expresses the load components corresponding to all active kinematic degrees of freedom (NDOF = 15 in this case, see Fig. 4.4-2). The load vector is in this case given by

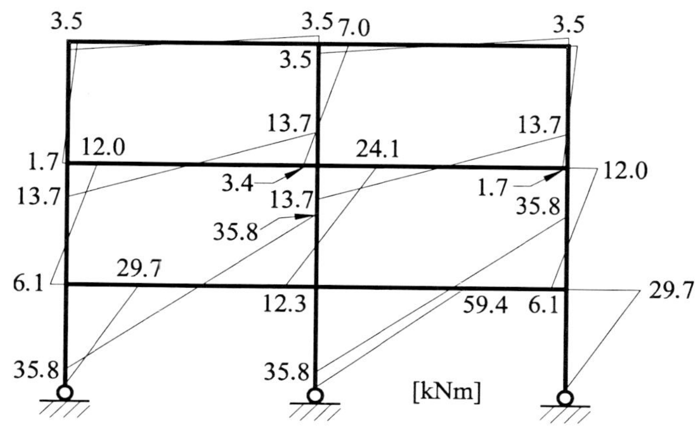

Fig. 4.5-5 : Bending moment diagram at time t = 0.61 s

```
0., 0., 0., 10., 0., 0., 0., 10., 0., 0., 0., 5., 0., 0., 0
```

with the point loads of 10.0, 10.0, and 5.0 kN associated with the degrees of freedom 4, 8 and 12. The pertinent part of the output file ARAHM is reproduced below:

```
ELEMENT NO.          1
   .0000E+00   .0000E+00  -.1687E-02   .4415E-02   .0000E+00  -.4108E-03
  -.6250E+01  -.3827E-15   .1032E-13   .6250E+01   .3827E-15   .2188E+02
 ELEMENT NO.         2
   .4415E-02   .0000E+00  -.4108E-03   .5862E-02   .0000E+00  -.1605E-03
  -.3750E+01  -.2296E-15   .4417E+01   .3750E+01   .2296E-15   .8708E+01
 ELEMENT NO.         3
   .5862E-02   .0000E+00  -.1605E-03   .6445E-02   .0000E+00  -.8783E-04
  -.1250E+01  -.7654E-16   .1565E+01   .1250E+01   .7654E-16   .2810E+01
 ELEMENT NO.         4
   .0000E+00   .0000E+00  -.1687E-02   .4415E-02   .0000E+00  -.4108E-03
  -.6250E+01  -.3827E-15   .1032E-13   .6250E+01   .3827E-15   .2188E+02
 ELEMENT NO.         5
   .4415E-02   .0000E+00  -.4108E-03   .5862E-02   .0000E+00  -.1605E-03
  -.3750E+01  -.2296E-15   .4417E+01   .3750E+01   .2296E-15   .8708E+01
 ELEMENT NO.         6
   .5862E-02   .0000E+00  -.1605E-03   .6445E-02   .0000E+00  -.8783E-04
  -.1250E+01  -.7654E-16   .1565E+01   .1250E+01   .7654E-16   .2810E+01
 ELEMENT NO.         7
   .0000E+00   .0000E+00  -.1687E-02   .4415E-02   .0000E+00  -.4108E-03
  -.1250E+02  -.7654E-15  -.9640E-14   .1250E+02   .7654E-15   .4375E+02
 ELEMENT NO.         8
   .4415E-02   .0000E+00  -.4108E-03   .5862E-02   .0000E+00  -.1605E-03
  -.7500E+01  -.4592E-15   .8834E+01   .7500E+01   .4592E-15   .1742E+02
 ELEMENT NO.         9
   .5862E-02   .0000E+00  -.1605E-03   .6445E-02   .0000E+00  -.8783E-04
  -.2500E+01  -.1531E-15   .3129E+01   .2500E+01   .1531E-15   .5621E+01
 ELEMENT NO.        10
   .4415E-02   .0000E+00  -.4108E-03   .4415E-02   .0000E+00  -.4108E-03
   .0000E+00   .8764E+01  -.2629E+02   .0000E+00  -.8764E+01  -.2629E+02
 ELEMENT NO.        11
   .4415E-02   .0000E+00  -.4108E-03   .4415E-02   .0000E+00  -.4108E-03
   .0000E+00   .8764E+01  -.2629E+02   .0000E+00  -.8764E+01  -.2629E+02
 ELEMENT NO.        12
   .5862E-02   .0000E+00  -.1605E-03   .5862E-02   .0000E+00  -.1605E-03
   .0000E+00   .3424E+01  -.1027E+02   .0000E+00  -.3424E+01  -.1027E+02
 ELEMENT NO.        13
   .5862E-02   .0000E+00  -.1605E-03   .5862E-02   .0000E+00  -.1605E-03
   .0000E+00   .3424E+01  -.1027E+02   .0000E+00  -.3424E+01  -.1027E+02
 ELEMENT NO.        14
   .6445E-02   .0000E+00  -.8783E-04   .6445E-02   .0000E+00  -.8783E-04
   .0000E+00   .9368E+00  -.2810E+01   .0000E+00  -.9368E+00  -.2810E+01
```

```
ELEMENT NO.          15
 .6445E-02   .0000E+00  -.8783E-04   .6445E-02   .0000E+00  -.8783E-04
 .0000E+00   .9368E+00  -.2810E+01   .0000E+00  -.9368E+00  -.2810E+01
```

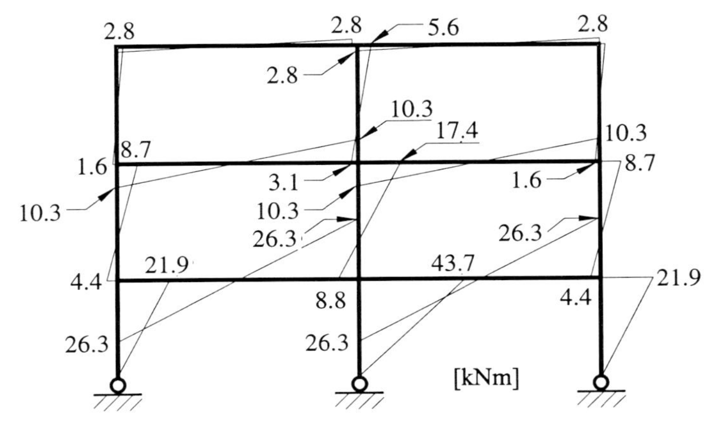

Fig. 4.5-6: Bending moment diagram for static loading

The comparison of the "dynamic" bending moments of Fig. 4.5-5 with the "static" values of Fig. 4.5-6 shows that the amplification factor $\frac{S_{dyn}}{S_{stat}}$ varies from section to section. A useful procedure for testing the plausibility of the computed results of a dynamic analysis is conducting an additional program run with the loading increasing slowly from zero to a (constant) maximum value and comparing the "dynamic" results with the output of a "static" program for this loading.

Results for the maximum and minimum values of the bending moment in the structure are given in the output file MAXMIN. The extreme values of the bending moment appear in the last column, the times at which they occurred in the first column and the internal forces occurring simultaneously are also given sequentially in the second and third columns (horizontal and vertical force component):

```
 Element No.            1
Max. pos., Beam end 1:      1.0700    .4161E+01   .2548E-15    .2644E-05
Max. neg., Beam end 1:       .6100   -.8480E+01  -.5192E-15   -.5365E-05
Max. pos., Beam end 2:       .6100    .8480E+01   .5192E-15    .2968E+02
Max. neg., Beam end 2:      1.0800   -.4161E+01  -.2548E-15   -.1456E+02
 Element No.            2
Max. pos., Beam end 1:       .6000   -.5197E+01  -.3182E-15    .6141E+01
Max. neg., Beam end 1:      1.0700    .2647E+01   .1620E-15   -.3146E+01
Max. pos., Beam end 2:       .6000    .5197E+01   .3182E-15    .1205E+02
Max. neg., Beam end 2:      1.0700   -.2647E+01  -.1620E-15   -.6117E+01
 Element No.            3
Max. pos., Beam end 1:       .5300   -.1485E+01  -.9096E-16    .1793E+01
Max. neg., Beam end 1:      1.0300    .5735E+00   .3512E-16   -.5597E+00
Max. pos., Beam end 2:       .6100    .1486E+01   .9101E-16    .3503E+01
Max. neg., Beam end 2:      1.0600   -.5707E+00  -.3495E-16   -.1483E+01
 Element No.            4
Max. pos., Beam end 1:      1.0700    .4161E+01   .2548E-15    .2644E-05
Max. neg., Beam end 1:       .6100   -.8480E+01  -.5192E-15   -.5365E-05
Max. pos., Beam end 2:       .6100    .8480E+01   .5192E-15    .2968E+02
Max. neg., Beam end 2:      1.0800   -.4161E+01  -.2548E-15   -.1456E+02
 Element No.            5
```

```
Max. pos., Beam end 1:     .6000  -.5197E+01  -.3182E-15   .6141E+01
Max. neg., Beam end 1:    1.0700   .2647E+01   .1620E-15  -.3146E+01
Max. pos., Beam end 2:     .6000   .5197E+01   .3182E-15   .1205E+02
Max. neg., Beam end 2:    1.0700  -.2647E+01  -.1620E-15  -.6117E+01
 Element No.          6
Max. pos., Beam end 1:     .5300  -.1485E+01  -.9096E-16   .1793E+01
Max. neg., Beam end 1:    1.0300   .5735E+00   .3512E-16  -.5597E+00
Max. pos., Beam end 2:     .6100   .1486E+01   .9101E-16   .3503E+01
Max. neg., Beam end 2:    1.0600  -.5707E+00  -.3495E-16  -.1483E+01
 Element No.          7
Max. pos., Beam end 1:    1.0700   .8321E+01   .5095E-15   .5288E-05
Max. neg., Beam end 1:     .6100  -.1696E+02  -.1038E-14  -.1073E-04
Max. pos., Beam end 2:     .6100   .1696E+02   .1038E-14   .5936E+02
Max. neg., Beam end 2:    1.0800  -.8322E+01  -.5096E-15  -.2913E+02
 Element No.          8
Max. pos., Beam end 1:     .6000  -.1039E+02  -.6365E-15   .1228E+02
Max. neg., Beam end 1:    1.0700   .5293E+01   .3241E-15  -.6292E+01
Max. pos., Beam end 2:     .6000   .1039E+02   .6365E-15   .2410E+02
Max. neg., Beam end 2:    1.0700  -.5293E+01  -.3241E-15  -.1223E+02
 Element No.          9
Max. pos., Beam end 1:     .5300  -.2971E+01  -.1819E-15   .3587E+01
Max. neg., Beam end 1:    1.0300   .1147E+01   .7023E-16  -.1119E+01
Max. pos., Beam end 2:     .6100   .2973E+01   .1820E-15   .7006E+01
Max. neg., Beam end 2:    1.0600  -.1141E+01  -.6989E-16  -.2966E+01
 Element No.         10
Max. pos., Beam end 1:    1.0700   .0000E+00  -.5903E+01   .1771E+02
Max. neg., Beam end 1:     .6100   .0000E+00   .1194E+02  -.3581E+02
Max. pos., Beam end 2:    1.0700   .0000E+00   .5903E+01   .1771E+02
Max. neg., Beam end 2:     .6100   .0000E+00  -.1194E+02  -.3581E+02
 Element No.         11
Max. pos., Beam end 1:    1.0700   .0000E+00  -.5903E+01   .1771E+02
Max. neg., Beam end 1:     .6100   .0000E+00   .1194E+02  -.3581E+02
Max. pos., Beam end 2:    1.0700   .0000E+00   .5903E+01   .1771E+02
Max. neg., Beam end 2:     .6100   .0000E+00  -.1194E+02  -.3581E+02
 Element No.         12
Max. pos., Beam end 1:    1.0700   .0000E+00  -.2206E+01   .6619E+01
Max. neg., Beam end 1:     .6000   .0000E+00   .4579E+01  -.1374E+02
Max. pos., Beam end 2:    1.0700   .0000E+00   .2206E+01   .6619E+01
Max. neg., Beam end 2:     .6000   .0000E+00  -.4579E+01  -.1374E+02
 Element No.         13
Max. pos., Beam end 1:    1.0700   .0000E+00  -.2206E+01   .6619E+01
Max. neg., Beam end 1:     .6000   .0000E+00   .4579E+01  -.1374E+02
Max. pos., Beam end 2:    1.0700   .0000E+00   .2206E+01   .6619E+01
Max. neg., Beam end 2:     .6000   .0000E+00  -.4579E+01  -.1374E+02
 Element No.         14
Max. pos., Beam end 1:    1.0600   .0000E+00  -.4944E+00   .1483E+01
Max. neg., Beam end 1:     .6100   .0000E+00   .1168E+01  -.3503E+01
Max. pos., Beam end 2:    1.0600   .0000E+00   .4944E+00   .1483E+01
Max. neg., Beam end 2:     .6100   .0000E+00  -.1168E+01  -.3503E+01
 Element No.         15
Max. pos., Beam end 1:    1.0600   .0000E+00  -.4944E+00   .1483E+01
Max. neg., Beam end 1:     .6100   .0000E+00   .1168E+01  -.3503E+01
Max. pos., Beam end 2:    1.0600   .0000E+00   .4944E+00   .1483E+01
Max. neg., Beam end 2:     .6100   .0000E+00  -.1168E+01  -.3503E+01
```

Fig. 4.5-7 shows the time history of the bending moment at the top of the middle column in the ground floor of the frame.

The programs introduced in this section are summarised in the following table:

Program name	Input files	Output files
JACOBI	KMATR MDIAG	AUSJAC OMEG PHI
INTERP	FKT AMPL	LASTV
MODAL	OMEG PHI MDIAG V0 VP0 AMAT LASTV	THIS.MOD THISDU THISDG
INTFOR	EKOND THISDU AMAT	FORSTA THHVM MAXMIN

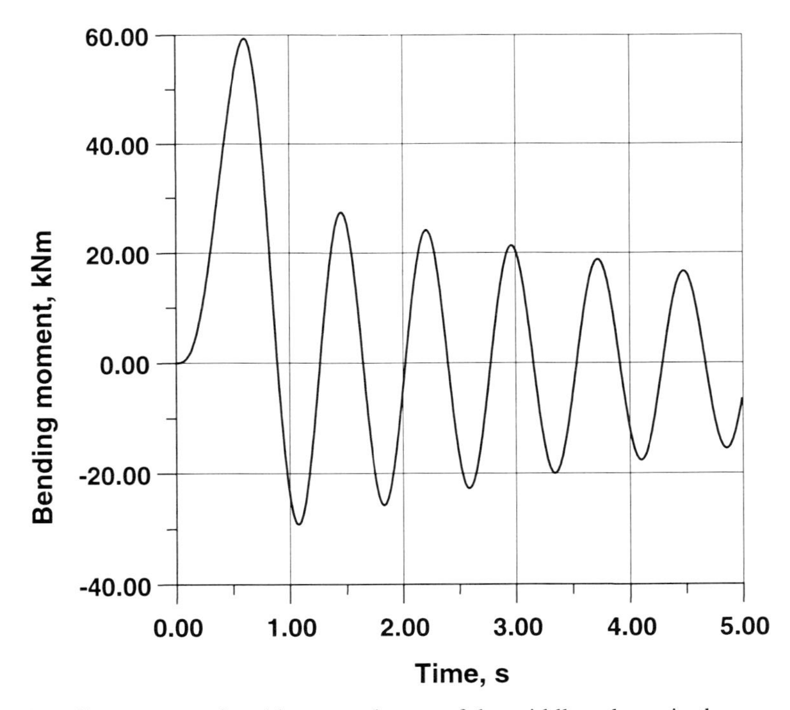

Fig. 4.5-7: Bending moment time history at the top of the middle column in the ground floor.

4.6 Solving the linear eigenvalue problem

Starting point is the general linear eigenvalue problem

$$\underline{K}\,\underline{\Phi} = \underline{M}\;\underline{\Phi}\;\underline{\omega}^2 \qquad (4.6.1)$$

which can be written as follows for the ith eigenvalue ω_i^2 and the corresponding eigenvector (mode shape vector) Φ_i :

$$\underline{K}\,\underline{\Phi}_i = \omega_i^2\,\underline{M}\,\underline{\Phi}_i \qquad (4.6.2)$$

Both matrices $\underline{M}$ and $\underline{K}$ are symmetric. The stiffness matrix $\underline{K}$ is positive definite after the support conditions of the (stable) structure have been considered, meaning that the following relationship holds for arbitrary vectors $\underline{V}$:

$$\underline{V}^T\,\underline{K}\,\underline{V} > 0 \qquad (4.6.3)$$

The mass matrix $\underline{M}$ as consistent mass matrix is positive definite; this is also the case for diagonal "lumped mass" matrices if the coefficients on the main diagonal are positive, $m_{ii} > 0$. Since generally both the stiffness and the mass matrix are real, symmetric and positive definite (n,n) matrices, the existence of n real and positive eigenvalues ω_i^2, $i = 1,2,\ldots,n$ is guaranteed; however, multiple or repeated eigenvalues may be among them. If all eigenvalues are different, the corresponding eigenvectors constitute a complete orthogonal reference system. If some of the eigenvalues are the same, any linear combination of their eigenvectors is also an eigenvector. Usually, it is only the eigenvalues and eigenvectors corresponding to the lower natural frequencies of the system that must be determined, because computed higher modes are necessarily inaccurate due to the unavoidable discrepancies between the numerical model, with its finite number of degrees of freedom, and the actual structure.

Eq. (4.6.2) can be written as a system of linear equations

$$(\underline{K} - \omega_i^2\,\underline{M})\,\underline{\Phi}_i = 0 \qquad (4.6.4)$$

which possesses a nontrivial solution only if the determinant of the system matrix is equal to zero:

$$\det(\underline{K} - \omega_i^2\,\underline{M}) = 0 \qquad (4.6.5)$$

Eq. (4.6.5) is referred to as the characteristic equation or frequency equation of the multi-degree-of-freedom system. It has n solutions ω_i^2, $i = 1,2,\ldots,n$ (corresponding to n eigenvectors $\underline{\Phi}_i$) which may be computed e.g. by finding the roots of the nth degree polynomial in ω_i^2 that results from evaluating the determinant. This classical approach is, of course, much too tedious and error-prone for all but the simplest systems, if done by hand.

The general linear eigenvalue problem may be tackled directly or reduced first to the special form

$$\underline{C}\,\underline{Y} = \lambda\,\underline{Y} \qquad (4.6.6)$$

with only one system matrix $\underline{C}$. The basic physical meaning of eigenvalue problems can be easily seen from this equation: We look for vectors $\underline{Y}$ which, in being pre-multiplied by $\underline{C}$, retain their direction and simply change their length by a factor λ. For structural dynamics applications it is important that the algebraically largest eigenvalue λ_1 (which is the simplest to determine) be defined in such a way that it corresponds to the lowest natural frequency of the system. To this effect, the eigenvalues are defined as reciprocals of the squares of the circular natural frequencies:

$$\lambda_i = \frac{1}{\omega_i^2} \qquad (4.6.7)$$

The general eigenvalue problem now reads:

$$\underline{M}\,\underline{\Phi}_i = \frac{1}{\omega_i^2}\ \underline{K}\,\underline{\Phi}_i \tag{4.6.8}$$

The so-called RAYLEIGH quotient is defined as the scalar quantity

$$R(\underline{V}_i) = \frac{\underline{V}_i^T \underline{K}\,\underline{V}_i}{\underline{V}_i^T\,\underline{M}\,\underline{V}_i} \tag{4.6.9}$$

for any arbitrary vector $\underline{V}$. It can be shown that the following inequality holds:

$$\omega_1^2 \le R(\underline{V}) \le \omega_n^2 \tag{4.6.10}$$

If a vector $\underline{V}$ which appears somehow similar to the structure's first mode-shape vector is used (having been derived, for example, by judicious application of a static loading pattern), the RAYLEIGH quotient furnishes an upper bound for ω_1^2 which usually yields a good approximation of its true value.

The easiest way to arrive at the special form Eq. (4.6.6) from Eq. (4.6.8) is by pre-multiplying the latter with the inverse stiffness matrix of the system, that is, its flexibility matrix $\underline{F}$, $\underline{F} = \underline{K}^{-1}$. We obtain

$$\underline{F}\,\underline{M}\,\underline{\Phi}_i = \frac{1}{\omega_i^2}\ \underline{\Phi}_i \tag{4.6.11}$$

or

$$\underline{D}\ \underline{\Phi}_i = \frac{1}{\omega_i^2}\ \underline{\Phi}_i \tag{4.6.12}$$

where the "dynamic matrix" $\underline{D} = \underline{F}\,\underline{M}$ is not symmetrical. A useful algorithm for solving the eigenvalue problem of Eq. (4.6.12) is the so-called power method. In it, an arbitrary vector $\underline{\varphi}_0$ is pre-multiplied by $\underline{D}$ yielding another vector $\underline{\varphi}_1$:

$$\underline{D}\,\underline{\varphi}_0 = \underline{\varphi}_1 = \frac{1}{\omega_1^2}\ \underline{\varphi}_0 \tag{4.6.13}$$

As seen from this equation, if $\underline{\varphi}_0$ already were the correct eigenvector, the corresponding eigenvalue $\frac{1}{\omega_1^2}$ could be computed as a quotient of the corresponding vector components of $\underline{\varphi}_0$ and $\underline{\varphi}_1$, with all n quotients of the n components yielding the same value. As it is, with $\underline{\varphi}_0$ only distantly related to the true eigenvector, the n quotients differ widely and only yield a rough approximation of the correct eigenvalue. However, the vector $\underline{\varphi}_1$ is a better approximation of the "first mode shape" (this being the eigenvector corresponding to the algebraically largest eigenvalue $\frac{1}{\omega_1^2}$) than $\underline{\varphi}_0$, and further iteration steps according to

$$\underline{D}\,\underline{\varphi}_i = \underline{\varphi}_{i+1} \tag{4.6.14}$$

eventually furnish the correct "largest mode" shape and the value $\frac{1}{\omega_1^2}$, corresponding to the largest natural period $T_1 = \frac{2\pi}{\omega_1}$. This algorithm is also suitable for hand calculation for small systems and, like all iterative schemes, it rewards a shrewd assumption for the starting vector $\underline{\varphi}_0$ by quick convergence. It can also serve for determining the algebraically smallest eigenvalue with the corresponding mode shape by first inverting $\underline{D}$ and applying the algorithm to $\underline{D}^{-1}$.

The computer program EIGVOL uses this algorithm for computing the first natural period of a system and the corresponding mode shape; it requires the full consistent mass and stiffness matrix of the structure, which, for plane frames, are determined by the program VOLLST (see Section 4.2).

Program name	Input files	Output files
EIGVOL	MVOLL KVOLL	AUSVOL

The program can be used for judging the quality of the idealisation of a plane frame system when only a few master degrees of freedom are retained.

A symmetrical system matrix offers many advantages for the computer-based solution of the eigenvalue problem. To that effect, a standard CHOLESKY decomposition of the stiffness matrix in the lower triangular matrix $\underline{L}$ and its transpose is carried out as follows:

$$\underline{K} = \underline{L}\,\underline{L}^T \tag{4.6.15}$$

Next, both sides of the general eigenvalue problem

$$\underline{M}\,\underline{\Phi}_i = \frac{1}{\omega_i^2}\underline{K}\,\underline{\Phi}_i \tag{4.6.16}$$

are pre-multiplied with $\underline{L}^{-1}$, leading to:

$$\underline{L}^{-1}\,\underline{M}\,\underline{\Phi}_i = \frac{1}{\omega_i^2}\,\underline{L}^T\,\underline{\Phi}_i \tag{4.6.17}$$

or

$$\underline{C}\,\underline{Y}_i = \frac{1}{\omega_i^2}\underline{Y}_i \tag{4.6.18}$$

with the transformed eigenvector

$$\underline{Y}_i = \underline{L}^T\,\underline{\Phi}_i \tag{4.6.19}$$

and the symmetrical system matrix

$$\underline{C} = \underline{L}^{-1}\,\underline{M}\,(\underline{L}^{-1})^T \tag{4.6.20}$$

The eigenvalues of $\underline{C}$ are identical to the ones of the original problem, while the eigenvectors of the original system can be easily obtained through Eq. (4.6.19). In deriving Eq. (4.6.20), use has been made of the relationship $(\underline{L}^{-1})^T = (\underline{L}^T)^{-1}$ which is valid for lower triangular matrices.

A number of powerful algorithms are available for solving the eigenvalue problem Eq. (4.6.18). Usually, eigenvectors and eigenvalues are determined from a special, easier to manipulate form of the system matrix (e.g. a tridiagonal matrix), into which the system matrix must be firstly transformed. Of course, the eigenvalues of the tridiagonal and the original matrix must be the same; this is the case, for instance, if the former is derived from the latter by similarity transformations, where the system matrix is repeatedly pre- and post-multiplied by another matrix and its inverse. This can be shown by introducing into the eigenvalue problem

$$\underline{A}\,\underline{\Phi} = \underline{\Phi}\,\underline{\Lambda} \tag{4. 6. 21}$$

a new modal matrix $\underline{Y}$ according to $\underline{Y} = \underline{T}\,\underline{\Phi}$, $\underline{\Phi} = \underline{T}^{-1}\,\underline{Y}$. This leads to

$$\underline{A}(\underline{T}^{-1}\,\underline{Y}) = (\underline{T}^{-1}Y)\,\underline{\Lambda} \tag{4. 6. 22}$$

or, by pre-multiplying both sides with $\underline{T}$:

$$\underline{T}\,\underline{A}\,\underline{T}^{-1}\,\underline{Y} = \underline{Y}\,\underline{\Lambda} \tag{4. 6. 23}$$

Comparing this expression with Eq. (4.6.21) shows that the eigenvalues $\underline{\Lambda}$ of the matrix $\underline{A}$ are identical to the ones of the matrix $\underline{T}\,\underline{A}\,\underline{T}^{-1}$ which has been derived from $\underline{A}$ by a similarity transformation. Special matrices $\underline{T}$ are normally used for similarity transformations, such as orthogonal matrices ($\underline{T}^T = \underline{T}^{-1}$) which preserve the symmetry of the system matrix. A classical algorithm is named after JACOBI and has been implemented in the program by the same name. Other well-known methods are the HOUSEHOLDER algorithm that also utilises similarity transformations and also methods relying on „simultaneous vector iteration“. The latter are, essentially, extensions of the simple power method in which the iteration is carried out with a whole group of mutually orthogonal vectors that eventually converge into the eigenvectors of the system.

4.7 The linear viscous damping model

The correct consideration of damping is of prime importance for the meaningfulness and accuracy of the results obtained in any structural dynamics problem. However, in view of the complexity of the different energy dissipation mechanisms contributing to damping, one is usually quite satisfied in being able to reproduce the correct order of magnitude of the overall energy dissipation, mostly by referring to past experience with similar structures. The mathematically simple linear viscous damping model is widely used for this purpose because of its ease of application, even if it cannot describe the damping phenomenon in all its complexity and heterogeneity.

Again, our starting point is the differential equation system

$$\underline{M}\,\ddot{\underline{V}} + \underline{C}\,\dot{\underline{V}} + \underline{K}\,\underline{V} = \underline{P}(t) \tag{4. 7. 1}$$

We are dealing with “proportional” damping if the viscous damping matrix $\underline{C}$ is diagonalised by a transformation with the modal matrix $\underline{\Phi}$ containing the mode shapes of the undamped system as columns:

$$\underline{\tilde{C}} = \underline{\Phi}^T \ \underline{C} \ \underline{\Phi} = \text{diag}[\tilde{c}_{ii}] = \text{diag}[2\, D_i \ \omega_i] \tag{4. 7. 2}$$

As for the single-degree-of-freedom system, D_i is the damping ratio (in percent of the critical damping) for the ith modal shape with the corresponding circular natural frequency ω_i. Because of the simplicity of this idealisation, other non-viscous damping mechanisms (such as material damping, see Section 4.9) are often approximated by equivalent linear viscous damping values. In addition to the damping ratio D (or ξ), the logarithmic decrement Λ introduced in Section 3.3 can also be used for quantifying viscous damping. If Eq. (4.7.2) holds, the system described by Eq. (4.7.1) is said to possess "classical" mode shapes, meaning that the eigenvectors as columns of the modal matrix $\underline{\Phi}$ are real and mutually orthogonal. A necessary and sufficient condition for the existence of such classical mode shapes is, after CAUGHEY [4.4]:

$$\underline{C}\, \underline{M}^{-1}\, \underline{K} = \underline{K}\, \underline{M}^{-1}\, \underline{C} \tag{4. 7. 3}$$

Physically, the existence of classical mode shapes implies that the damping mechanisms all over the structure are somehow similar. Strong discrepancies in the distribution of the energy dissipation rate (e.g. if soil-structure interaction effects are considered) lead to a distribution of damping forces that varies significantly from those of the restoring and inertia forces. If Eq. (4.7.3) is not satisfied, we have the general case of non-proportional viscous damping.

Usually, no explicit viscous damping matrix is known, and one can only assume modal damping ratios D_i (i=1, ..., r) for the r modal contributions which could be considered in the context of a modal analysis approach. For the modal analysis itself, it is sufficient to supply the damping ratio D_i alone and no explicit damping matrix $\underline{C}$ is needed. If, on the other hand, $\underline{C}$ itself is needed, e.g. for solving Eq. (4.7.1) by Direct Integration, two methods are recommended for setting up $\underline{C}$ based on one or more known modal damping ratios:

- **RAYLEIGH damping:**

Here, $\underline{C}$ is computed as a linear combination of the system's stiffness and mass matrix according to

$$\underline{C} = \alpha\, \underline{M} + \beta\, \underline{K} \tag{4. 7. 4}$$

Analogously to the single-degree-of-freedom system where $\frac{c}{m} = 2\, D\, \omega_1$, we obtain:

$$2\, D_i\, \omega_i = \underline{\Phi}_i^{\ T}\, \underline{C}\, \underline{\Phi}_i = \underline{\Phi}_i^{\ T}(\alpha\, \underline{M} + \beta\, \underline{K})\underline{\Phi}_i = \alpha + \beta\, \omega_i^{\ 2} \tag{4. 7. 5}$$

and further

$$D_i = \frac{\alpha}{2\,\omega_i} + \beta\, \frac{\omega_i}{2} = \alpha \frac{T_i}{4\pi} + \beta \frac{\pi}{T_i} \tag{4. 7. 6}$$

The two coefficients α and β can be determined in such a manner that prescribed damping ratios D_1 and D_2 appear at two period values T_1 and T_2 (which are not necessarily natural periods of the structure). The coefficients α and β can be determined from the expressions:

$$\alpha = 4\pi\, \frac{T_1 D_1 - T_2 D_2}{T_1^{\ 2} - T_2^{\ 2}} \tag{4. 7. 7}$$

$$\beta = T_1\, T_2\, \frac{T_1 D_2 - T_2 D_1}{\pi\left(T_1^2 - T_2^2\right)} \qquad (4.7.8)$$

In the special case of stiffness proportional damping, we have $\alpha = 0$ and β is obtained from Eq. (4.7.6) as

$$\beta = \frac{D_1 T_1}{\pi} \qquad (4.7.9)$$

For mass proportional damping $(\beta = 0)$, α is given by

$$\alpha = \frac{4\pi\, D_1}{T_1} \qquad (4.7.10)$$

For these special cases (mass or stiffness proportional damping), only one damping ratio D_1 for the period T_1 can be prescribed. Generally, the stiffness proportional assumption is more realistic than the mass proportional one, because the latter yields smaller damping values for higher natural frequencies (lower periods), contrary to experience. To illustrate, Fig. 4.7-1 shows curves for the damping ratio D against period for all three RAYLEIGH damping variants discussed, these being the stiffness proportional ($\beta \underline{K}$), the mass proportional ($\alpha \underline{M}$) and the general ($\alpha \underline{M} + \beta \underline{K}$) variant. The computer program ALFBET supplies the coefficients α and β and also the curve D(T) for an arbitrary period range. No input file is required (interactive prompts) and the output file is APERD. The example shown in Fig. 4.7-1 refers to the plane frame of Fig. 4.4-1, the natural periods of which have been determined in Section 4.5 to be equal to 0.7551, 0.1817 and 0.1055 s. A modal damping ratio $D_1 = 0.02$ was chosen for the first natural period T_1=0.7551 s. For a purely mass- or stiffness proportional damping, the coefficients α and β can be computed from Eqs. (4.7.10) and (4.7.9) as α=0.3328 and β=0.004807. If, additionally, a damping value $D_3 = 0.06$ is assumed at $T_3 = 0.1055$ s, α and β become equal to $\alpha = 0.1972$ and $\beta = 0.001959$. Once α and/or β have been determined, the damping ratio D is fixed for all other periods, as shown in Fig. 4.7-1.

The damping matrix $\underline{C}$ itself can be computed by the program CRAY. It furnishes $\underline{C}$ for mass proportional, stiffness proportional or general RAYLEIGH damping, with the stiffness matrix (file KMATR) and the diagonal of the lumped-mass matrix (file MDIAG) as input files. The necessary data for damping ratio(s) and corresponding period(s) are entered interactively, and the resulting damping matrix is written in the output file CMATR.

If more than two given modal damping ratios must be taken into consideration in computing the viscous damping matrix $\underline{C}$, the following approach according to CLOUGH/PENZIEN [4.10] is recommended:

- **Complete modal damping**

Eq. (4.7.2) may be written as

$$\underline{\Phi}^T\, \underline{C}\, \underline{\Phi} = \text{diag}\left[2\, D_i\, \omega_i\right] \qquad (4.7.11)$$

$$\underline{C}\, \underline{\Phi} = \left(\underline{\Phi}^T\right)^{-1}\, \text{diag}\left[2\, D_i\, \omega_i\right] \qquad (4.7.12)$$

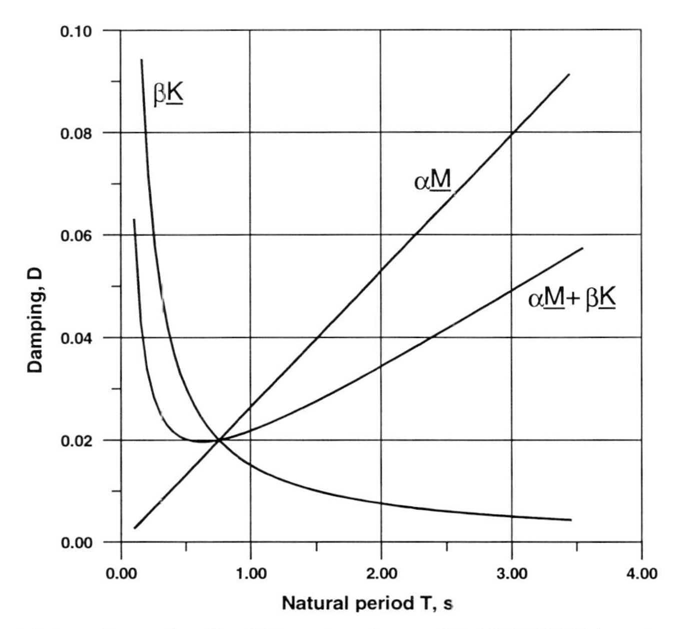

Fig. 4.7-1: Modal damping ratios for different variants of RAYLEIGH damping

$$\underline{C} = \left(\underline{\Phi}^T\right)^{-1} \operatorname{diag}\left[2\, D_i\, \omega_i\right]\, \underline{\Phi}^{-1} \tag{4. 7. 13}$$

and considering that

$$\left(\underline{\Phi}^T\right)^{-1} = \underline{M}\, \underline{\Phi} \tag{4. 7. 14}$$

we obtain

$$\underline{C} = \underline{M}\, \underline{\Phi}\, \operatorname{diag}\left[2\, D_i\, \omega_i\right]\, \underline{\Phi}^T\, \underline{M} \tag{4. 7. 15}$$

All known modal damping ratios $D_i, i=1,\ldots,r$ may now be included. As an example, the damping matrix of the three-story frame of Fig. 4.4-1 is evaluated. Assuming damping ratios of $D_1 = 0.02$, $D_2 = 0.04$ and $D_3 = 0.06$ for the three natural periods given above, the program CMOD writes the following damping matrix in the file CMATR:

```
 .77036E+02 -.58677E+02  .76702E+01
-.58677E+02  .98537E+02 -.40996E+02
 .76702E+01 -.40996E+02  .35133E+02
```

CMOD needs MDIAG, PHI and OMEG as input files and also (in the file DAEM) the r damping values D_i. To check this result, the transformation $\underline{\Phi}^T\, \underline{C}\, \underline{\Phi}$ is carried out by means of the program FTCF which needs PHI and CMATR as input files. The output file CCMAT of FTCF contains the matrix

```
0.3328    0.0000    0.0000
0.0000    2.7660    0.0000
0.0000    0.0000    7.1452
```

that is, $\mathrm{diag}\left[2\ D_i\ \omega_i\right]$ with modal damping ratios of 2%, 4% and 6% , as assumed.

The single elements of a building structure (columns, girders, shear walls, load-carrying and non load-carrying partition walls) exhibit different damping characteristics which usually depend on the vibration amplitude. If the single damping ratios of all elements and subassemblages are known, the damping matrix $\underline{C}$ of the overall structure can be set up by adding the contributions of each element as already done for the overall stiffness matrix. However, the resulting damping matrix containing the contribution of all element damping matrices is generally "non-proportional", in the sense that it cannot be diagonalised by the mode shapes of the undamped system. Apart from tackling Eq. (4.7.1) straightforwardly by Direct Integration, the following possibilities exist for solving a reduced equation system in the time domain:

1. Modal Analysis with complex eigenvectors,

2. Approximate reduction to the proportionally damped case:

    2.1 By equivalent modal damping

    2.2 By somehow diagonalising $\underline{\Phi}^T\ \underline{C}\ \underline{\Phi}$

3. Direct Integration of a subset of coupled modal equations.

Some remarks to the single approaches follow:

ad 1: As already mentioned in Section 4.5, complex eigenvalues and eigenvectors can be used to decouple a non-proportionally damped system [4.1][4.5][4.6]. However, the numerical effort is noticeable and not always compatible with the additional gain in accuracy considering the suitability of the viscous damping model in the first place.

ad 2.1: An equivalent modal damping ratio can be derived by somehow "weighing" the damping ratios D_i of the single structural elements [4.7]. The weighing can be carried out according to the energy stored in the elements in single modal shapes:

$$D_j = \frac{\sum_i D_i\ W_{i,j}}{\sum_i W_{i,j}} \tag{4.7.16}$$

Here, $W_{i,j}$ is the work stored in the ith structural element in the jth mode shape, given by

$$W_{i,j} = \frac{1}{2}\ \underline{\Phi}_j^{\ T}\ \underline{k}_i\ \ \underline{\Phi}_j \tag{4.7.17}$$

ad 2.2: It is also possible to simply disregard the non-diagonal elements of the product $\underline{\Phi}^T\ \underline{C}\ \underline{\Phi}$ and thus decouple the system. A serious drawback of this approach is the impossibility of realistically assessing the error margin involved.

ad 3: According to MOJTAHEDI [4.8], a reduced coupled differential equation system in modal Co-ordinates as variables may be treated by Direct Integration in the time domain. The reduction results from simply omitting higher mode contributions. A drawback of this approach is that the influence of a higher-mode contribution might be important because of the damping coupling.

If all damping ratios of the structural elements are known (which is normally not the case), the program CALLG can be used to compute the corresponding non-proportional damping matrix in all active kinematic degrees of freedom (that is, without condensation to a few master degrees of freedom). In CALLG, the stiffness matrix of each of the NELEM beam elements is multiplied by a factor β according to Eq. (4.7.9) for the stiffness-proportional RAYLEIGH damping, and the resulting element damping matrix is added to the overall system damping matrix $\underline{C}$. The input file BETA contains the NELEM β-factors for the elements; if additional dashpots or springs are present, their damping matrices must be available in the input file FEDCMT and their incidence vectors in INZFED. The output file CVOLL contains the square system damping matrix $\underline{C}$.

The programs introduced in this section are summarised below:

Program name	Input files	Output files
ALFBET	-	APERD
CRAY	MDIAG KMATR	CMATR
CMOD	MDIAG PHI OMEG DAEM	CMATR
FTCF	PHI CMATR	CCMAT
CALLG	EKOND BETA INZFED FEDCMT	CVOLL

4.8 Direct integration methods

Direct numerical integration methods require no modal analysis and are also applicable to non-linear systems. They generate numerically approximate solutions for a response process $\{\underline{V}, \dot{\underline{V}}, \ddot{\underline{V}}\}$ at discrete time points $t = 1 \cdot \Delta t,\ 2 \cdot \Delta t,\ \ldots, n \cdot \Delta t$. As a starting point, it is assumed that the response process is known at a defined time t (usually t = 0) and from these known initial values (and the known excitation process $\underline{P}$) the system response at time $t + \Delta t$ must be calculated. The expressions derived for the single degree of freedom system in section 3.4 retain their validity for multi-degree-of-freedom systems, if, instead of the mass m, damping coefficient c, stiffness k, load function F(t) and system response u(t), the corresponding matrices or vectors $\underline{M}$, $\underline{C}$, $\underline{K}$, $\underline{P}$ and $\underline{V}$ are introduced.

The importance of using an unconditionally stable integration scheme has already been stressed in section 3.4 for the single-degree-of-freedom oscillator; with multi-degree-of-freedom systems this choice is even more important because the time step Δt is normally larger than the highest natural period of the structure. If "blow up" occurred, not only would the higher mode contributions to the solution be lost, but also the important low-frequency solution components would be rendered useless. Therefore, in structural dynamic applications, unconditionally stable implicit single-step integration algorithms such as the NEWMARK method [4.9] are used very often. In the NEWMARK algorithm, the solution at time $t+\Delta t$ is given as follows:

$$\underline{\dot{V}}_{t+\Delta t} = \underline{\dot{V}}_t + \int_t^{t+\Delta t} \underline{\ddot{V}}(\tau)\, d\tau \tag{4. 8. 1}$$

$$\underline{V}_{t+\Delta t} = \underline{V}_t + \underline{\dot{V}}_t\, \Delta t + \int_t^{t+\Delta t} \underline{\ddot{V}}(\tau)(t+\Delta t-\tau)\, d\tau \tag{4. 8. 2}$$

with the integrals being evaluated numerically. This leads to the expressions

$$\underline{\dot{V}}_{t+\Delta t} = \underline{\dot{V}}_t + \Delta t\cdot(1-\gamma)\underline{\ddot{V}}_t + \Delta t\cdot\gamma\cdot\underline{\ddot{V}}_{t+\Delta t} \tag{4. 8. 3}$$

$$\underline{V}_{t+\Delta t} = \underline{V}_t + \underline{\dot{V}}_t\, \Delta t + (\Delta t)^2\cdot\left(\frac{1}{2}-\beta\right)\cdot\underline{\ddot{V}}_t + (\Delta t)^2\cdot\beta\cdot\underline{\ddot{V}}_{t+\Delta t} \tag{4. 8. 4}$$

where the parameters β und γ for the unconditionally stable „constant acceleration pattern" have the values,

$$\beta=\frac{1}{4},\ \gamma=\frac{1}{2} \tag{4. 8. 5}$$

The acceleration within the time step ($t\le\tau\le t+\Delta t$) is given by

$$\underline{\ddot{V}}(\tau) = \frac{1}{2}\left(\underline{\ddot{V}}_t + \underline{\ddot{V}}_{t+\Delta t}\right) \tag{4. 8. 6}$$

For parameter values

$$\beta=\frac{1}{6},\ \gamma=\frac{1}{2} \tag{4. 8. 7}$$

the acceleration varies linearly within the time step according to

$$\underline{\ddot{V}}(\tau) = \underline{\ddot{V}}_t + \frac{\tau}{\Delta t}\left(\underline{\ddot{V}}_{t+\Delta t} - \underline{\ddot{V}}_t\right) \tag{4. 8. 8}$$

This „linear acceleration" variant is, however, only conditionally stable and should be used with some caution. A meaningful choice of the time step for the unconditionally stable „constant acceleration" variant amounts to about one quarter of the smallest system natural period T_n , but also with larger values of Δt one can obtain sufficiently accurate solutions when these mainly contain low-frequency components.

The computer program NEWMAR requires the square system stiffness and damping matrices $\underline{K}$ and $\underline{C}$ as input, to be read from the files KMATR and CMATR, while the diagonal of the mass matrix $\underline{M}$ is read from the file MDIAG. The determination of CMATR can be carried out, for example, by means of the programs CRAY or CMOD as described in section 4.7. The single steps of the algorithm are as follows:

1st step : Assigning values to the parameters ß and γ and evaluating the constants c_1 to c_6 :

$$c_1 = \frac{1}{\beta(\Delta t)^2} \tag{4. 8. 9}$$

$$c_2 = \frac{1}{\beta\,\Delta t} \tag{4. 8. 10}$$

$$c_3 = \frac{1}{2\beta} \tag{4. 8. 11}$$

$$c_4 = \frac{\gamma}{\beta\,\Delta t} = c_2\,\gamma \tag{4. 8. 12}$$

$$c_5 = \frac{\gamma}{\beta} \tag{4. 8. 13}$$

$$c_6 = \Delta t\left(\frac{\gamma}{2\beta} - 1\right) = \Delta t\left(\frac{c_5}{2} - 1\right) \tag{4. 8. 14}$$

2nd step : Evaluation of

$$\underline{K}^* = \underline{K} + c_1\,\underline{M} + c_4\,\underline{C} \tag{4. 8. 15}$$

and calculation of the reciprocal matrix $\underline{F}^*$ of $\underline{K}^*$.

3rd step : Evaluation of the generalised load as

$$\underline{P}^* = \Delta\underline{P} + \underline{M}\left(c_2\,\dot{\underline{V}}_t + c_3\,\ddot{\underline{V}}_t\right) + \underline{C}\left(c_5\,\dot{\underline{V}}_t + c_6\,\ddot{\underline{V}}_t\right) \tag{4. 8. 16}$$

where $\Delta\underline{P} = \underline{P}_{t+\Delta t} - \underline{P}_t$.

4th step: Determination of the displacement increment $\Delta\underline{V}$ from

$$\Delta\underline{V} = \underline{V}_{t+\Delta t} - \underline{V}_t = \underline{F}^* \cdot \underline{P}^* \tag{4. 8. 17}$$

5th step : Determination of the solution vectors at time $t + \Delta t$:

$$\ddot{\underline{V}}_{t+\Delta t} = \ddot{\underline{V}}_t + c_1\,\Delta\underline{V} - c_2\,\dot{\underline{V}}_t - c_3\,\ddot{\underline{V}}_t \tag{4. 8. 18}$$

$$\dot{\underline{V}}_{t+\Delta t} = \dot{\underline{V}}_t + c_4\,\Delta\underline{V} - c_5\,\dot{\underline{V}}_t - c_6\,\ddot{\underline{V}}_t \tag{4. 8. 19}$$

$$\underline{V}_{t+\Delta t} = \underline{V}_t + \Delta\underline{V} \tag{4. 8. 20}$$

For linear systems no updating of system properties and evaluation of unbalanced forces is necessary and the algorithm proceeds by evaluating the next generalised load as in step 3.

A summary of the input and output files of the program NEWMAR follows:

Program name	Input files	Output files
NEWMAR	KMATR MDIAG CMATR LASTV V0 VP0	THIS.NEW THISDU

In the file THIS.NEW, the computed values of (alternatively) displacement, velocity or acceleration in the NDU master degrees of freedom are given together with the corresponding time points, with the latter in the first column. The file THISDU, containing the displacements in the NDU master degrees of freedom without the corresponding time points, is used as an input file by the program INTFOR which evaluates the internal forces in the structure.

As an example, once again we consider the frame of Fig. 4.4-1, which, in section 4.5, has already been analysed for the loading vector $\underline{P}^T = f(t) \cdot (10.0 \;\; 10.0 \;\; 5.0)$ acting at the 1st, 2nd and 3rd story level, respectively, by modal analysis . The program CMOD provides the following damping matrix for 2% damping ratio in all three modes:

```
  .35673E+02 -.19323E+02  .29939E-01
 -.19323E+02  .37126E+02 -.11639E+02
  .29939E-01 -.11639E+02  .13368E+02
```

The stiffness and mass matrix are already known from section 4.4:

$$\underline{K} = \begin{bmatrix} .3429090E+05 & -.2933414E+05 & .4739090E+04 \\ -.2933414E+05 & .4707862E+05 & -.2116977E+05 \\ .4739090E+04 & -.2116977E+05 & .1678259E+05 \end{bmatrix} \quad (4.8.21)$$

$$\text{diag}\,\underline{M} = [30.0 \;\; 30.0 \;\; 8.0] \quad (4.8.22)$$

Time histories for the displacements, velocities and accelerations of the 1st and 3rd story of the frame are shown in Figs. 4.8-1, 4.8-2 and 4.8-3. The time step chosen was 0.01s and the results agree completely with the ones obtained in section 4.5 by modal analysis.

As already explained in section 4.5, the further determination of the deformations in the slave degrees of freedom and the internal forces can be carried out by means of the program INTFOR, which obtains its data from the output file THISDU of NEWMAR. Additionally, INTFOR requires the transformation matrix $\underline{A}$ which is supplied by the program KONDEN as file AMATR, and also the input file EKOND of KONDEN. More details on the output of INTFOR, especially regarding the options for determining extreme values and time histories of given internal forces, are given in section 4.5.

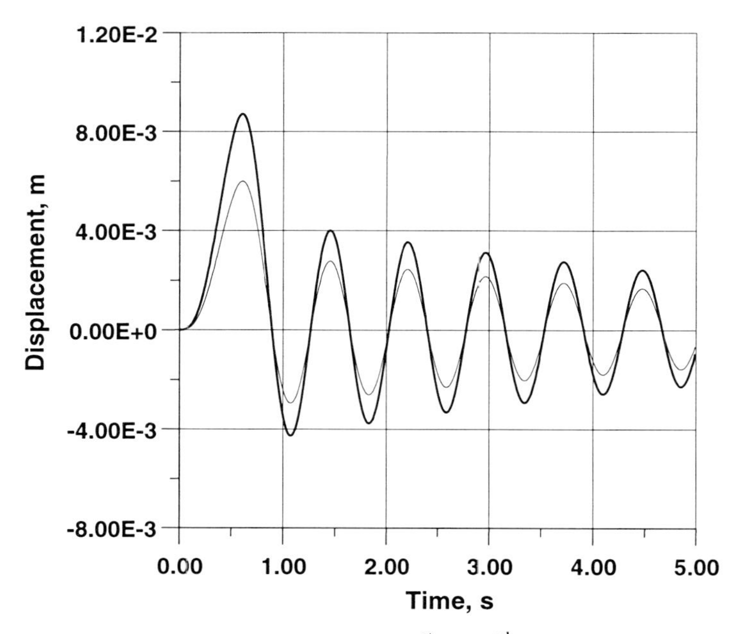

Fig. 4.8-1: Horizontal displacement time histories, 1st and 3rd story

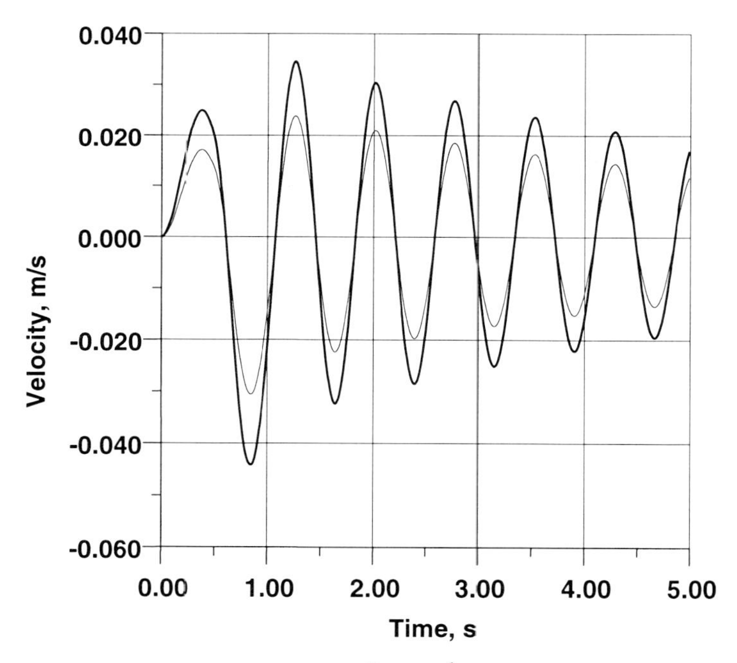

Fig. 4.8-2: Horizontal velocity time histories, 1st and 3rd story

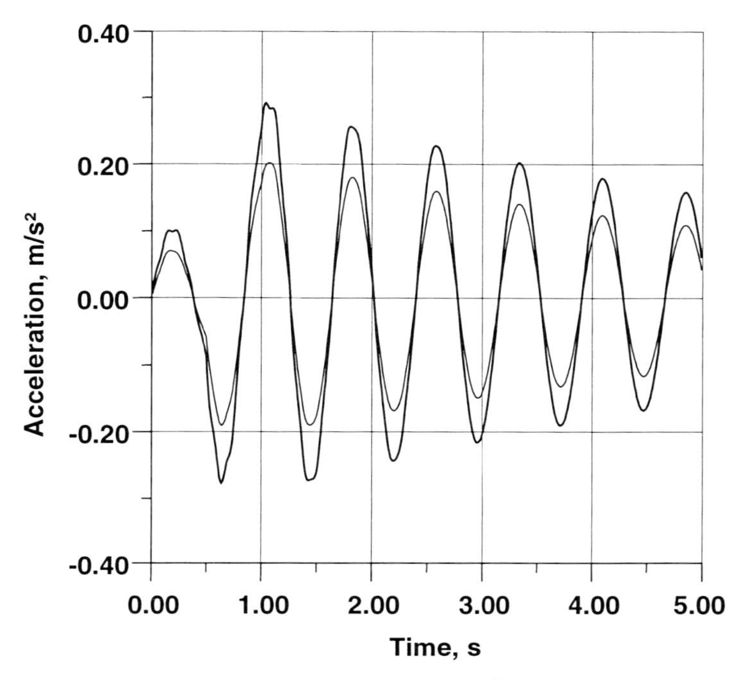

Fig. 4.8-3: Horizontal acceleration time histories, 1st and 3rd story

4.9 Frequency domain methods

The transformation of the differential equation system to an algebraic equation system in the frequency domain by means of FOURIER or LAPLACE integral transformations is yet another way for solving the differential equation system (4.7.1). Physically, this approach corresponds to a description of the structure through transfer functions (frequency response functions), which reproduce its response for stationary harmonic excitation. The pertinent transfer functions are multiplied by the frequency spectrum of the loading, and the resulting system response in the frequency domain can then be transformed back to the time domain by means of an inverse integral transformation. The method is, in principle, only applicable to linear systems, since it makes use of the principle of superposition in the frequency domain. It is especially suitable when elastic or visco-elastic continua form part of the system (e.g. an elastic half-space as foundation medium) or when soil-structure interaction effects are important.

We again consider the differential equation system (4.7.1) with a stationary harmonic excitation of the form

$$\underline{P}(t)=\underline{\tilde{P}}(i\Omega)\, e^{i\Omega t} \tag{4.9.1}$$

Assuming an analogous solution

$$\underline{V}(t)= \underline{\tilde{V}}(i\Omega)\, e^{i\Omega t} \tag{4.9.2}$$

leads to the complex equation system

$$\left(-\Omega^2 \underline{M} + i\Omega \underline{C} + \underline{K}\right) \underline{\tilde{V}}(i\Omega) = \underline{\tilde{P}}(i\Omega) \tag{4.9.3}$$

or

$$\underline{\tilde{K}}(i\Omega)\, \underline{\tilde{V}}(i\Omega) = \underline{\tilde{P}}(i\Omega) \tag{4.9.4}$$

with

$$\underline{\tilde{K}}(i\Omega) = -\Omega^2 \underline{M} + i\Omega \underline{C} + \underline{K} \tag{4.9.5}$$

The system response to a harmonic excitation (or the FOURIER components of non-harmonic excitations) is thus given by

$$\underline{\tilde{V}}(i\Omega) = \left[\underline{\tilde{K}}(i\Omega)\right]^{-1} \underline{\tilde{P}}(i\Omega) = \underline{H}(i\Omega)\, \underline{\tilde{P}}(i\Omega) \tag{4.9.6}$$

Here, $\underline{H}(i\Omega)$ is the complex matrix containing the frequency response functions of the system (4.7.1):

$$\underline{H}(i\Omega) = \left[\underline{\tilde{K}}(i\Omega)\right]^{-1} = \left[-\Omega^2 \underline{M} + i\Omega \underline{C} + \underline{K}\right]^{-1} \tag{4.9.7}$$

The numerical effort necessary for determining $\underline{H}(i\Omega)$ should not be underestimated, because the inversion of the expression in brackets shown in Eq. (4.9.7) must be carried out for a large number of frequencies Ω. It is usually easier to first solve the eigenvalue problem in order to determine the frequency response functions of the modal co-ordinates. Due to the orthogonality property of the mode shapes, the frequency response functions of modal co-ordinates can be obtained directly, without having to solve an equation system. Furthermore, the consideration of only a few modal contributions is usually sufficient for solutions of quite acceptable accuracy. For modal co-ordinates $\underline{\tilde{\eta}}$ defined according to

$$\underline{\tilde{V}}(i\Omega) = \underline{\Phi}\, \underline{\tilde{\eta}}(i\Omega) \tag{4.9.8}$$

we obtain from Eq. (4.9.4):

$$\underline{\tilde{K}}(i\Omega)\, \underline{\Phi}\, \underline{\tilde{\eta}}(i\Omega) = \underline{\tilde{P}}(i\Omega) \tag{4.9.9}$$

and pre-multiplying with the transpose of the modal matrix yields

$$\underline{\Phi}^T \underline{\tilde{K}}(i\Omega)\, \underline{\Phi}\, \underline{\tilde{\eta}}(i\Omega) = \underline{\Phi}^T \left[-\Omega^2 \underline{M} + i\Omega \underline{C} + \underline{K}\right] \underline{\Phi}\, \underline{\tilde{\eta}}(i\Omega) = \underline{\Phi}^T\, \underline{\tilde{P}}(i\Omega) \tag{4.9.10}$$

Carrying out the multiplication leads to the following expression for the ith mode shape:

$$\left[-\Omega^2 + i\Omega\, \underline{\Phi}_i^T \underline{C}\, \underline{\Phi}_i + \omega_i^2\right] \underline{\tilde{\eta}}_i(i\Omega) = \underline{\Phi}_i^T\, \underline{\tilde{P}}(i\Omega) \tag{4.9.11}$$

For the special case of proportional damping we obtain:

$$\left[-\Omega^2 + i\Omega \cdot 2 D_i\, \omega_i + \omega_i^2\right] \underline{\tilde{\eta}}_i(i\Omega) = \underline{\Phi}_i^T\, \underline{\tilde{P}}(i\Omega) \tag{4.9.12}$$

Finally, the expression for the ith modal co-ordinate is given by

$$\underline{\tilde{\eta}}_i(i\Omega) = \underline{H}_i\, \underline{\Phi}_i^T\, \underline{\tilde{P}}(i\Omega) = \frac{1}{\omega_i^2 - \Omega^2 + i\Omega \cdot 2 D_i\, \omega_i} \underline{\Phi}_i^T\, \underline{\tilde{P}}(i\Omega) \tag{4.9.13}$$

with $\underline{H}_i$ as frequency response function. The modal frequency response matrix is then given by the diagonal matrix

$$\underline{H}_{\eta}(i\Omega) = \operatorname{diag}(H_i) = \operatorname{diag}\left[\frac{1}{\omega_i^2 - \Omega^2 + 2i\,D_i\omega_i\,\Omega}\right] \qquad (4.9.14)$$

The solution for the original co-ordinates in the frequency domain is

$$\underline{\tilde{V}}(i\Omega) = \underline{\Phi}\;\underline{H}_{\eta}\,\underline{\Phi}^T\,\underline{\tilde{P}}(i\Omega) \qquad (4.9.15)$$

By carrying out the inverse transform

$$\underline{V}(t) = \int_{-\infty}^{\infty} \underline{\tilde{V}}\,e^{i\omega t}\,d\omega \qquad (4.9.16)$$

we finally obtain the components of the solution vector in the form of time functions.

While in the time domain the viscous damping model $\underline{F}_D = \underline{C}\,\dot{\underline{V}}$ is the most popular, mostly due to its simplicity, the so-called hysteretic damping is often used in the frequency domain. This damping model, also known as material damping, exhibits frequency independent damping properties in agreement with experimental results, while viscous damping is strongly frequency dependent. The damping force due to linear material damping is proportional to the displacement with a phase lag of $\pi/2$, according to

$$\underline{F}_D = i \cdot d \cdot \underline{K}\,\underline{V} \qquad (4.9.17)$$

Here, i is the imaginary unit, d the damping coefficient ("loss factor"), $\underline{K}$ the stiffness matrix and $\underline{V}$ the displacement vector. A general complex matrix including stiffness and material damping is given by

$$\underline{K}_{ges} = \underline{K} + i\,\underline{K}_d \qquad (4.9.18)$$

yielding the differential equation system

$$\underline{M}\,\ddot{\underline{V}} + \underline{K}_{ges}\,\underline{V} = \underline{P}(t) \qquad (4.9.19)$$

Eqs. (4.9.1) and (4.9.2) lead to

$$\left(-\Omega^2\,\underline{M} + \underline{K} + i\,\underline{K}_d\right)\underline{\tilde{V}} = \underline{\tilde{P}} \qquad (4.9.20)$$

For the special case of Eq. (4.9.17) with $\underline{K}_d = d\,\underline{K}$ the modal analytical approach leads to

$$\underline{\Phi}^T\left[-\Omega^2\,\underline{M} + +\underline{K} + i\,\underline{K}_d\right]\underline{\Phi}\,\underline{\tilde{\eta}}(i\Omega) = \underline{\Phi}^T\,\underline{\tilde{P}}(i\Omega) \qquad (4.9.21)$$

$$\left[-\Omega^2 + \omega_i^2(1 + i\,d_i)\right]\,\underline{\tilde{\eta}}_i(i\Omega) = \underline{\Phi}_i^T\,\underline{\tilde{P}}(i\Omega) \qquad (4.9.22)$$

and further to the following expression which corresponds to Eq. (4.9.13):

$$\underline{\tilde{\eta}}_i(i\Omega) = \frac{1}{\omega_i^2 - \Omega^2 + i d_i\,\omega_i^2}\,\underline{\Phi}_i^T\,\underline{\tilde{P}}(i\Omega) \qquad (4.9.23)$$

In the case of resonance, $\Omega = \omega_i$, the condition of equal amplification factors for viscous and material damping leads, through the comparison of Eqs. (4.9.23) and (4.9.13), to the expression

$$D_i = \frac{d_i}{2} \tag{4.9.24}$$

As shown by this derivation, this expression is, in a strict sense, only valid for the case of resonance. Furthermore, the material damping model of Eq. (4.9.18) itself is actually only valid for stationary harmonic vibrations, although in practice it is also applied to transient vibrations.

5 Systems with distributed mass and stiffness properties

5.1 General

Frame structures with continuously distributed mass and elasticity have an infinite number of degrees of freedom, which may be thought of as corresponding to an infinite number of infinitesimal lumped masses. Their motion can be described by partial differential equations in space and time, the solution of which is more time-consuming than the treatment of systems such as discrete multi-degree-of-freedom (MDOF) systems, which can be described by ordinary differential equations. Closed-form solutions are only available for special cases, e.g. a simple beam in flexure with uniformly distributed mass and stiffness. If the natural frequencies and vibration modes of these structures are known, their forced vibration response can be investigated using modal analysis, providing the system behaves linearly. Analogous to previously, we obtain the following ordinary differential equation for the ith modal co-ordinate Y_i

$$\ddot{Y}_i(t) + 2\xi_i\omega_i\dot{Y}_i(t) + \omega_i^2 Y_i(t) = \frac{P_i(t)}{M_i} \tag{5. 1. 1}$$

with M_i as generalised mass and P_i as generalised load. As an example, for a beam of length ℓ and mass distribution m(x) these values are given by

$$M_i = \int_0^\ell m(x)\,\varphi_i^2(x)\,dx, \quad P_i(t) = \int_0^\ell P(x,t)\,\varphi_i(x)\,dx \tag{5. 1. 2}$$

in analogy to $M_i = \underline{\Phi}_i^T \underline{M}\,\underline{\Phi}_i$ and $P_i = \underline{\Phi}_i^T \underline{P}$ of the discrete multi-degree-of-freedom idealisation. In Eq. (5.1.2) $\varphi_i(x)$ represents the ith mode shape and P(x,t) the loading, which is variable in space and time. Usually, only relatively few mode shapes need to be considered.

A useful way of determining the response values in a dynamically excited system with distributed mass and stiffness is to examine its response to stationary forced harmonic vibration with a loading function $P(x, t) = P(x)\sin\omega t$, since in this case, the time function for all force and deformation quantities is already known. After the transient response at the onset of the loading has been damped out, the generalised deflection V(x, t) as a result of the harmonic loading $P(x, t) = P(x)\sin\omega t$ is given by

$$V(x,t) = V(x)\,\sin\omega t \tag{5. 1. 3}$$

This allows the variables of time and space to be separated, and ordinary differential equations for the deflection curve V(x) with $\sin\omega t = 1$ are obtained. As in the general deformation method of statics, the solution of these differential equations leads to the equation system

$$\underline{P} = \underline{K}_{dyn} \cdot \underline{V} \tag{5. 1. 4}$$

which describes the relationship between the harmonically varying nodal forces $\underline{P}$ and the likewise harmonically varying nodal displacements $\underline{V}$. The "dynamic" coefficient k_{ij} in $\underline{K}_{dyn}$ physically represents the harmonic (with circular frequency ω) force component acting along the degree of freedom i which causes a harmonically varying deflection of unit amplitude in the degree of freedom j. It can be shown (PAZ, [5.1]) that a series decomposition of these dynamic influence coefficients yields the coefficients of the "static" stiffness matrix and the consistent mass matrix. Damping can be introduced e.g. by the linear viscous model, in which case the energy dissipated by damping is proportional to the square of the amplitude of vibration. Since the work done by the restoring forces is also proportional to the square of the vibration amplitude, the loss factor d defined in Eq. (3.3.30) is independent of the vibration amplitude and the principle of superposition retains its validity. Setting

$$\sigma = E(\varepsilon + \vartheta\dot{\varepsilon}) \tag{5. 1. 5}$$

and $\varepsilon = \varepsilon_0 \, e^{i\omega t}$ leads to

$$\sigma = \varepsilon \hat{E}, \quad \hat{E} = E\left(1 + id\right) \tag{5. 1. 6}$$

with the loss factor $d = \omega\vartheta$. This result shows that linear-viscous damping can be considered simply by replacing E with the complex modulus of elasticity $\hat{E}$.

In the following sections 5.2 to 5.6, some frequently used dynamic stiffness matrices of common structural elements will be derived.

5.2 Longitudinal vibration of straight bars

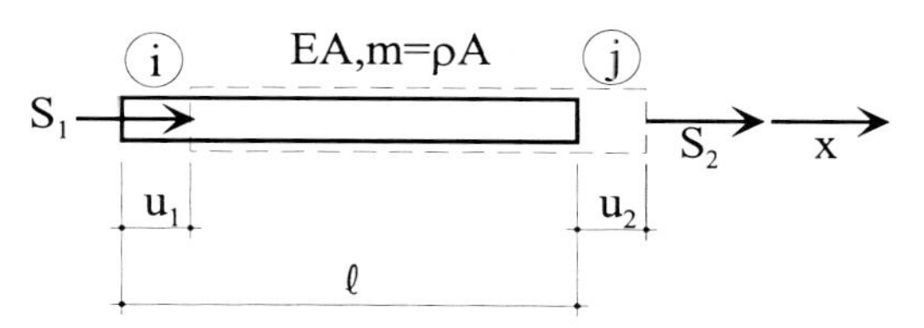

Fig. 5.2-1: Longitudinally vibrating bar

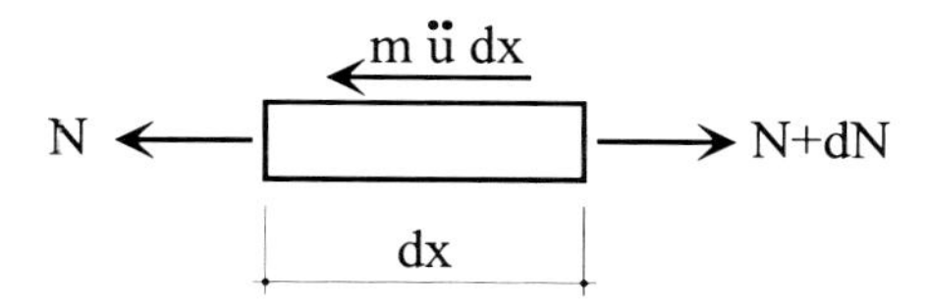

Fig. 5.2-2: Equilibrium of an infinitesimal element

Consider the bar in Fig. 5.2-1 of constant mass distribution $m = \rho A$, length ℓ and axial stiffness EA. The longitudinal forces S_1 and S_2 act on the end sections i and j of the element; they vary harmonically with a circular frequency ω and they cause displacements u_1 and u_2 respectively. The equilibrium equation (see Fig. 5.2-2) is given by

$$\frac{dN}{dx} = N' = m\ddot{u} \tag{5. 2. 1}$$

Here, as in the following, the dots represent the derivative with respect to time, and the primes the derivative with respect to the x co-ordinate. The constitutive law is as follows

$$N = EA\,\varepsilon \tag{5. 2. 2}$$

EA is the axial stiffness and ε the longitudinal strain which is related to the longitudinal displacement by

$$\varepsilon = \frac{du}{dx} = u' \tag{5.2.3}$$

By substituting the derivative u' for the strain ε in Eq. (5.2.2) and differentiating it again with respect to x we obtain

$$\frac{dN}{dx} = EA\,u'' \tag{5.2.4}$$

The term on the left-hand side may be written as $m\ddot{u}$ according to Eq. (5.2.1), and for a harmonic vibration $u(x, t) = u(x) \sin \omega t$, $\ddot{u} = -\omega^2 u$ we obtain the differential equation for the deflected shape u(x):

$$EAu'' + m\omega^2 u = 0. \tag{5.2.5}$$

With the substitution

$$\psi = \ell \sqrt{\frac{m\omega^2}{EA}} \tag{5.2.6}$$

the general solution of Eq. (5.2.5) can be written as:

$$u(x) = C_1 \cos\left(\frac{\psi x}{\ell}\right) + C_2 \sin\left(\frac{\psi x}{\ell}\right). \tag{5.2.7}$$

For the general boundary conditions $u(0) = u_1$, $u(\ell) = u_2$, the constants emerge as

$$\begin{aligned} C_1 &= u_1, \\ C_2 &= \frac{u_2}{\sin\psi} - u_1 \cot\psi. \end{aligned} \tag{5.2.8}$$

The forces normal to the end sections may be written as:

$$\begin{aligned} S_1 &= -N(0) = -EA\,u'(0), \\ S_2 &= N(\ell) = EA\,u'(\ell), \end{aligned} \tag{5.2.9}$$

and this leads to:

$$\begin{bmatrix} S_1 \\ S_2 \end{bmatrix} = \frac{EA}{\ell} \begin{bmatrix} \psi\cot\psi & -\dfrac{\psi}{\sin\psi} \\ -\dfrac{\psi}{\sin\psi} & \psi\cot\psi \end{bmatrix} \begin{bmatrix} u_1 \\ u_2 \end{bmatrix} \tag{5.2.10}$$

There is an important difference between this "dynamic" case and the static case: In the "static" expression corresponding to Eq. (5.2.10)

$$\begin{bmatrix} S_1 \\ S_2 \end{bmatrix} = \frac{EA}{\ell} \begin{bmatrix} 1 & -1 \\ -1 & 1 \end{bmatrix} \begin{bmatrix} u_1 \\ u_2 \end{bmatrix} \tag{5.2.11}$$

the stiffness matrix $\underline{K}$ is always singular, if the rigid body displacements are not eliminated by support conditions present. This is not the case with the dynamic matrix $\underline{K}_{dyn}$ in Eq. (5.2.10) (leading to the so-called free-free vibration mode), which, however, includes the static rigid body displacement mode (for $\omega = 0$).

5.3 Torsional vibration of straight bars

We consider the bar depicted in Fig. 5.3-1 with length ℓ, torsional stiffness GI_T and mass moment of inertia Θ per unit length with reference to the longitudinal bar axis.

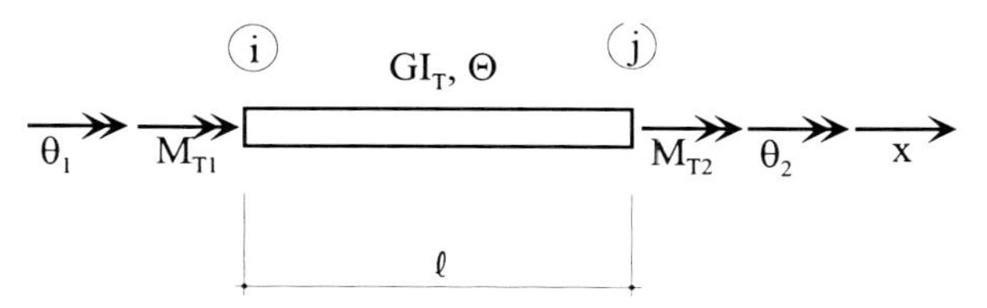

Fig. 5.3-1: Bar element in torsion

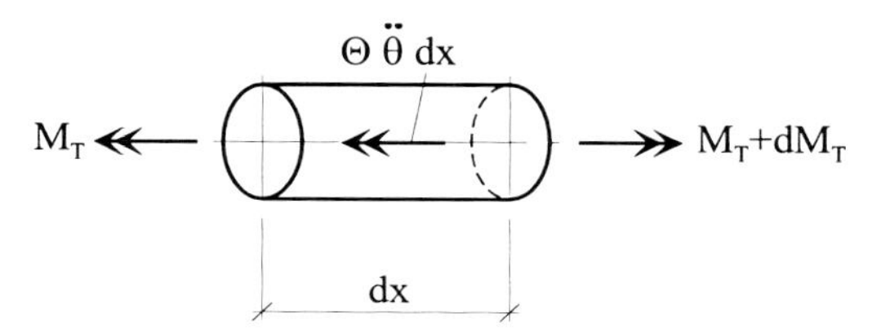

Fig. 5.3-2: Equilibrium of an infinitesimal element

The torsional moments acting on the end sections of the element are

$$M_{T,1}(t) = M_{T,1}\sin\omega t \tag{5. 3. 1}$$

and

$$M_{T,2}(t) = M_{T,2}\sin\omega t \tag{5. 3. 2}$$

They cause end rotations ϑ_1 and ϑ_2 about the element axis. The consideration of equilibrium of the differential element (see Fig. 5.3-2) yields

$$\frac{dM_T}{dx} = \Theta\,\ddot{\vartheta} \tag{5. 3. 3}$$

The constitutive law and compatibility are given by

$$M_T = G\,I_T\,\vartheta' \tag{5. 3. 4}$$

with ϑ' as torsional rotation per unit length. This leads to

$$\frac{dM_T}{dx} = G\,I_T\,\vartheta'' \tag{5. 3. 5}$$

and with Eq. (5.3.3) for harmonic vibrations $\ddot{\vartheta} = -\omega^2\vartheta$:

$$\Theta\left(-\omega^2\vartheta\right) = G\,I_T\vartheta'' \tag{5. 3. 6}$$

This produces the desired differential equation for the "torsion curve":

$$G I_T \vartheta'' + \Theta \omega^2 \vartheta = 0 \tag{5. 3. 7}$$

This equation can be solved analogously to the differential equation for the element vibrating longitudinally as presented in section 5.2. By inserting the characteristic parameter λ according to

$$\lambda = \ell \sqrt{\frac{\Theta\omega^2}{GJ_T}} \tag{5. 3. 8}$$

the following relationships between internal forces and displacements at the end sections of the bar are established:

$$\begin{bmatrix} M_{T,1} \\ M_{T,2} \end{bmatrix} = \frac{GJ_T}{\ell} \begin{bmatrix} \lambda \cot \lambda & -\frac{\lambda}{\sin \lambda} \\ -\frac{\lambda}{\sin \lambda} & \lambda \cot \lambda \end{bmatrix} \begin{bmatrix} \varphi_1 \\ \varphi_2 \end{bmatrix} \tag{5. 3. 9}$$

5.4 Flexural vibration of EULER-BERNOULLI beams

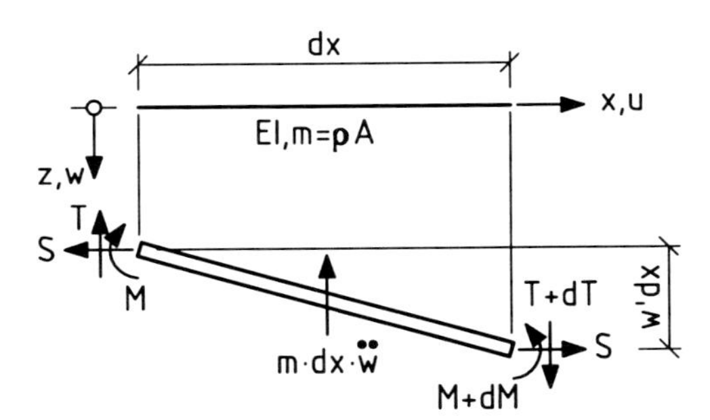

Fig. 5.4-1: Internal and inertia forces for the EULER-BERNOULLI beam

The vibration of slender beams in a predominantly flexural mode, with negligible effects of shear deformations and rotational inertia, can be described by the classical EULER-BERNOULLI theory. The differential element shown in Fig. 5.4-1 has a constant bending stiffness EI and mass m per unit length; the transversal force T acting perpendicular to the initial beam axis is equal to the shear force of the respective section, T = Q, only if 2nd order theory effects are neglected. Consideration of vertical equilibrium gives

$$\frac{dT}{dx} = m\,\ddot{w} \tag{5. 4. 1}$$

and the condition of moment equilibrium

$$M + dM - M - T dx - m\,\ddot{w}\frac{(dx)^2}{2} + S\,w'\,dx = 0 \tag{5. 4. 2}$$

yields, after neglecting higher order terms and differentiating once more with respect to x:

$$\frac{d^2M}{dx^2}-\frac{dT}{dx}+S\,w''=0\,. \tag{5.4.3}$$

The kinematic relationship for a beam element in flexure is given by

$$\frac{dw}{dx}=-\varphi \tag{5.4.4}$$

or, in terms of the radius of curvature κ of the cross-section

$$\frac{d\varphi}{dx}=\varphi'=\kappa \tag{5.4.5}$$

The constitutive law for beams in flexure is

$$M=EI\,\kappa=-EI\,w'' \tag{5.4.6}$$

By combining the expressions from constitutive law, kinematics and equilibrium we determine the differential equation for the deflected shape w(x) as

$$EI\,w^{IV}-Sw''+m\ddot{w}=0 \tag{5.4.7}$$

or

$$w^{IV}-\frac{S}{EI}w''+\frac{m}{EI}\ddot{w}=0 \tag{5.4.8}$$

Neglecting the influence of the axial force, S = 0, furnishes the following differential equation for the deflected shape w(x) of the beam in harmonic vibration with w(x, t) = w(x) sin ωt:

$$w^{IV}-\left(\frac{\lambda^4}{\ell^4}\right)w=0. \tag{5.4.9}$$

Here, λ stands for

$$\lambda=\ell\sqrt[4]{\frac{m\omega^2}{EI}} \tag{5.4.10}$$

The general solution of Eq. (5.4.9) is given by

$$w=C_1\sinh\lambda\xi+C_2\cosh\lambda\xi+C_3\sin\lambda\xi+C_4\cos\lambda\xi, \tag{5.4.11}$$

with the dimensionless co-ordinate $\xi=\frac{x}{\ell}$. The constants C_1 to C_4 are found from the four boundary conditions $w(0)=w_1$, $w'(0)=-\varphi_1$, $w(\ell)=0$ and $w'(\ell)=-\varphi_2$. Using the constitutive law Eq. (5.4.6) and the corresponding expression for the shear force

$$T(x)=Q(x)=-EI\,w'''(x) \tag{5.4.12}$$

the end forces T_1, M_1, T_2, M_2 can be expressed in terms of the end displacements w_1, φ_1, w_2 and φ_2. For the vectors

$$\underline{P}^T=(T_1\,,M_1\,,T_2\,,M_2) \tag{5.4.13}$$

and

$$\underline{V}^T=(w_1,\varphi_1,w_2,\varphi_2) \tag{5.4.14}$$

the dynamic stiffness matrix $\underline{K}_{dyn}$ according to Eq. (5.1.4) becomes

$$\underline{K}_{dyn} = \frac{EI}{1-f_8}\begin{bmatrix} k_{11} & & & \\ k_{21} & k_{22} & \text{symm.} & \\ k_{31} & k_{32} & k_{33} & \\ k_{41} & k_{42} & k_{43} & k_{44} \end{bmatrix} \tag{5. 4. 15}$$

with the coefficients

$$\begin{aligned}
k_{11} &= \frac{\lambda^3}{\ell^3}(f_6 + f_7) \\
k_{21} &= -\frac{\lambda^2}{\ell^2} f_5 \\
k_{22} &= \frac{\lambda}{\ell}(f_6 - f_7) \\
k_{31} &= -\frac{\lambda^3}{\ell^3}(f_1 + f_3) \\
k_{32} &= -\frac{\lambda^2}{\ell^2}(f_2 - f_4) \\
k_{33} &= k_{11} \\
k_{41} &= -k_{32} \\
k_{42} &= \frac{\lambda}{\ell}(f_3 - f_1) \\
k_{43} &= -k_{21} \\
k_{44} &= k_{22}.
\end{aligned} \tag{5. 4. 16}$$

and the auxiliary functions

$$\begin{aligned}
f_1 &= \sin\lambda \\
f_2 &= \cos\lambda \\
f_3 &= \sinh\lambda \\
f_4 &= \cosh\lambda \\
f_5 &= f_1 f_3 \\
f_6 &= f_1 f_4 \\
f_7 &= f_2 f_3 \\
f_8 &= f_2 f_4
\end{aligned} \tag{5. 4. 17}$$

If the circular frequency of the excitation ω coincides with a natural frequency of the beam, its stiffness becomes equal to zero:

$$\det(\underline{K}(\omega)) = 0 \tag{5. 4. 18}$$

This fact can be used for determining the natural frequencies of single beams or of plane frame structures (see section 5.7) by finding the roots ω_i of this transcendental equation. This procedure furnishes also the natural frequencies in the case of free-free vibrations, that is if no support reactions are there. Of course, the natural frequency $\omega = 0$ is also present in this case, corresponding to a rigid body displacement. A pertinent example is given in section 5.8.

The first three natural circular frequencies for some common cases are shown in Table 5.4-1; for plane frames see sections 5.7 and 5.8.

Case	k_1	k_2	k_3
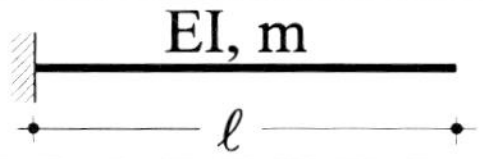 EI, m; ℓ	3.52	22.05	61.70
	2.47	22.18	61.70
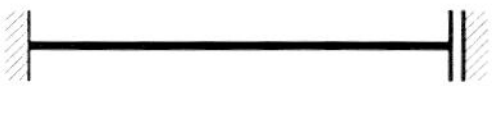	5.59	30.22	74.77
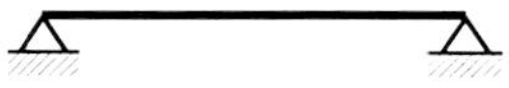	9.86	39.48	88.59
	15.39	49.95	104.3
	22.37	61.70	120.6
Natural circular frequencies $\omega_i = k_i \cdot \sqrt{\frac{EI}{m\ell^4}}$ [rad/s], $i = 1,2,3$			

Table 5.4-1: Natural circular frequencies of simple beams

5.5 Flexural vibration considering axial forces (2nd order theory)

The influence of axial force needs to be considered, for safety reasons, in the case of columns subjected to large compressive forces [5.2]. Referring once again to Fig. 5.4-1 we consider the axial load D = -S (assumed to be constant), with positive compressive force D. The transversal force T and the shear force Q are no longer identical; the expressions relating T, Q, N and S can be written as follows (see Fig. 5.5-1):

$$S = N + Q\varphi \approx N \tag{5. 5. 1}$$

$$T = Q - N\varphi \approx Q - S\varphi = Q + S w' \tag{5. 5. 2}$$

The differential equation for the deflected shape w(x) resulting from a harmonic excitation ($\ddot{w} = -\omega^2 w$) is, according to Eq. (5.4.8)

$$w^{IV} + \frac{D}{EI} w'' - \frac{m\omega^2}{EI} w = 0 \tag{5. 5. 3}$$

Using the substitutions

$$\lambda_1 = \sqrt{-\frac{D}{2EI} + \sqrt{\frac{D^2}{(2EI)^2} + \frac{m\omega^2}{EI}}}$$
$$\lambda_2 = \sqrt{\frac{D}{2EI} + \sqrt{\frac{D^2}{(2EI)^2} + \frac{m\omega^2}{EI}}} \qquad (5.5.4)$$

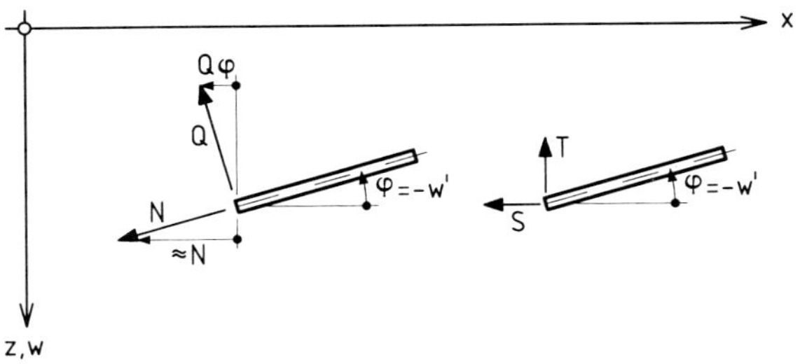

Fig. 5.5-1: Relationship between element end forces (S, T) and (N, Q)

its general solution becomes:

$$w(x) = C_1 \sinh(\lambda_1 x) + C_2 \cosh(\lambda_1 x) + C_3 \sin(\lambda_2 x) + C_4 \cos(\lambda_2 x) \qquad (5.5.5)$$

with $\xi = \frac{x}{\ell}$ as a dimensionless longitudinal co-ordinate. The dynamic stiffness matrix is now given by

$$\underline{K}_{dyn} = \frac{EI}{2\lambda_1\lambda_2(1-f_8) + f_5(\lambda_1^2 - \lambda_2^2)} \begin{bmatrix} k_{11} & & & \\ k_{21} & k_{22} & \text{symm.} & \\ k_{31} & k_{32} & k_{33} & \\ k_{41} & k_{42} & k_{43} & k_{44} \end{bmatrix} \qquad (5.5.6)$$

with the coefficients

$$\begin{aligned}
k_{11} &= f_7(\lambda_1^2\lambda_2^3 + \lambda_1^4\lambda_2) + f_6(\lambda_1\lambda_2^4 + \lambda_1^3\lambda_2^2) \\
k_{21} &= \lambda_1^3\lambda_2 - \lambda_1\lambda_2^3 + f_8(\lambda_1\lambda_2^3 - \lambda_1^3\lambda_2) - 2f_5\lambda_1^2\lambda_2^2 \\
k_{22} &= f_6(\lambda_1\lambda_2^2 + \lambda_1^3) - f_7(\lambda_1^2\lambda_2 + \lambda_2^3) \\
k_{31} &= -f_3(\lambda_1^4\lambda_2 + \lambda_1^2\lambda_2^3) - f_1(\lambda_1^3\lambda_2^2 + \lambda_1\lambda_2^4) \\
k_{32} &= (f_4 - f_2)(\lambda_1\lambda_2^3 + \lambda_1^3\lambda_2) \\
k_{33} &= k_{11} \\
k_{41} &= -k_{32} \\
k_{42} &= f_3(\lambda_1^2\lambda_2 + \lambda_2^3) - f_1(\lambda_1\lambda_2^2 + \lambda_1^3) \\
k_{43} &= -k_{21} \\
k_{44} &= k_{22}
\end{aligned} \qquad (5.5.7)$$

and the auxiliary functions

$$\begin{aligned} f_1 &= \sin \lambda_2 \ell \\ f_2 &= \cos \lambda_2 \ell \\ f_3 &= \sinh \lambda_1 \ell \\ f_4 &= \cosh \lambda_1 \ell \\ f_5 &= f_1 \, f_3 \\ f_6 &= f_1 f_4 \\ f_7 &= f_2 f_3 \\ f_8 &= f_2 f_4 \end{aligned} \tag{5. 5. 8}$$

Natural frequencies can be determined here as earlier by setting the determinant of the stiffness matrix equal to zero. By introducing the constant compressive force D, the natural frequency of the beam decreases in comparison to the case with no axial compression; if D becomes equal to the EULER buckling load, we obtain a first natural frequency of zero, since the element no longer possesses any stiffness. If the vibration modes for buckling and flexural vibration are the same (e.g. as in the case of a simply supported beam, EULER case II), the lowest vibration eigenvalue ω_1^2 is linearly dependent on the ratio of the compressive force to the EULER buckling load D_{ki} (DUNKERLEY line):

$$\left(\frac{\omega^2{}_{II}}{\omega^2{}_{I}} \right) = 1 - \frac{D}{D_{ki}} \tag{5. 5. 9}$$

For the EULER case II (simply supported beam), the buckling load of an element of length ℓ and bending stiffness EI is given by

$$D_{ki} = \frac{EI \, \pi^2}{\ell^2} \tag{5. 5. 10}$$

In cases where the mode shapes for buckling and natural vibration are not the same (e.g. in all other EULER cases as well as in general plane frames), the relationship expressed by Eq. (5.5.9) is only approximate and should be written in the form

$$\left(\frac{\omega^2{}_{II}}{\omega^2{}_{I}} \right) + \frac{D}{D_{ki}} > 1 \tag{5. 5. 11}$$

Assuming the validity of Eq. (5.5.9), the buckling load of the structure can be determined [5.3] by measuring the first natural frequencies ω_1, ω_2 under two different compressive load levels D_1, D_2 and using Eq. (5.5.9) to obtain

$$D_{ki} = \frac{D_2 \cdot \omega_1^2 - D_1 \cdot \omega_2^2}{\omega_1^2 - \omega_2^2} \tag{5. 5. 12}$$

An example is given in section 5.8.

5.6 Flexural vibration of the TIMOSHENKO beam

Shear deformations should be considered in the case of short beams and the rotary inertia of deep cross-sections may also be significant in some cases.

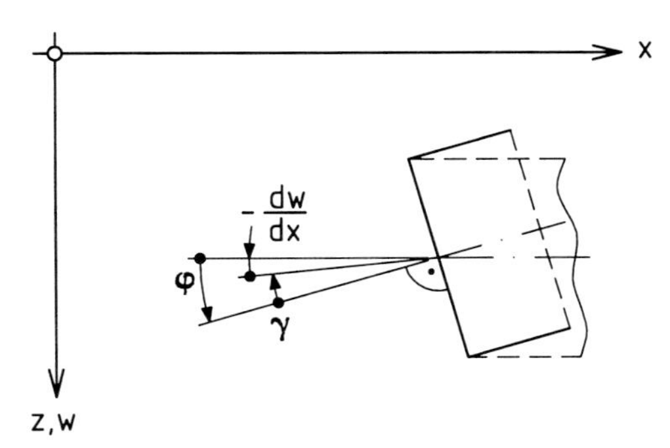

Fig. 5.6-1: Basic relationship for a cross-section subjected to shear

The basic relationship for an element in shear is illustrated by Fig. 5.6-1; the slope of the element axis $\frac{dw}{dx}$ is given by the difference of the angle φ resulting from pure bending and the angle γ due to shear deformation. We obtain

$$\frac{dw}{dx} = w' = \gamma - \varphi \tag{5.6.1}$$

The constitutive law for shear is given by

$$\gamma = \frac{\tau}{G} = \frac{Q}{\alpha_s A} \frac{1}{G} = \frac{Q}{GA_s} \tag{5.6.2}$$

with the shear modulus G, the shear force Q, the average shear stress τ over the cross-section, and the effective shear area A_s, which is calculated by multiplying the actual cross-sectional area A by α_s. The factor α_s has a value of approximately $\frac{5}{6}$ for rectangular cross-sections; for I-sections, where the shear force is only carried by the web, α_s is equal to the ratio of the web area to the total area of the profile. Combining Eqs. (5.6.1) and (5.6.2) yields

$$\varphi' = -w'' + \frac{Q'}{GA_s} \tag{5.6.3}$$

As shear deformation does not contribute to the bending angle, we can write

$$\varphi' = \frac{M}{EI} \Rightarrow M = EI\left(-w'' + \frac{Q'}{GA_s}\right) \tag{5.6.4}$$

With deep cross-sections, harmonic vibration of a differential element causes not only inertia forces $-m\,\ddot{w}\,dx$, but also bending moments equal to $-m\frac{I}{A}\ddot{\varphi}\,dx = -\rho I \ddot{\varphi}\,dx$ resulting from the angular acceleration $\ddot{\varphi}$ (see Fig. 5.6-2). These bending moments are proportional to the radius of gyration $i = \sqrt{\frac{I}{A}}$, where I is the cross-section moment of inertia and A the area of the cross-section. Here, the transversal force T is equal to the shear force Q (1st order theory).

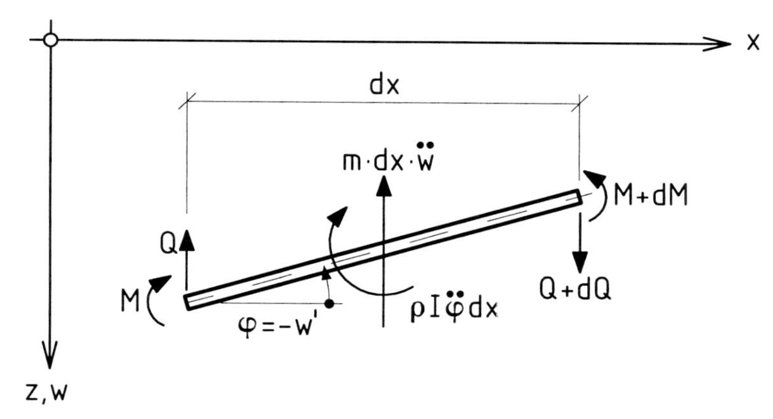

Fig. 5.6-2: Internal forces and inertia forces for a TIMOSHENKO beam

Consideration of vertical equilibrium of the differential element yields

$$\frac{dQ}{dx} - m\ddot{w} = 0 \tag{5. 6. 5}$$

and from moment equilibrium we get

$$\frac{dM}{dx} - Q - \rho I \ddot{\varphi} = 0 \tag{5. 6. 6}$$

Inserting Eq. (5.6.5) into (5.6.4) leads to the following equation for harmonic vibration with $\ddot{w} = -\omega^2 w$:

$$M = EI\left(-w'' + \frac{Q'}{GA_s}\right) = -EI\,w'' + \frac{EI}{GA_s}\left(-m\omega^2 w\right) \tag{5. 6. 7}$$

or, after differentiating twice

$$M'' = -EI\,w^{IV} + \frac{EI}{GA_s}\left(-m\omega^2 w''\right) \tag{5. 6. 8}$$

For harmonic vibration, the condition for moment equilibrium Eq. (5.6.6) is given by

$$M' = Q - \rho I \omega^2 \varphi \tag{5. 6. 9}$$

Differentiating Eq. (5.6.9) once with respect to x and taking the derivative of the shear force from Eq. (5.6.5) leads to

$$M'' = Q' - \rho I \omega^2 \varphi' = -m\omega^2 w - \rho I \omega^2 \varphi' \tag{5. 6. 10}$$

and with Eq. (5.6.3)

$$M'' = -m\omega^2 w - \rho I \omega^2 \left(-w'' + \frac{-m\omega^2 w}{GA_s}\right) \tag{5. 6. 11}$$

By comparing Eq. (5.6.11) and Eq. (5.6.8), the differential equation for harmonic vibration results:

$$\mathrm{EI}\,w^{IV} + w'' \cdot m\omega^2 i^2 \left(1 + \alpha_s \frac{E}{G}\right) + w \cdot m\omega^2 \left(-1 + \frac{i^2\omega^2 m}{GA_s}\right) = 0 \qquad (5.6.12)$$

The stiffness matrix $\underline{K}_{dyn}$, which links the vectors $\underline{P}$ and $\underline{V}$ containing the internal forces and displacements of the end sections according to Eqs. (5.4.13) and (5.4.14), is given by

$$\underline{K}_{dyn} = B \begin{bmatrix} k_{11} & & & \\ k_{21} & k_{22} & \text{symm.} & \\ k_{31} & k_{32} & k_{33} & \\ k_{41} & k_{42} & k_{43} & k_{44} \end{bmatrix} \qquad (5.6.13)$$

The coefficients of the matrix have the following values:

$$\begin{aligned}
k_{11} &= -\frac{1}{\ell^3} h_1 h_2 \left(\alpha^2 + \beta^2\right)\left(h_1 f_7 + h_2 f_6\right) \\
k_{21} &= \frac{1}{\ell^2} h_1 h_2 \left[\left(\beta h_2 - \alpha h_1\right)\left(f_5 - 1\right) + h_1 h_2 f_8 \left(h_1 \beta + h_2 \alpha\right)\right] \\
k_{22} &= -\frac{1}{\ell}\left(\alpha^2 + \beta^2\right)\left(h_2 f_7 - h_1 f_6\right) \\
k_{31} &= \frac{h_1 h_2}{\ell^3}\left(\alpha^2 + \beta^2\right)\left(h_2 f_4 + h_1 f_2\right) \\
k_{32} &= \frac{1}{\ell^2} h_1 h_2 \left(\alpha^2 + \beta^2\right)\left(f_1 - f_3\right) \\
k_{33} &= k_{11} \\
k_{41} &= -k_{32} \\
k_{42} &= -\frac{1}{\ell}\left(\alpha^2 + \beta^2\right)\left(h_1 f_4 - h_2 f_2\right) \\
k_{43} &= -k_{21} \\
k_{44} &= k_{22}
\end{aligned} \qquad (5.6.14)$$

with the abbreviations

$$\begin{aligned}
&\lambda = \ell \sqrt[4]{\frac{m\omega^2}{EI}}, \quad i = \sqrt{\frac{I}{A}}, \quad \zeta = \frac{EI}{GA_s \ell^2}, \\
&\alpha = \lambda \sqrt{0.50\,\lambda^2 \left(\frac{i^2}{\ell^2} + \zeta\right) + \sqrt{0.25\,\lambda^4 \left(\frac{i^2}{\ell^2} - \zeta\right)^2 + 1}} \\
&\beta = \lambda \sqrt{-0.50\lambda^2 \left(\frac{i^2}{\ell^2} + \zeta\right) + \sqrt{0.25\,\lambda^4 \left(\frac{i^2}{\ell^2} - \zeta\right)^2 + 1}} \\
&h_1 = \alpha - \lambda^4 \frac{\zeta}{\alpha} \\
&h_2 = \beta + \lambda^4 \frac{\zeta}{\beta} \\
&B = \frac{EI}{2 h_1 h_2 \left(f_5 - 1\right) + f_8 \left(h_1^2 - h_2^2\right)}
\end{aligned} \qquad (5.6.15)$$

and the auxiliary functions

$$
\begin{aligned}
f_1 &= \cos\alpha \\
f_2 &= \sin\alpha \\
f_3 &= \cosh\beta \\
f_4 &= \sinh\beta \\
f_5 &= f_1 f_3 \\
f_6 &= f_1 f_4 \\
f_7 &= f_2 f_3 \\
f_8 &= f_2 f_4
\end{aligned}
\tag{5. 6. 16}
$$

While both shear deformation and rotary inertia are considered in the case of the TIMOSHENKO beam, the so-called RAYLEIGH beam idealisation includes only the influence of rotary inertia. It can be regarded as a special case of the TIMOSHENKO beam for very high shear stiffness.

5.7 Programming aspects

The stiffness matrices of all single beam elements must refer to the global kinematic degrees of freedom u_1, w_1, φ_1, u_2, w_2, φ_2 according to Fig. 5.7-1 before they can be added into the system stiffness matrix.

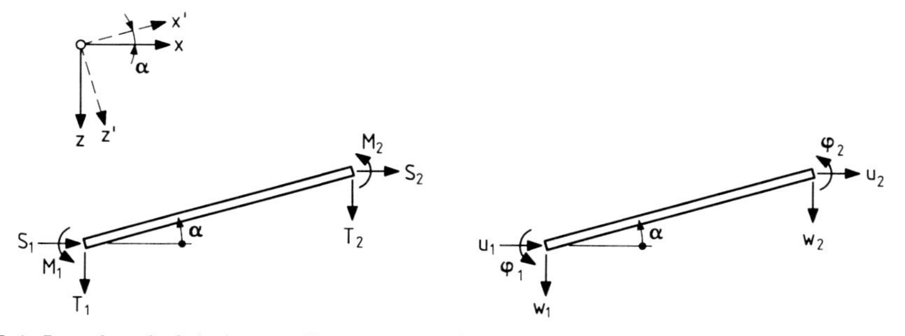

Fig. 5.7-1: Local and global co-ordinate system for beam elements

The (6,6) element stiffness matrix of the beam in the local (x', z') co-ordinate system is given by

$$
\underline{k} = \begin{bmatrix}
s_{11} & & & & & \\
0 & k_{11} & & \text{symm.} & & \\
0 & k_{21} & k_{22} & & & \\
s_{21} & 0 & 0 & s_{22} & & \\
0 & k_{31} & k_{32} & 0 & k_{33} & \\
0 & k_{41} & k_{42} & 0 & k_{43} & k_{44}
\end{bmatrix}
\tag{5. 7. 1}
$$

The coefficients s_{ij} can be extracted from Eq. (5.2.10) while the coefficients k_{ij} are given in Eqs. (5.4.15), (5.5.6) or (5.6.13) according to the theory employed. The element stiffness matrix in the global kinematic degrees of freedom $(u_1, w_1, \varphi_1, u_2, w_2, \varphi_2)$ is then given by

$$\underline{k}_g = \underline{T}^T \underline{k} \underline{T} \tag{5.7.2}$$

Here $\underline{T}$ is the orthogonal (inverse equal to its transpose) transformation matrix

$$\underline{T} = \begin{bmatrix} c & -s & 0 & 0 & 0 & 0 \\ s & c & 0 & 0 & 0 & 0 \\ 0 & 0 & 1 & 0 & 0 & 0 \\ 0 & 0 & 0 & c & -s & 0 \\ 0 & 0 & 0 & s & c & 0 \\ 0 & 0 & 0 & 0 & 0 & 1 \end{bmatrix} \tag{5.7.3}$$

with the abbreviations s = sin α, c = cos α. The following table summarises the names and input/output files of the three computer programs which evaluate the internal forces and displacements of harmonically excited plane frames without damping. The loading consists of sinusoidal nodal forces and moments with a common excitation circular frequency ω, and the internal forces and displacements are understood to be the maximum values for $\sin \omega t = 1$. The single programs are based on different idealisations, with the program EULBER utilising the EULER/BERNOULLI 1st order theory, the program EUBER2 the EULER/BERNOULLI 2nd order theory and the program TIMOSH the TIMOSHENKO theory.

Program name	Input file	Output file
EULBER	EEBN	AEBN
EUBER2	EEB2	AEB2
TIMOSH	ETIM	ATIM

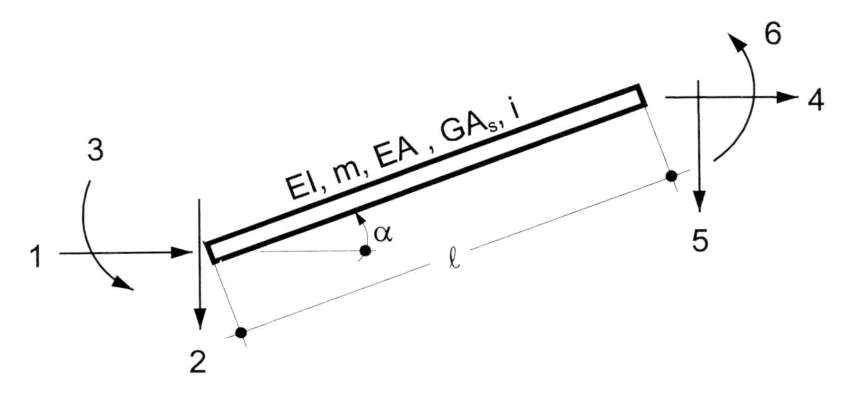

Fig. 5.7-2: General beam element with numbering of the active kinematic degrees of freedom

As already mentioned, the loading consists of forces and moments with common circular frequency ω acting at the nodes in the directions of the active kinematic degrees of freedom. The general case is a plane frame structure consisting of NELEM beam elements and having NDOF active kinematic degrees of freedom, the input file of which is formed as follows: In the first of three data blocks, NELEM rows contain the stiffness, mass and geometry data of each beam element, followed in the second block by another NELEM rows with the incidence vectors of all elements (that is, the numbers of system degrees of freedom corresponding to the global element degrees of freedom 1 to 6 according to Fig. 5.7-2) and, in the final block,

NDOF amplitudes of the loads acting along the system degrees of freedom. The second and third data block are the same for all three programs, while the data to be entered in the first block vary according to the following table:

Program name	Input data for each beam element
EULBER	EI, ℓ, EA, m, α
EUBER2	EI, ℓ, EA, m, α, D
TIMOSH	EI, ℓ, EA, m, α, i, GA_s

Here, EI is the bending stiffness and EA the axial stiffness of the beam (in kNm^2 and kN, respectively), ℓ the element length (in m), m the constant mass per unit length in t/m, α the angle shown in Fig. 5.7-1 (in degrees, positive counter-clockwise), D the constant compressive force (in kN), i the section's radius of gyration in m ($i=\sqrt{I/A}$) and GA_s the shear stiffness of the section in kN. Tensile axial forces which would produce a stiffening of the system are not considered in the program EUBER2; for elements with tensile axial forces or no axial forces at all D should be assigned a small value (say D = 0.01). More details regarding the input files and the interpretation of the output are given in the context of the examples presented in section 5.8.

If the excitation circular frequency ω coincides with a natural frequency of the structure, the determinant of the system stiffness matrix $\underline{K}$ vanishes and no solution of the system $\underline{P}=\underline{K}\cdot\underline{V}$ can be found. It is thus possible to determine the natural frequencies of the plane frame structures by computing the value of the determinant of $\underline{K}(\omega)$ for several ω values and determining the roots of the equation det $(\underline{K}(\omega)) = 0$. A stumbling block of this approach are the numerically large values of the determinants, which easily exceed the range available to a PC user; using logarithms has its own problems. It is recommended instead to evaluate the reciprocal of some displacement value and plot it versus frequency, the roots of this function being the natural frequencies. For easier localisation of these roots, it is further recommended to compute and plot the square root of the reciprocal of a suitable displacement value rather than the reciprocal value itself. This is actually done in the computer programs mentioned below, in which a unit vector is used as loading vector, thus applying unit loads to all active kinematic degrees of freedom. Their names and input/output files are:

Program name	Input file	Output files
EUBFRQ	EEBN	ECHO AEBNFR
EB2FRQ	EEB2	ECHO AEB2FR
TIMFRQ	ETIM	ECHO ATIMFR

The input files of these programs are the same as those used for the determination of internal forces and displacements. The input data are written by the programs in the file ECHO for checking purposes; the computed values (two columns with circular frequencies in the first and corresponding square roots of the reciprocals of a specified displacement in the second) are written in the files AEBNFR, AEB2FR or ATIMFR ready to be plotted. From the plot it is readily discernible if a sign change corresponds to a root or to a vertical asymptote [5.3].

While roots denote natural frequencies of the system, vertical asymptotes correspond to zero values of the flexibility matrix $\underline{F}(\omega)$, with

$$\underline{F}(\omega)=\underline{K}^{-1}(\omega) \qquad (5.7.4)$$

5.8 Examples

This section illustrates by a series of examples the computation of internal forces and displacements of harmonically excited undamped plane frames and also the determination of their natural frequencies as roots of frequency-dependent reciprocal values of suitable displacement components. In practice, the computation of natural frequencies by means of a discrete lumped-mass model is much quicker and numerically much more robust; nevertheless, the approach illustrated in this section is more precise for higher natural frequencies, especially if the number of degrees of freedom employed in the discrete model is not as large as it should have been in order to accurately describe higher modes.

The first example refers to the two-story frame shown in Fig. 5.8-1, the members of which have the following properties:

Element no.	EI	EA	m	$i=\sqrt{\frac{I}{A}}$	GA_s
	kNm^2	kN	t/m	m	kN
1, 2, 3, 4	270112.0	$6.620\ 10^6$	0.788	0.200	$2.297\ 10^6$
8	294000.0	$10.71\ 10^6$	4.35	0.166	$1.968\ 10^6$
6, 7	170100.0	$9.765\ 10^6$	5.14	0.132	$1.640\ 10^6$
5	71761.0	$4.252\ 10^6$	0.506	0.130	$1.641\ 10^6$

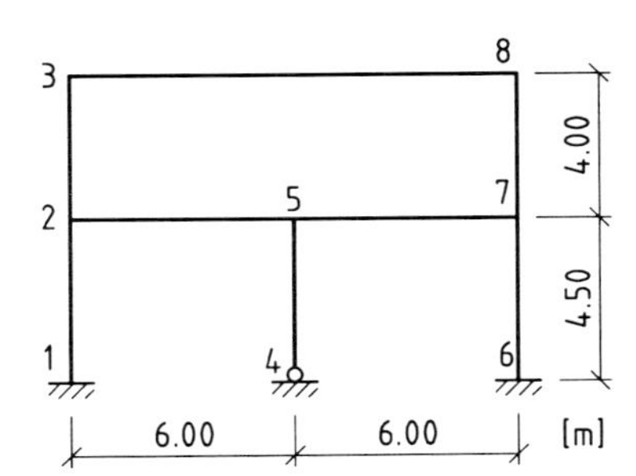

Fig. 5.8-1: Plane frame

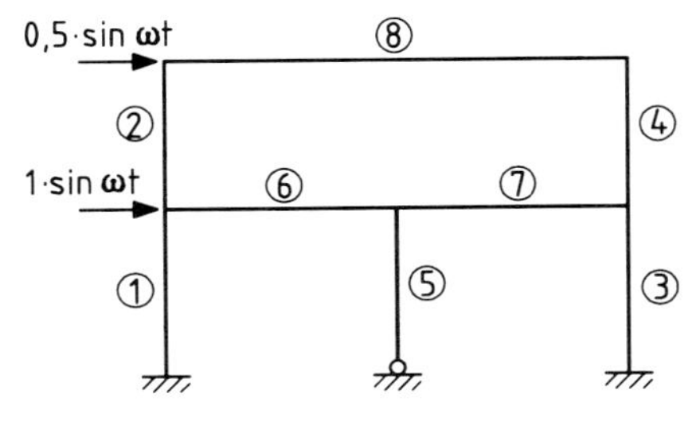

Fig. 5.8-2: Element numbering and loading

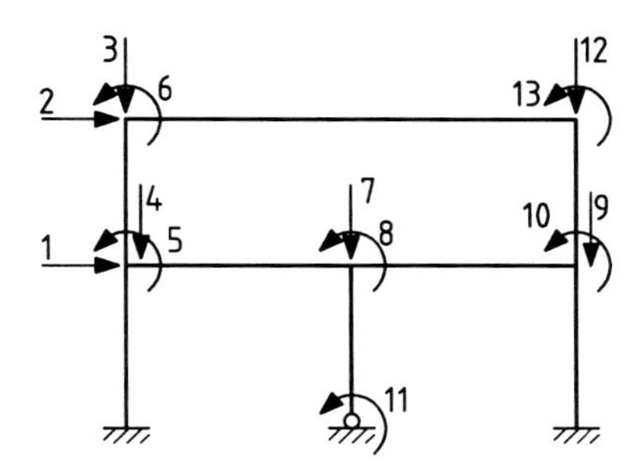

Fig. 5.8-3: Discretised system

The loading consists of two harmonic horizontal loads as shown in Fig. 5.8-2 and the discretised system (Fig. 5.8-3) is made up of 8 beam elements and 13 active kinematic degrees of freedom (NDOF = 13, NELEM = 8) . The input file EEBN for the program EULBER is given by:

```
270112., 4.5, 6.62e6,  0.788, 90.
270112., 4.0, 6.62e6,  0.788, 90.
270112., 4.5, 6.62e6,  0.788, 90.
270112., 4.0, 6.62e6,  0.788, 90.
71761.,  4.5, 4.252e6, 0.506, 90.
170100., 6.,  9.765e6, 5.14,  0.
170100., 6.,  9.765e6, 5.14,  0.
294000., 12., 10.71e6, 4.35,  0.
0,0,0,  1,4,5,
1,4,5,  2,3,6,
0,0,0,  1,9,10
1,9,10, 2,12,13,
0,0,11, 1,7,8,
1,4,5,  1,7,8,
1,7,8,  1,9,10,
2,3,6,  2,12,13,
1., 0.5, 0., 0., 0.,
0., 0., 0., 0., 0.,
0., 0., 0.,
```

For an excitation circular frequency $\omega = 10$ rad/s the program EULBER yields the bending moment diagram depicted in Fig. 5.8-4. For comparison, Fig. 5.8-5 shows the “static” bending moment diagram that has also been computed by EULBER for $\omega \approx 0$. The circular frequency ω cannot be input as zero since this would lead to numerical problems, so its value is given as $\omega = 1$ or 2 rad/s, which approximates the static case well enough.

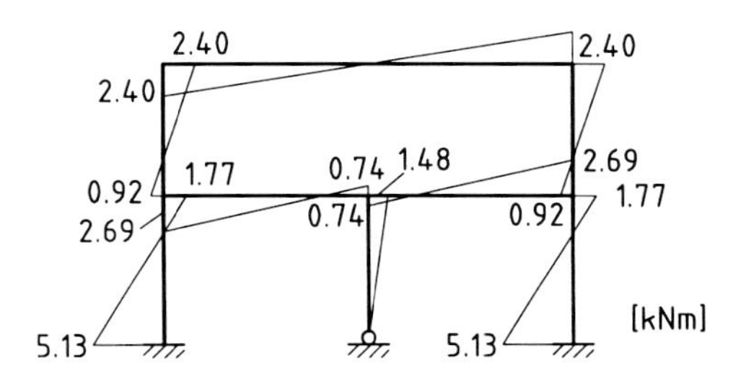

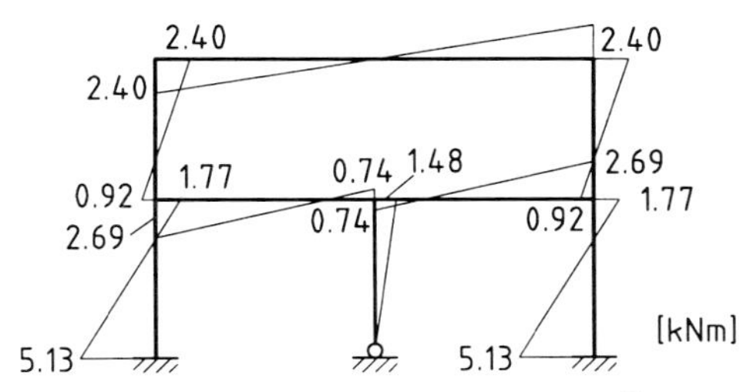

Fig. 5.8-4: Bending moment diagram, EULER/BERNOULLI 1st order theory

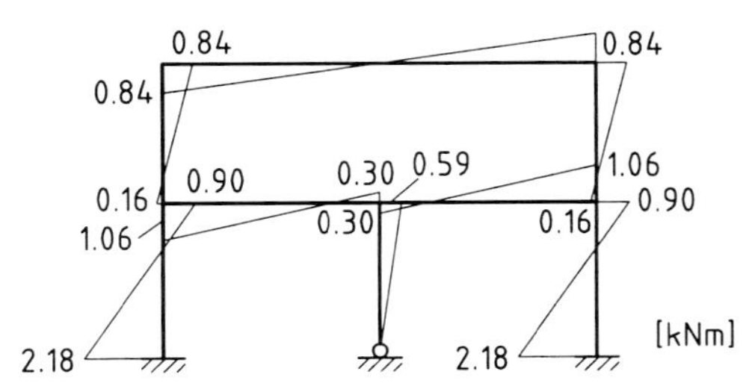

Fig. 5.8-5 Static bending moment diagram

The input files for the programs EUBER2 (for the computation according to 2nd order theory with compressive forces in the range of 277 to 449 kN in the columns) and TIMOSH (considering shear deformation and rotary inertia) are given below. Only the first NELEM rows of the input files are given, since the remainder does not change. The relevant data for EUBER2 (input file EEB2) are:

```
270112., 4.5, 6.62e6,  0.788, 90., 449.,
270112., 4.0, 6.62e6,  0.788, 90., 277.,
270112., 4.5, 6.62e6,  0.788, 90., 449.0,
270112., 4.0, 6.62e6,  0.788, 90., 277.0,
 71761., 4.5, 4.252e6, 0.506, 90., 320.,
170100., 6.,  9.765e6, 5.14,  0.,  0.001
170100., 6.,  9.765e6, 5.14,  0.,  0.001
294000., 12., 10.71e6, 4.35,  0.,  0.001
```

For the program TIMOSH, the first NELEM rows of its input file ETIM are as follows:

```
270112., 4.5, 6.62e6,  0.788, 90., 0.20,  2.297e6
270112., 4.0, 6.62e6,  0.788, 90., 0.20,  2.297e6
270112., 4.5, 6.62e6,  0.788, 90., 0.20,  2.297e6
270112., 4.0, 6.62e6,  0.788, 90., 0.20,  2.297e6
 71761., 4.5, 4.252e6, 0.506, 90., 0.13,  1.641e6
170100., 6.,  9.765e6, 5.14,  0.,  0.132, 1.64e6
170100., 6.,  9.765e6, 5.14,  0.,  0.132, 1.64e6
294000., 12., 10.71e6, 4.35,  0.,  0.166, 1.968e6
```

The computed bending moment diagrams are depicted in Figs. 5.8-6 and 5.8-7.

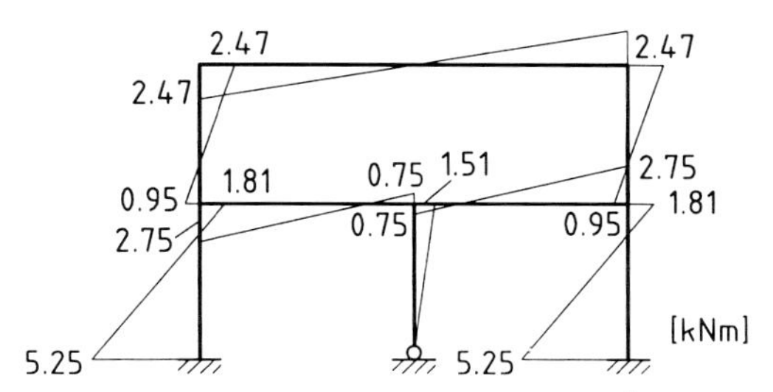

Fig. 5.8-6: Bending moment diagram, EULER/BERNOULLI 2nd order theory

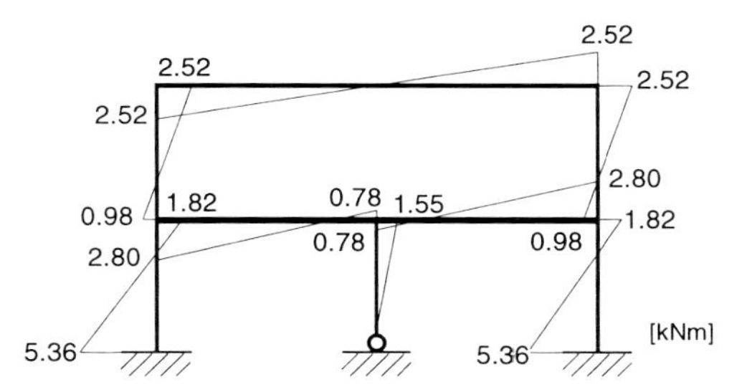

Fig. 5.8-7: Bending moment diagram, TIMOSHENKO theory

For determining the system's natural frequencies, the square root of the reciprocal of the horizontal roof displacement v (corresponding to the degree of freedom no. 2) is plotted versus the excitation circular frequency ω. The program EUBFRQ (EULER/BERNOULLI 1st order theory) yields the curve shown in Fig. 5.8-8.

A vertical asymptote is present at about 66 rad/s, corresponding to a root of the determinant of the flexibility matrix, while the actual natural circular frequencies are located at 12.85, 46.50, 73.26 and 89.96 rad/s. The following table compares these results to those obtained by the EULER/BERNOULLI 2nd order theory and the TIMOSHENKO theory:

Theory	ω_1 (rad/s)	ω_2 (rad/s)	ω_3 (rad/s)	ω_4 (rad/s)
EULER/BERN.	12.85	46.50	73.26	89.96
2nd order	12.74	46.40	73.23	89.95
TIMOSHENKO	12.63	45.31	70.67	86.31

A second example deals with the determination of circular natural frequencies of the built-in beam shown in Fig. 5.8-9.

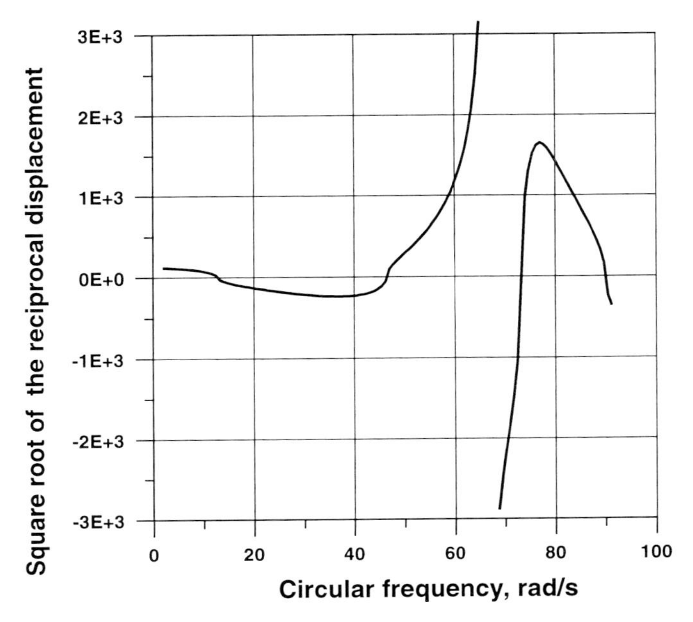

Fig. 5.8-8: Determination of natural frequencies

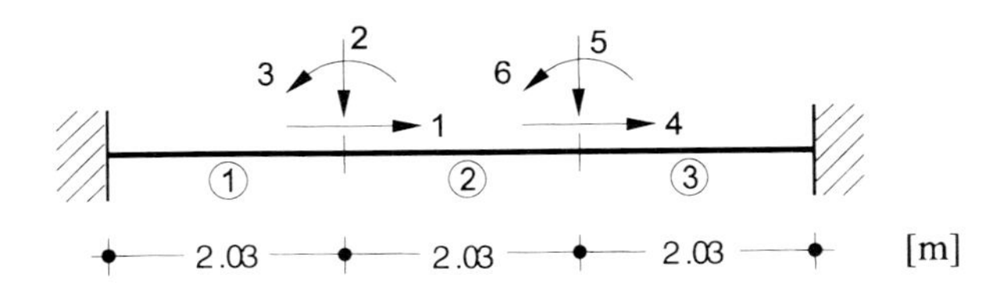

Fig. 5.8-9: Built-in beam with discretisation

Its mass is given as 0.0804 t/m, its bending stiffness EI = 36170 kNm^2, axial stiffness EA = 2.06 10^6 kN, radius of gyration i = 0.132 m and its shear stiffness GA_s = 2.152 10^5 kNm^2. The following table compares results for the first three circular natural frequencies in rad/s computed by the EULER/BERNOULLI 1st order theory and the TIMOSHENKO theory; it is evident that the differences are quite noticeable, especially for higher harmonics.

	ω_1	ω_2	ω_3
EULER/BERNOULLI	405	1115	2187
TIMOSHENKO	365	907	1596

The third example considers the natural circular frequency of the column shown in Fig. 5.8-10, considering the influence of the static compressive axial force D.

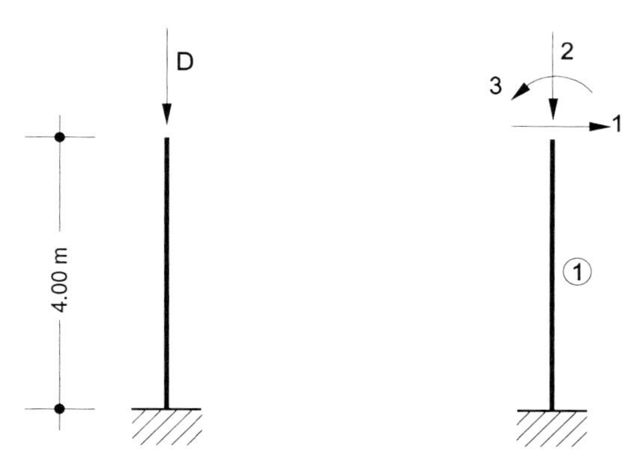

Fig. 5.8-10: Column under compression with discretisation

The 1st order EULER/BERNOULLI theory (D = 0) yields a first natural circular frequency of $\omega_1 = 3.52\sqrt{EI/m\ell^4}$ according to the table 5.4-1; for numerical values of EI=44800 kNm^2, m=0.40 t/m, $\ell = 4.00$ m we obtain ω_1 = 73.6 rad/s. The evaluation of the first natural circular frequency for several values of the compressive force D, normalised by the EULER buckling load D_{ki} of the column

$$D_{ki} = \frac{\pi^2 EI}{(2\ell)^2} = 6908.7\text{kN} \tag{5. 8. 1}$$

yields the following results for ω_D and for the ratio $\left(\frac{\omega_D}{\omega_{D=0}}\right)^2$:

$\frac{D}{D_{ki}}$	ω_D rad/s	$\left(\frac{\omega_D}{\omega_{D=0}}\right)^2$
0	73.60	1.0
0,1	70.02	0.9051
0,2	66.27	0.8107
0,3	62.23	0.7149
0,4	57.84	0.6176
0,5	53.01	0.5188
0,6	47.61	0.4184
0,7	41.41	0.3166
0,8	33.96	0.2129
0,9	24.12	0.1074
1,0	0.0	0.0

The decrease of the natural frequency with increasing compressive force is best illustrated by the interaction diagram shown in Fig. 5.8-11.

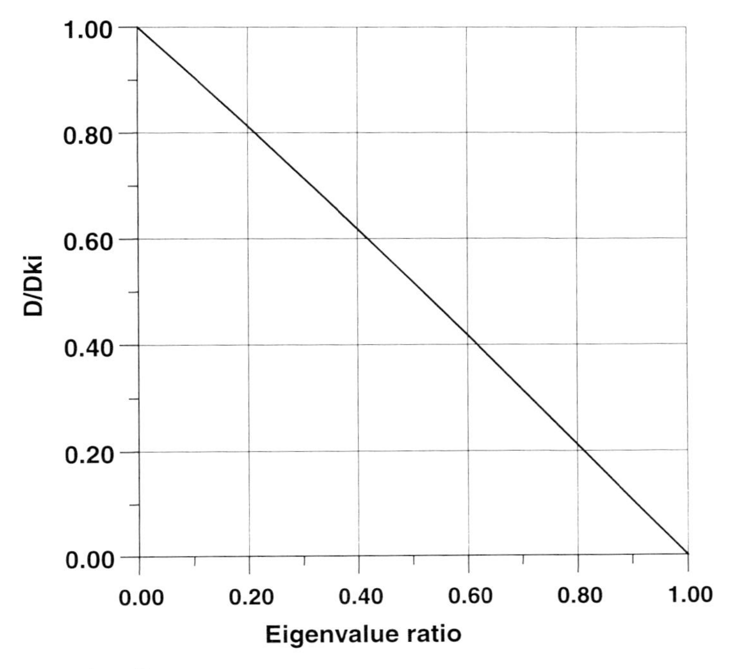

Fig. 5.8-11: Interaction diagram

In the present case, compressive forces of D_1 = 1000 kN and D_2 = 5000 kN correspond to natural circular frequencies of 68.4 rad/s and 39.8 rad/s. Here, the relationship (5.5.12) is only valid as an approximation, because the buckling and vibration mode shapes are not identical, as in the case of the simply supported beam. This leads to

$$D_{ki} \approx \frac{D_2 \cdot \omega_1^{\ 2} - D_1 \cdot \omega_2^{\ 2}}{\omega_1^{\ 2} - \omega_2^{\ 2}} \tag{5. 8. 2}$$

and for the values given above, (1000 kN, 68.4 rad/s) and (5000 kN, 39.8 rad/s), the approximate value for the buckling load becomes

$$D_{ki} \approx \frac{5000 \cdot 68.4^{\ 2} - 1000 \cdot 39.8^{\ 2}}{68.4^2 - 39.8^2} = 7047.5\text{kN} \tag{5. 8. 3}$$

This is merely 2% higher than the correct value of 6908.7 kN.

In the last example the "free-free" natural circular frequencies of a beam are determined. Its properties are given as follows: Bending stiffness EI = 20 000 kNm^2, length $\ell = 10$ m, axial stiffness EA = 2.0 10^6 kN, mass 2.0 t/m. We obtain values of ω_2=22.4 and ω_3 = 61.7 rad/s in addition to the rigid-body natural frequency of $\omega_1 = 0$. Fig. 5.8-12 shows the curve that has been used for determining these values; the rotation φ of a beam end section has been chosen here as characteristic displacement.

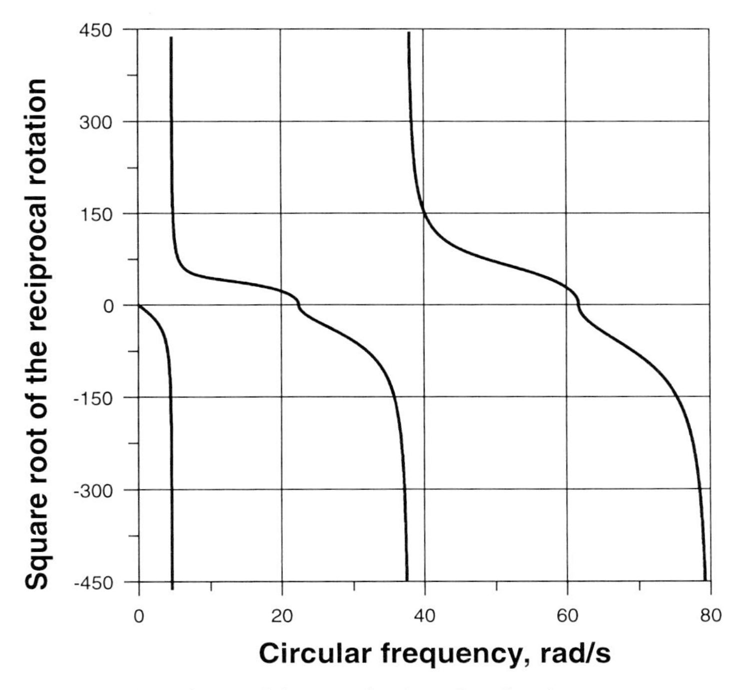

Fig. 5.8-12: Determination of natural frequencies for a free-free beam

6 Bell tower response to bell-induced forces

6.1 Analytical preliminaries

Bell towers are relatively tall structures, the main purpose of which is to safely convey the dynamic forces associated with the ringing of the church bells to the foundation. In this section the basic relationships for describing the response of a tower under such an excitation are presented. The tower's natural frequencies and damping are recognised as the most important parameters; section 6.2 deals with their experimental determination while section 6.3 is devoted to the discussion of two practical examples.

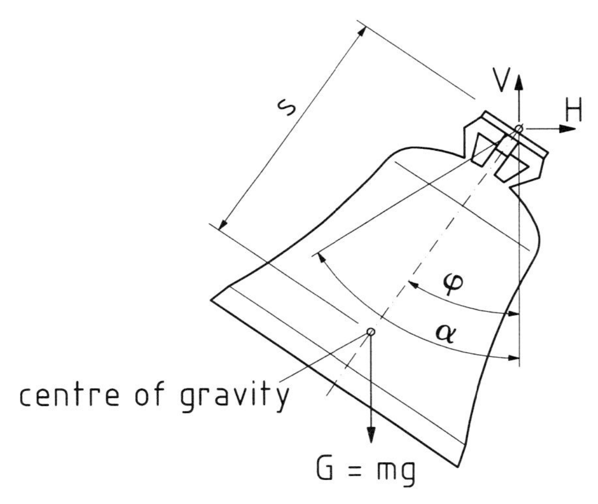

Fig. 6.1-1: Church bell idealised as a pendulum

Referring to Fig. 6.1-1, the maximum angle of rotation is denoted by α; its actual value depends on size and form as well as on the musical note and the frequency of the bell. The following quantities with their corresponding notation are introduced in DIN 4178, which deals with the analysis and design of bell towers:

- The number of bell swings per minute, denoted by n. It is one-half the strike number a which expresses how many times the clapper strikes the bell in a minute, n = a/2. Transformed to a circular excitation frequency Ω, this yields

$$\Omega = \frac{\pi n}{30} = \frac{\pi a}{60} \quad \left[\frac{\text{rad}}{\text{s}}\right] \tag{6. 1. 1}$$

- The natural circular frequency of the tower is denoted by ω; the relationship between ω and the tower natural cycle number n_e is given by

$$\omega = \frac{\pi n_e}{30} \quad \left[\frac{\text{rad}}{\text{s}}\right] \tag{6. 1. 2}$$

- A characteristic parameter or shape factor for the bell, c, is introduced as a function of the bell mass moment of inertia Θ_S about its centre of gravity, the bell mass m (bell weight G = mg) and the distance s between the bell's centre of gravity and the axis of rotation.

$$c = \frac{m \cdot s^2}{\Theta_s + m \cdot s^2} \qquad (6.\ 1.\ 3)$$

If no values are known for these parameters, approximate values can be obtained from a table given in DIN 4178. The mass of the clapper is normally neglected, being much smaller than the bell's mass.

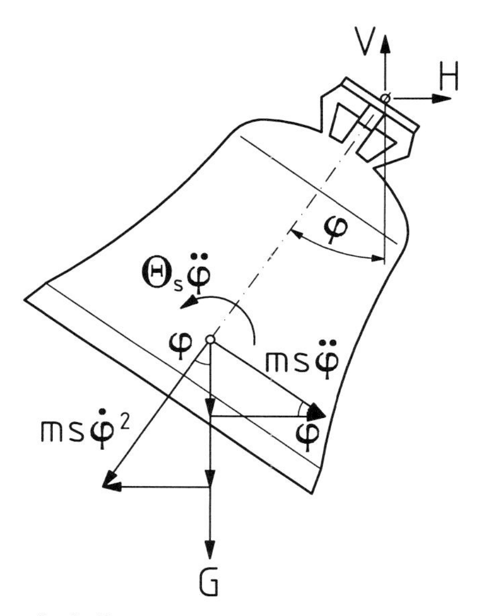

Fig. 6.1-2: Forces acting on the bell

As shown in Fig. 6.1-2, acting forces include the self-weight G, the inertia force $m\,s\,\ddot{\varphi}$ and moment $\Theta_S\,\ddot{\varphi}$, both acting in the opposite direction to the bell's movement, and finally the centripetal force $m\,s\,\dot{\varphi}^2$ [6.1]. The condition that the sum of the moments about the axis of rotation must vanish leads to the pertinent differential equation of motion:

$$\sum M = 0\!: \;\; G \cdot s \cdot \sin\varphi + m\,s\,\ddot{\varphi} \cdot s + \Theta_S\,\ddot{\varphi} = 0 \qquad (6.\ 1.\ 4)$$

The equilibrium equations furnish the vertical and horizontal component of the support reaction as a function of the rotation angle φ:

$$\sum H = 0\!: \;\; H + m\,s\,\ddot{\varphi}\cos\varphi - m\,s\,\dot{\varphi}^2 \sin\varphi = 0 \qquad (6.\ 1.\ 5)$$

$$\sum V = 0\!: \;\; V - G - m\,s\,\ddot{\varphi}\sin\varphi - m\,s\,\dot{\varphi}^2 \cos\varphi = 0 \qquad (6.\ 1.\ 6)$$

According to the law of energy conservation, the sum of kinetic and potential energy present at any time (angle of rotation equal to φ) must be the same as the potential energy when the maximum angle of rotation $\varphi = \alpha$ is reached, since the corresponding kinetic energy is zero:

$$G \cdot s \cdot (1 - \cos\varphi) + \frac{m s^2 \dot{\varphi}^2}{2} + \frac{\Theta_s \dot{\varphi}^2}{2} = G \cdot s \cdot (1 - \cos\alpha) \qquad (6.\,1.\,7)$$

These relationships lead to the following expressions for the components H and V of the support reaction as functions of the rotation angle φ, with Θ_0 as mass moment of inertia of the bell about its axis of rotation, $\Theta_0 = \Theta_S + m s^2$:

$$V(\varphi) = \frac{G \cdot m \cdot s^2}{\Theta_s + m \cdot s^2}\left(3\cos^2\varphi - 2\cos\varphi\,\cos\alpha - 1\right) + G \qquad (6.\,1.\,8)$$

$$H(\varphi) = \frac{G \cdot m \cdot s^2}{\Theta_s + m \cdot s^2}\left(\frac{3}{2}\sin 2\varphi - 2\sin\varphi\,\cos\alpha\right) \qquad (6.\,1.\,9)$$

The non-linear differential equation of motion of the pendulum Eq. (6.1.4) can be written as:

$$\ddot{\varphi} + \frac{G \cdot s}{\Theta_s + m \cdot s^2}\sin\varphi = 0 \qquad (6.\,1.\,10)$$

Its solution $\varphi(t)$ contains Jacobian elliptic functions [6.2], which have been expressed by trigonometric series in DIN 4178 as follows:

$$H(t) = c \cdot G \sum_i \gamma_i \sin\Omega_i t\,,\quad i = 1,3,5,\ldots \qquad (6.\,1.\,11)$$

$$V(t) = c \cdot G \sum_i \beta_i \cos\Omega_i t\,,\quad i = 2,4,6,\ldots \qquad (6.\,1.\,12)$$

In these expressions, c is the shape factor according to Eq. (6.1.3). The excitation circular frequencies Ω_i are given as

$$\Omega_i = i \cdot \Omega = i\frac{\pi n}{30} \qquad (6.\,1.\,13)$$

The γ_i and β_i coefficients can be obtained from the diagrams included in DIN 4178 (see for example Fig. 6.1-3), where they are plotted as functions of the maximum rotation angle α. The series expansion of the vertical component contains only even (cosine) terms, the one for the horizontal component only odd (sine) terms.

As an example, the time history of the horizontal support reaction component of a 2.5 kN heavy bell is investigated. Further parameters are given as follows:

Shape factor	$c = 0.76$
Bell swings per minute	$n = 32\ \text{min}^{-1}$ ($a = 64$ 1/ min)
	$\rightarrow \Omega = 2\pi\, 32/60 = 3.35$ rad/s
	$\rightarrow$ Period T = 1.875 s
Max. rotation angle	$\alpha = 70°$

H(t) in kN is given by Eq. (6.1.11) as:

$$\begin{aligned} H(t) &= \sum_i H_i(t) = G \cdot c \cdot \sum_i \gamma_i \cdot \sin\Omega_i \cdot t \quad \text{for } i = 1,3,5 \\ &= 2.5 \cdot 0{,}76 \cdot \left(\gamma_1 \cdot \sin\Omega_1 \cdot t + \gamma_3 \cdot \sin\Omega_3 \cdot t + \gamma_5 \cdot \sin\Omega_5 \cdot t\right) \end{aligned}$$

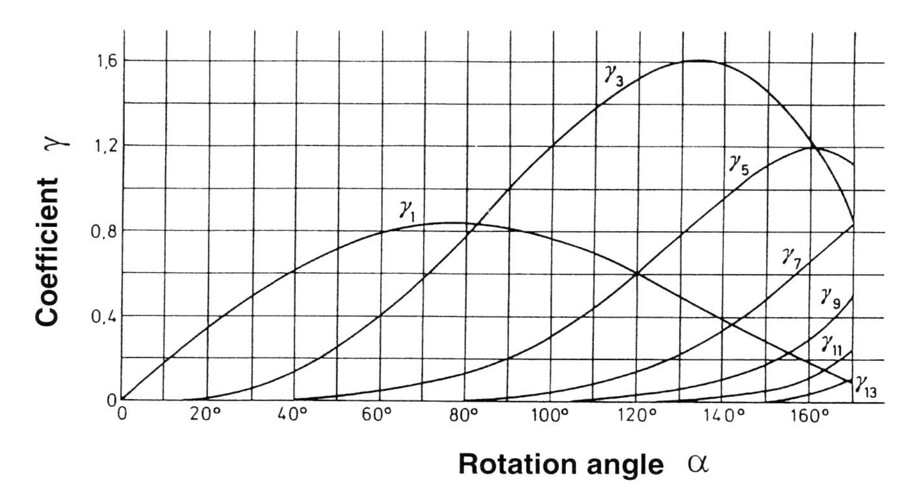

Fig. 6.1-3: Coefficients γ for the horizontal support reaction component after DIN 4178

Fig. 6.1-3 yields the following coefficients for $\alpha = 70°$:

$$\gamma_1 = 0.84$$
$$\gamma_3 = 0.57 \qquad (6.\ 1.\ 14)$$
$$\gamma_5 = 0.08$$

This leads to

$$H(t) = 2.5 \cdot 0.76 \cdot (0.84 \cdot \sin 3.35 \cdot t + 0.57 \cdot \sin 3.35 \cdot 3 \cdot t + 0.08 \cdot \sin 3.35 \cdot 5 \cdot t) \qquad (6.\ 1.\ 15)$$

The time history H(t) is shown in Fig. 6.1-4. If more than one bell is present, a conservative assumption is that they all swing in phase, so that their maximum support reactions may be summed algebraically.

Of course, it is also possible to solve the non-linear differential equation (6.1.10) directly by means of a simple direct integration method. This has been implemented in the program GLOCKE which directly yields the support reaction components H(t) and V(t) beginning with $\varphi(t = 0) = \alpha$. Input data are the bell self-weight in kN, the shape factor c, angle α, the circular excitation frequency Ω and also the distance s between the rotation axis and the bell's centre of gravity; the latter can be assumed to be approximately 70% of the bell's diameter. All these data are input interactively; the output file AGLO contains four columns with the time points, the rotation angle φ, the horizontal component H and the vertical component V. The maximum values of H and V are also output on screen.

Program name	Input file	Output file
GLOCKE	-	AGLO

Fig. 6.1-5 shows the time history H(t) determined by the numerical solution of the differential equation as compared to the solution according to DIN 4178; the difference is negligible.

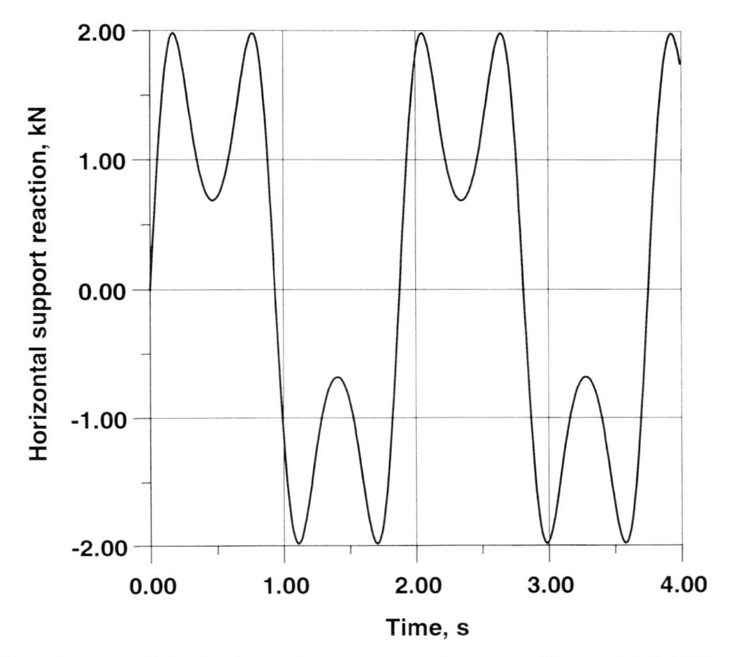

Fig. 6.1-4: Time history of the horizontal support reaction according to DIN 4178

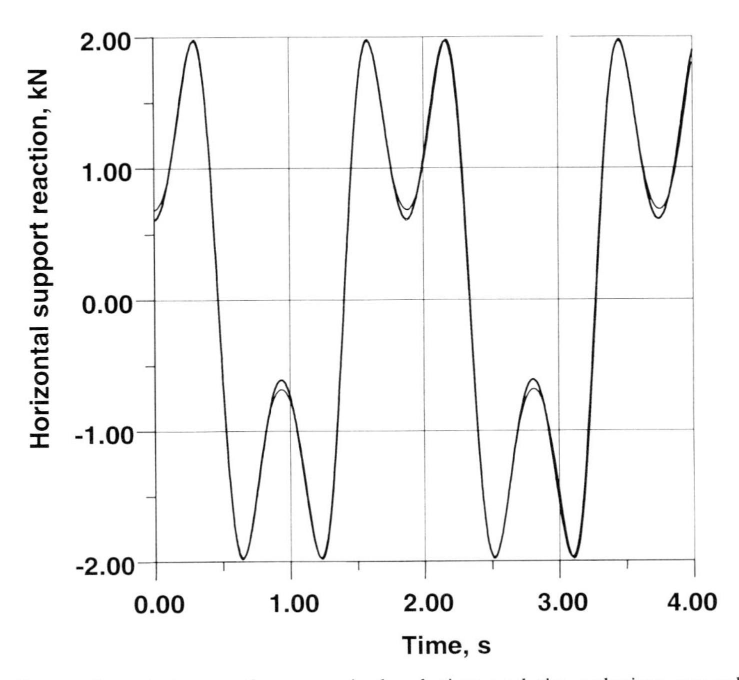

Fig. 6.1-5: Comparison between the numerical solution and the solution according to DIN 4178

Now to the investigation of the bell tower itself according to DIN 4178:

DIN 4178 states that the tower must be checked for the horizontal support reactions of the bells. If the bells are located in the upper third of the tower, an approximate method using equivalent static loads may be used. These equivalent horizontal forces H_{equiv} are given by

$$H_{equiv} = \max\left[c \cdot G \sum_i \gamma_i V_i \sin \Omega_i t\right], \quad i = 1,3,5,\ldots \qquad (6.\,1.\,16)$$

In this expression, the amplification factor V_i is defined as

$$V_i = \frac{1}{\sqrt{(1-\eta_i^2)^2 + (2D\eta_i)^2}} \qquad (6.\,1.\,17)$$

It is a function of the critical damping ratio D of the tower and of the frequency ratio

$$\eta_i = \frac{\Omega_i}{\omega} \qquad (6.\,1.\,18)$$

between the excitation frequency and the first natural frequency of the tower. This equivalent static load must be applied at the height of the bell support; if this point is located in the lower two-thirds of the tower, a true dynamic analysis must be carried out, e.g. by modal analysis. In view of the availability of the necessary computer programs, it is recommended that a dynamic analysis be carried out anyway, even if the bells are suspended in the upper third of the tower.

The following approximate damping ratios are recommended for the tower's damping, if no experimental results are available:

Structure type/Material	Damping ratio in %
Steel structure, welded or with high strength friction grip bolts	0.2-0.3%
Steel structure, bolted	0.5-0.6%
Reinforced concrete structure	1-1.5%
Masonry	1.5-2%

Additionally, DIN 4178 requires that the first circular natural frequency of the tower, ω_1, must exceed three times the bell excitation circular frequency by at least 20% (if the tower natural frequency has been determined computationally) or 10% (if the tower natural frequency has been measured):

$$\omega_1 \geq 1.2\ \Omega_3 \text{ (computed)}$$

$$\omega_1 \geq 1.1\ \Omega_3 \text{ (measured)} \qquad (6.\,1.\,19)$$

The influence of the foundation flexibility on the tower's natural frequency is quite noticeable and should be considered. For rectangular foundations, DIN 4178 introduces the dynamic rocking foundation modulus C_k as

$$C_k = \frac{E_{S,dyn}}{f \cdot \sqrt{A}} \qquad (6.\ 1.\ 20)$$

In this expression, A is the plan area of the foundation block, f a coefficient which may be approximated by 0.35 and $E_{S,dyn}$ is the dynamic stress-strain modulus for which DIN 4178 suggests the following values:

Soil type	$E_{S,dyn}$ in kN/m^2
Sand of average density	60 000 to 150 000
Heterogeneous sand/gravel	200 000 to 400 000
Glacial fill	300 000 to 600 000
Clay, medium	70 000 to 140 000
Clay, sandy (medium)	75 000 to 120 000
Clay, hard	35 000 to 70 000

The dynamic rocking foundation modulus C_k is measured in kN/m^3, since it expresses the stress which would cause a unit settlement. The corresponding rotational spring constant k_φ in kNm/rad for the elastic support is given by

$$k_\varphi = C_k \cdot I \qquad (6.\ 1.\ 21)$$

Here, I is the moment of inertia of the rectangular foundation about its rocking axis. For the investigation of the bell support construction, it is sufficiently accurate to apply a dynamic amplification factor of 1.3 to the maximum horizontal force component; a check of the bell support construction's natural frequency is not necessary. It is also unnecessary to investigate the effects of the vertical bell support reaction, since the high stiffness of the tower in the vertical direction implies a favourite frequency ratio η.

At the planning stage, it is recommended to design the foundation area larger than strictly necessary, in order to reduce the rocking flexibility which might lead to a low natural frequency. In the case of poor subgrade quality, it is also recommended to design the foundation with a hollow (ring or cellular) plan shape, so that a rocking motion with the geometrical centre as pivot is avoided.

6.2 Experimental investigations

The appearance of cracks or other damage signs in bell towers is often a motive for conducting an experimental investigation of the dynamic properties of the structure. The most important parameters to be determined are the natural frequencies of the tower (especially the lowest) and its damping characteristics. This section discusses some details of the following procedures for conducting such measurements:

1. Measuring the acceleration or velocity structural response due to ambient vibration,

2. Measuring the acceleration or velocity structural response due to impulsive loading (e.g. induced by a hammer stroke),

3. Measuring the acceleration or velocity structural response due to harmonic excitation (this approach is especially suitable for determining damping properties).

Generally, modern portable devices allow not only the storage of large data fields, but have additionally powerful on-line spectral analysis capabilities. Fig. 6.2-1 shows the equipment employed by us for conducting the analyses just mentioned, consisting of a four-canal recording and analysing device, an impulse hammer and a manually operated apparatus to generate harmonic horizontal forces.

Fig. 6.2-1: Equipment for experimental investigations

Fig. 6.2-2: Three-channel acceleration pick-up

An acceleration pick-up as shown in Fig. 6.2-2 is used for measuring acceleration time histories in three orthogonal directions.

The measurement of the dynamic structural response due to ambient vibration is discussed first. Even in non-seismic areas, the ambient vibration level, sometimes including wind-induced vibrations, is often strong enough to produce useful acceleration or velocity signals in bell towers, if their height is not too small. The excitation is typically broad-band noise, from which the structure filters out and converts into kinetic energy the harmonic components adjacent to its natural frequencies (predominantly the lowest). Therefore, a spectral analysis of the measured acceleration or velocity time histories yields a series of more or less well-defined peaks corresponding to the structure's natural frequencies. An illustrative example is given in Fig. 6.2-3 showing results of vibration measurements conducted at 100m height in the southern tower of the Cologne cathedral. The plot in the upper part of the figure depicts the measured acceleration time history, the amplitude spectrum of which is reproduced directly below.

Such ambient-vibration based dynamic structural investigations are relatively easy to conduct, since no extra equipment is necessary for exciting the structure. However, they normally yield only information on natural frequencies.

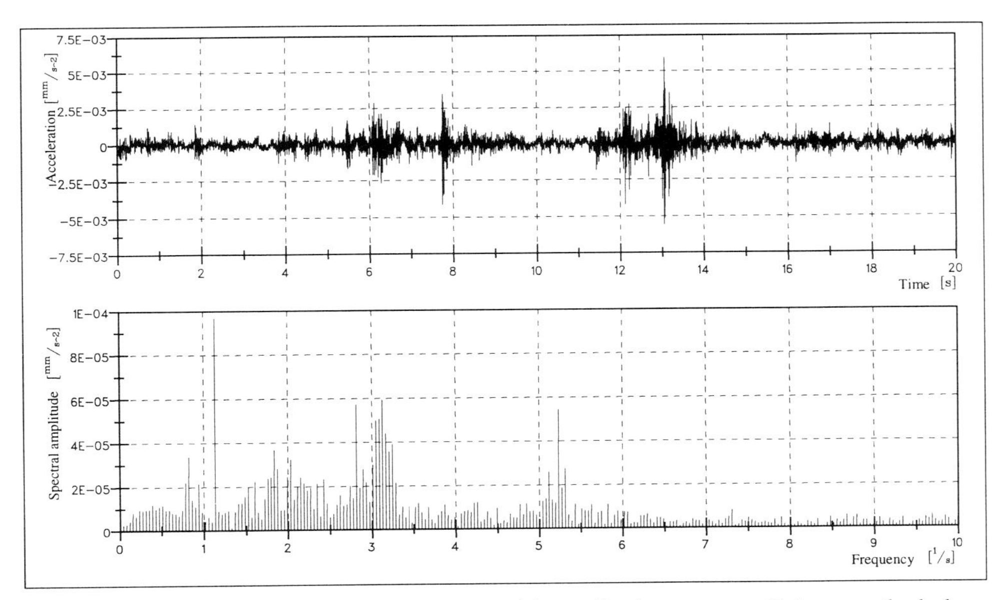

Fig. 6.2-3: Measured acceleration time history with amplitude spectrum, Cologne cathedral

The second approach mentioned utilises short transient loads for exciting the structure. This method is especially suitable for a quick determination of the first natural frequency of not-too-massive bell towers. A stroke with the special impulse hammer is applied to the tower (Fig. 6.2-4) as near the top as possible, and the acceleration or velocity time history is measured. An example is presented in section 6.3.

Fig. 6.2-4: Using an impulse hammer

The third approach, that is the vibration excitation of the tower by means of harmonically varying horizontal forces, is not as simple as the two methods already discussed. Its great advantage is that it yields very accurate results not only for the natural frequencies (and, if so desired, for the corresponding mode shapes) but also for the damping values. The necessary harmonic horizontal forces can be generated by means of electromagnetic or hydraulic shakers or by rotary devices powered by various types of motors or even by hand. Once a natural frequency has been detected, it is very easy to produce large structural deformations with small energy input. By stopping the device and letting the tower's free vibration slowly come to an end, the logarithmic viscous damping decrement can be easily derived from the measured time history, as described in section 3.3. A pertinent example is given in section 6.3.

If the frequencies of the tower bells are not known, they can be easily determined by carrying out a spectral analysis of measured acceleration or velocity time histories of the bell tower subjected to bell ringing.

6.3 Examples

Fig. 6.3-1: 30m bell tower

As first example we consider the masonry bell tower shown in Fig. 6.3-1. It is 30m and square in plan, with sides of 4.20 m. Its bells are described in the following table:

Bell No.	Tone	Diameter [mm]	Mass [kg]	Frequency [1/min]	Max. angle [deg]
I	E'+1	1237	1180	53	62
II	fis'+1	1102	820	56	64
III	a'+2	914	460	59	66
IV	h'+2	812	320	62	68

An impulse excitation by means of a hammer stroke yields the acceleration amplitude spectrum reproduced in Fig. 6.3-2.

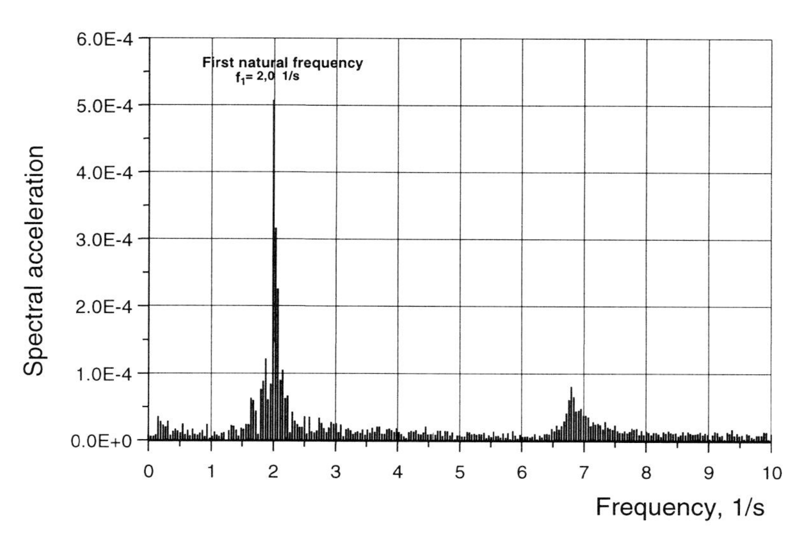

Fig. 6.3-2: Amplitude spectrum due to impulse hammer excitation

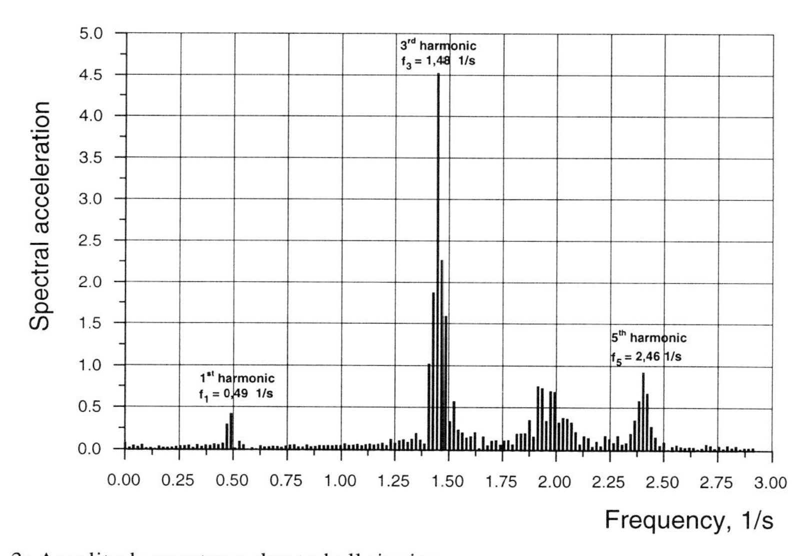

Fig. 6.3-3: Amplitude spectrum due to bell ringing

From the amplitude spectrum, the first natural frequency of the tower can be seen to be equal to $f_1 = 2.0$ Hz. The spectrum of the measured horizontal acceleration due to ringing of bell III is shown in Fig. 6.3-3. Considering the bell's ringing frequency of 59 /min, corresponding to an excitation frequency of $\frac{59 \cdot \pi}{60} = 3.09 \frac{\text{rad}}{\text{s}} = 0.49\,\text{Hz}$ according to Eq. (6.1.13), the peaks shown correspond to the 1st, 3rd and 5th harmonic. The distance between the natural frequency of the tower and the 3rd harmonic of this bell is large enough to preclude resonance effects.

7 Structural response to earthquakes

7.1 Seismological background

The outer layer of the earth's mantle, the crust, is only about 10 km thick under the oceans and about 50 km thick on the continents [7.1] [7.2]. The so-called Mohorovicic discontinuity (the discontinuity here referring to the seismic wave velocity), separates the crust from the underlying mantle. The latter is the layer between the crust and the (liquid) outer core; the inner core is again solid. According to plate tectonics, the earth's crust and upper mantle, the so-called lithosphere, consists of a number of plates which are constantly moving relative to each other. As an example, in subduction zones a plate is descending beneath another plate, while in regions where two plates are being pulled apart, new molten material is forced upward from the earth's interior. The plate boundaries, where the plates forming the earth's crust meet, are also the zones where the most severe earthquakes are to be expected (Fig. 7.1-1). However, earthquakes also may occur in the plates themselves (intraplate events); these are typically of shorter duration and higher frequency content than interplate earthquakes. Apart from tectonic earthquakes, caused by the sudden rupture of rock formations with correspondingly high energy release, earthquakes may also stem from volcanic activity or from the collapse of subterranean (natural or man-made) cavities; their significance is however much smaller than that of tectonic events.

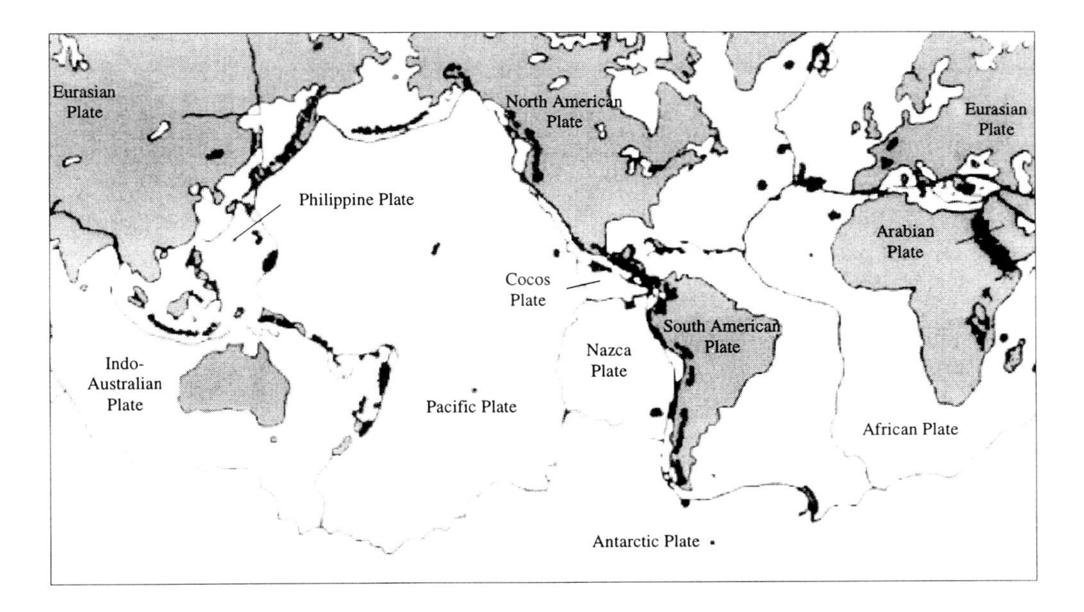

Fig. 7.1-1: The major continental plates

Rupturing and faulting occurs in the crust when the stresses that have been slowly building up during decades or even centuries exceed the strength of the local rock material, which is especially low along pre-existing faults. The energy released propagates in the form of seismic waves in all directions, eventually reaching the surface after multiple reflections and refractions. The most important physical parameters determining the destructiveness of an earthquake are the magnitude of the fractured area or rupture zone, which may be tens or even hundreds of km long and tens of km wide, the strength of the rock material and also the relative displacement or offset at the fault. The fastest waves propagating from the rupture or focal zone are the P (primary, longitudinal, push-pull, irrotational) waves, followed by the S (secondary, transverse, shear, rotational) waves. The shear wave velocity v_s of the S waves is given by

$$v_s = \sqrt{\frac{E}{\rho}}\sqrt{\frac{1}{2(1+\nu)}} \tag{7. 1. 1}$$

and is much less than the propagation velocity v_p of the compression waves

$$v_p = \sqrt{\frac{E}{\rho}}\sqrt{\frac{1-\nu}{(1+\nu)(1-2\nu)}} \tag{7. 1. 2}$$

In these expressions, ν is the POISSON's ratio of the medium, E its YOUNG's modulus and ρ its density. Typical v_p values at the level of the upper mantle lie between 7 and 8 km/s, compared with 4 to 5 km/s for S waves. Both P and S waves are body waves; at the earth's surface and also due to the existence of internal boundaries, so-called surface waves are generated, the amplitude of which decreases rapidly with depth. We further distinguish between RAYLEIGH and LOVE waves; the former, named after Lord RAYLEIGH, are quite slow and exhibit a "retrograde elliptical" motion, with the material points actually moving opposite to the wave's propagation direction in ellipses lying in planes perpendicular to the surface. LOVE waves, on the other hand, exhibit a horizontal motion transverse to their direction of propagation and parallel to the surface.

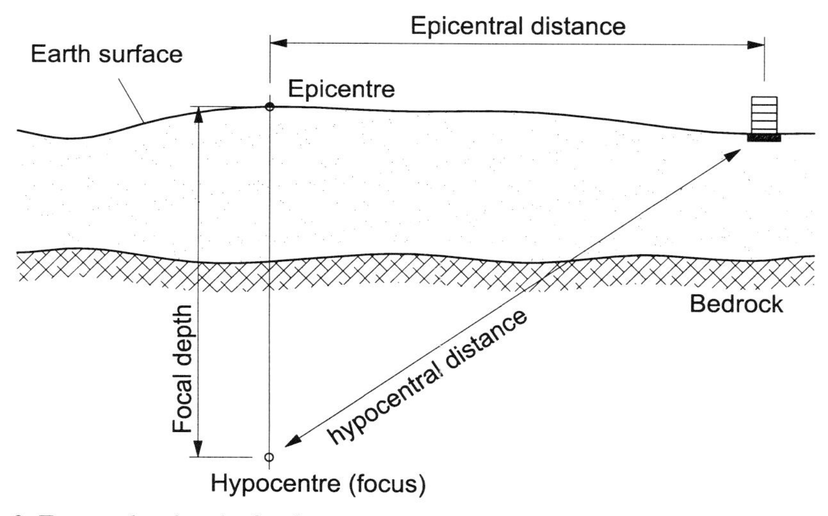

Fig. 7.1-2: Terms related to the focal zone

Fig. 7.1-2 summarises the most important terms characterising the position of the focal zone relative to the earth's surface. The epicentral distance (Δ or D) is defined as distance between the epicentre (the point on the surface directly above the hypocenter) and the reference point (e.g. building site). For large epicentral distances, Δ may also be measured in degrees. The focal depth is denoted by h and the hypocentral distance between the focus and the reference point by s. Most earthquakes are generated in relatively shallow depths of up to h = 60 km; earthquakes with deep foci (say h > 300 km) are unimportant from the structural engineering viewpoint, while shallow earthquakes with h about 10 km can cause very heavy damage. The seismically induced ground acceleration arriving at the foundation of a structure is a vector process normally described by its two horizontal (NS and EW) and the vertical translational components. The latter is usually not as strong as the former, although records do exist that exhibit higher vertical than horizontal peak acceleration values. The two horizontal components are usually of more or less equal strength.

A series of geophysical characteristic functions are employed for expressing the strength of a seismic event, the most important of which are explained below; more details can be readily found in specialised geophysical monographs such as [7.1] and [7.2]. The parameters discussed below are the magnitude (M), the intensity (I), the seismic moment M_0 and the moment magnitude M_W.

Magnitude:

The magnitude of an earthquake is a measure of the energy released by it and is thus independent of the observer's standpoint. The original definition by the seismologist Charles RICHTER in 1935 of the so-called local magnitude M_L involves a maximum displacement amplitude A (measured with a WOOD-ANDERSON seismometer), a normalising factor A_0 and a site-dependent term S to account for the local geological situation:

$$M_L = \log \frac{A}{A_0} + S \tag{7. 1. 3}$$

Eq. (7.1.3) can also be expressed as

$$M_L = \log A - \log A_0 + S \tag{7. 1. 4}$$

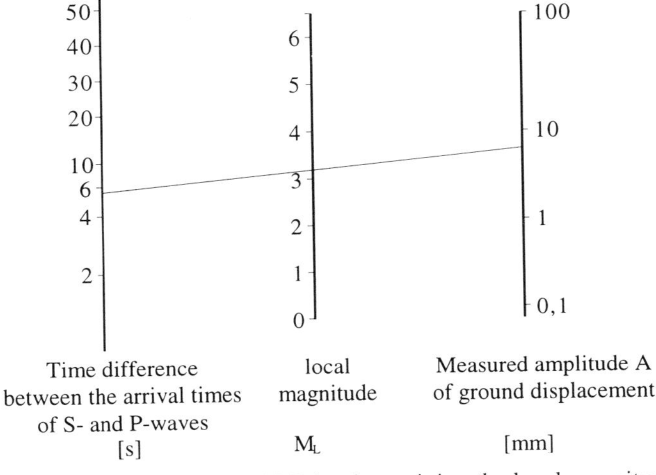

Fig. 7.1-3: Nomogram according to BOLT for determining the local magnitude (schematic)

Physically, the local magnitude can be regarded as the decadic logarithm of the measured displacement signal of a WOOD-ANDERSON-type seismometer located at an epicentral distance of 100 km. Even if registrations of seismic signals are no longer being carried out by WOOD-ANDERSON seismometers, the measured amplitudes are numerically transformed to simulate this instrument type in order to compute M_L. BOLT [7.3] has developed nomograms which permit a quick determination of the local magnitude; input data are the time lag between the arrival of the P and S waves (this being proportional to the distance Δ) and the S-wave amplitude (see Fig. 7.1-3 after [7.4]).

The determination of local magnitudes is only reasonable for short to medium epicentral distances, say up to 700 km.

Another magnitude type, the so-called surface wave magnitude M_S , is used for shallow earthquakes (h < 60 km) at distances over 20°. It is defined by an empirical expression of the type

$$M_S = \log A + C_1 \log \Delta + C_2 \tag{7. 1. 5}$$

or

$$M_S = \log \frac{A}{T} + C_3 \log \Delta + C_4 \tag{7. 1. 6}$$

and is independent of the instrument type; the variables present in these expressions are explained below. For deep earthquakes, a body wave magnitude m_b is introduced by the expression

$$m_b = \log \frac{A}{T} + Q(\Delta, h) + C_5 \tag{7. 1. 7}$$

Here, Q is a correction term considering the non-linear relationship between $\log \frac{A}{T}$ and the epicentral distance. In the Eqs. (7.1.5) to (7.1.7), A is the maximum amplitude of the surface waves in the period range from 18 to 22 s, Δ the epicentral distance, T the period of the surface waves and C_1 to C_5 various correction terms. Empirical (mostly linear) expressions have been given for transforming the different magnitude types into one another.

The most important point to remember in the context of earthquake magnitude values is that their logarithmic character implies a 10-fold increase in wave amplitude for an increase of one unit. The RICHTER scale has in principle no upper limit, but magnitudes in excess of, say, 9 to 9.5 are quite improbable due to the finite strength of the rock material in the rupture zone.

Intensity:

Contrary to the magnitude, earthquake intensity values always refer to a specific site and describe the effects of the earthquake at that site in more or less qualitative terms. Well-known intensity scales are the MSK (MEDVEDEV-SPONHEUER-KARNIK) scale included in DIN 4149 and the Modified MERCALLI (MM) scale. Both are 12-point scales ranging from I=1 (not felt) to I=12 (total destruction). Table 7.1-1 presents a simplified version of the MSK scale. There exist various empirical expressions for correlating intensity values with, say, ground accelerations, such as the relationships of MURPHY and O'BRIEN [7.28] . Such expressions are, for example, of the form

$$\log a_{g,h} = 0.25\ I_{MM} + 0.25$$
$$\log a_{g,v} = 0.30\ I_{MM} - 0.54 \quad (7.1.8)$$

Here, $a_{g,h}$ is the horizontal and $a_{g,v}$ the vertical acceleration component. To illustrate, according to these expressions, a Modified Mercalli intensity of 6 corresponds to a horizontal acceleration of 56 cm/s^2 and a vertical acceleration of 18 cm/s^2. Generally, the current trend is to describe the local intensity at a site by means of measured quantities („instrumental intensity"), rather than relying exclusively on the observation of the effects of the earthquake.

Intensity	Characteristics
1	Only registered by instruments
2	Noticed by very few persons
3	Noticed by some persons
4	Noticed by many; rattling of window panes and crockery
5	Suspended objects swing; sleeping persons are awaken
6	Slight damage to buildings, some barely visible plaster cracks
7	Noticeable plaster cracks, cracks in masonry walls and chimneys
8	Large masonry cracks, collapse of cornices and mouldings
9	Collapse of walls and roofs of some buildings, some ground sliding
10	Several buildings collapse, ground fissures of up to 1 m width
11	Many fissures and landslides
12	Marked landscape changes

Table 7.1-1: Simplified MSK scale according to DIN 4149 part 1

Seismic moment:

The seismic moment M_0 is defined by the relationship

$$M_0 = \mu D A \quad (7.1.9)$$

with D as mean displacement of the rupture surface A and μ as the value of the local rock strength; if these values are not constant, M_0 is evaluated as integral over a series of sub-regions. The seismic moment is a physically meaningful value that corresponds directly to the energy involved in the process; it is also used for defining the so-called moment magnitude M_W after KANAMORI [7.5]:

$$M_W = \frac{2}{3} \log M_0 - 10.73 \quad (7.1.10)$$

Here, the seismic moment M_0 must be input in dyn cm units, with 1 dyn = 10^{-8} kN. An advantage of the moment magnitude compared to other magnitude definitions is that it is free of the saturation effect observed in the latter for very strong earthquakes.

7.2 Characteristic functions for describing ground motions

For engineering purposes, measured time histories of the seismically induced ground acceleration, so-called accelerograms, are the most important data for describing the seismic hazard at a certain location. The three spatial ground motion components (two mutually perpendicular translational horizontal components and the vertical component) are measured by special "strong motion instruments" which are, in fact, single-degree-of-freedom oscillators with a natural frequency of about 25 Hz and a damping ratio of about 60%. The raw ground motion acceleration ordinates measured by these instruments must normally be first subjected to additional numerical processing in order to eliminate long-wave components. These become obvious as unrealistic velocity or displacement values at the end of the ground motion; the velocity and/or displacement time histories are computed by simply integrating the measured accelerogram once or twice with respect to time. Normally, the actual difference of the acceleration values between the corrected and the raw version of the accelerogram is quite small. Such numerical procedures for reducing residual velocities and unrealistic ground displacements should also be applied to synthetically generated accelerograms.

While the standard processing of raw acceleration data is a discipline in itself, involving multiple filtering by customised digital filters, a very simple procedure for such an enhancement of an accelerogram is the so-called baseline correction, which aims at minimising the mean square of the ground velocity. For a linear baseline correction, the corrected acceleration ordinates are expressed as

$$a_{cor} = a_{unc} - (c_1 + c_2 \cdot t) \tag{7. 2. 1}$$

and the velocity and displacement values, respectively, as

$$v_{cor} = v_{unc} - \left(c_1 \cdot t + c_2 \cdot \frac{t^2}{2} \right) \tag{7. 2. 2}$$

$$d_{cor} = d_{unc} - \left(c_1 \cdot \frac{t^2}{2} + c_2 \cdot \frac{t^3}{6} \right) \tag{7. 2. 3}$$

Minimising the square of the velocity integrated over time is expressed as

$$INT = \int_{t=0}^{t=s} v_{cor}^{\,2}\, dt = \int_{t=0}^{t=s} \left(v_{unc} - \left(c_1 \cdot t + c_2 \cdot \frac{t^2}{2} \right) \right)^2 dt \rightarrow Min \tag{7. 2. 4}$$

which yields a linear equation system for the unknown coefficients c_1 and c_2:

$$\frac{\partial}{\partial c_1}(INT) = 0, \quad \frac{\partial}{\partial c_2}(INT) = 0 \tag{7. 2. 5}$$

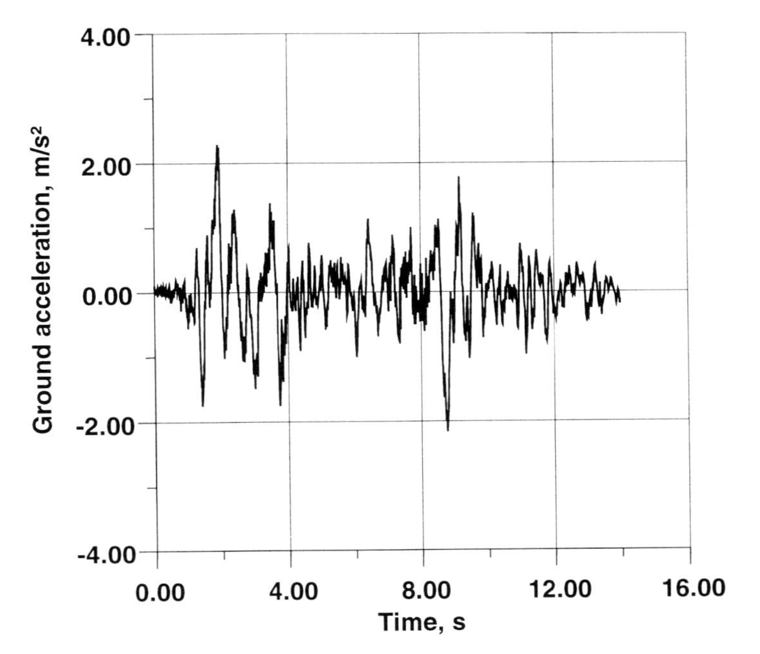

Fig. 7.2-1: „Albatros“ record

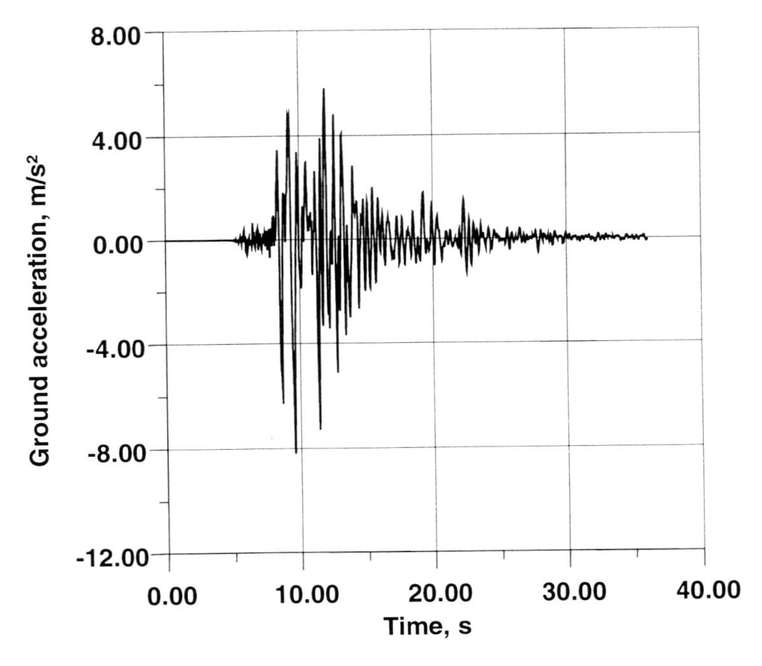

Fig. 7.2-2: „JMA“ record

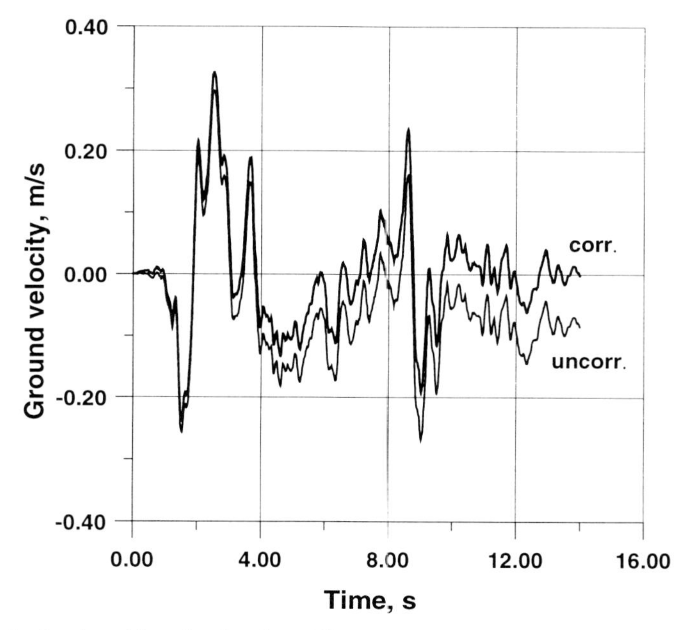

Fig. 7.2-3: Velocity time histories for the „Albatros" record, corrected/uncorrected

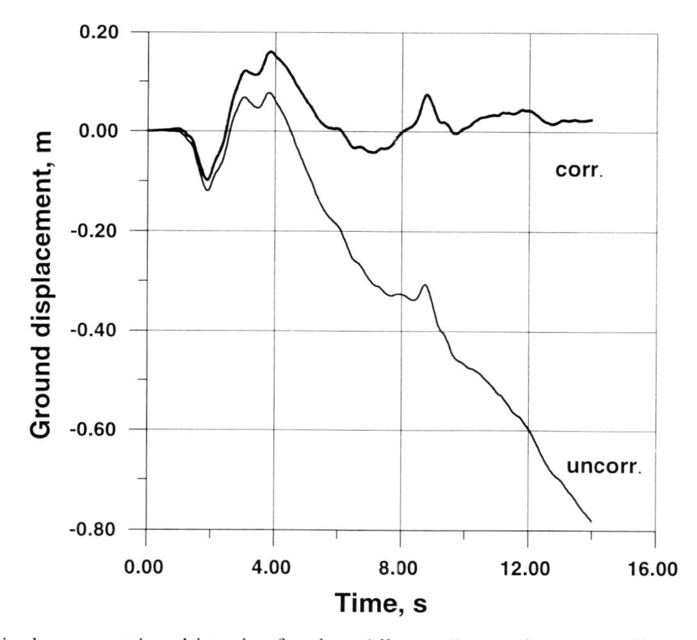

Fig. 7.2-4: Displacement time histories for the „Albatros" record, corrected/uncorrected

This linear baseline correction algorithm has been implemented in the computer program BASKOR. A simple high-pass filtering as described in section 3.5 can also suppress the unwanted long-wave components and lead to acceptable results.

Figs. 7.2-1 and 7.2-2 show some measured accelerograms. The record shown in Fig. 7.2-1 is the east-west (EW) component of the 1979 Montenegro earthquake measured at the Albatros Hotel in Ulcinj; Fig. 7.2-2 shows the north-south (NS) component of the 1995 Hyogoken-Nanbu event, measured at the Japan Meteorological Agency (JMA) in Kobe.

Figs. 7.2-3 to 7.2-5 show time histories for the ground velocities and displacements of these records before and after linear baseline corrections carried out by the program BASKOR.

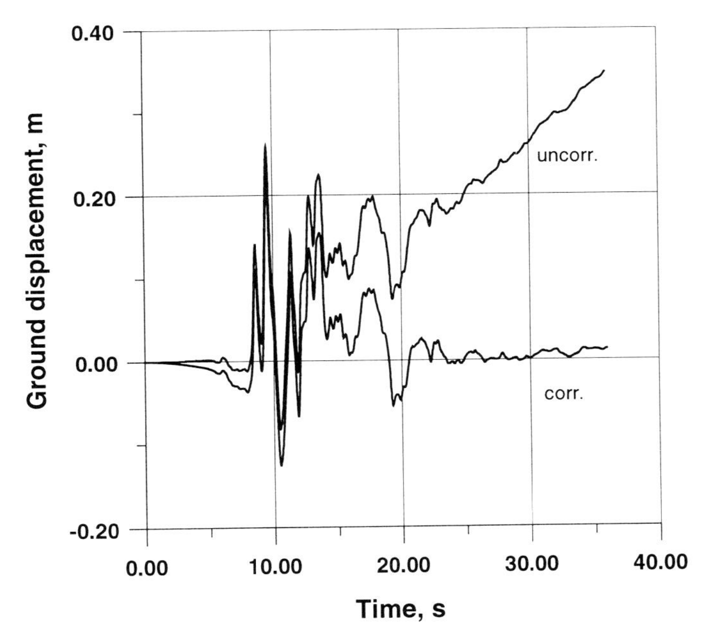

Fig. 7.2-5: Displacement time histories for the „JMA“ record, corrected/uncorrected

The evaluation of the velocity and displacement time histories from the acceleration record is done by means of the program INTEG; its input file ACC contains the accelerogram in a 2E14.7 format, with the equidistant time points in the first and the acceleration values in the second column. The results are written in the output files VEL (for velocities) and DIS (for displacements), also in a 2E14.7 format with the time points in the first column. The program requires the interactive input of a factor for scaling the given acceleration ordinates in order to transform them in units of m/s^2 ; in VEL and DIS velocities and displacements are output in m/s and m units, respectively. The program BASKOR also supplies time histories for the ground velocities and displacements for both the corrected and the uncorrected accelerogram.

The two following subsections 7.2.1 and 7.2.2 discuss some useful parameters and characteristic functions for describing the properties of accelerograms both in the time and in the frequency domain.

7.2.1 Time domain parameters

1 Peak ground acceleration (PGA)

Since the peak ground acceleration is an isolated value mostly related to the high-frequency components of the accelerogram, it does not correlate very well with the damage potential of the ground motion. It is, however, an important parameter for characterising the local intensity of a seismic event and it serves also as scaling factor for other characteristic functions like response spectra. The possibility of defining „effective“ acceleration values is discussed in section 7.2.2 in the context of determining response spectra. As examples, the PGA values of the JMA record and for the “Albatros” record were equal to 8.2 m/s^2 and 2.3 m/s^2, respectively; in Germany, PGA values typically lie below 1 m/s^2.

2 Effective (RMS) acceleration values

RMS values are defined as square roots of the integral of the square of the acceleration values over the nominal duration s of the shaking, divided by s:

$$\mathrm{RMS} = \sqrt{\frac{1}{s}\int_{t=0}^{t=s} a^2 \, dt} \tag{7. 2. 6}$$

Here, $a = \ddot{u}_g(t)$ is the ground acceleration component. The RMS value is obviously directly related to the energy of the accelerogram.

3 HUSID diagram H(t)

It is defined as the time integral of the square of the ground acceleration a and is usually normalised to a unit value at t=s, s being the nominal duration of the shaking:

$$H(t) = \frac{\int_{t=0}^{t} a^2 \, dt}{\int_{t=0}^{t=s} a^2 \, dt} \tag{7. 2. 7}$$

Since the HUSID diagram mirrors the temporal distribution of the accelerogram's energy, it can be used for defining effective strong motion duration values. It can be evaluated by means of the program HUSID, which requires the same input file ACC as the program INTEG; the results are written in the output file HUS. Fig. 7.2-7 shows the HUSID diagram for the record depicted in Fig. 7.2-6, which also stems from the 1979 Montenegro event, having been measured at the “Oliva”-Hotel in Petrovac. Fig. 7.2-8 depicts the extremely long SCT record of the 1985 Mexico City earthquake, the HUSID diagram of which is plotted in Fig. 7.2-9 together with the HUSID diagram of the JMA record.

4 ARIAS intensity AI

It may be defined as the normalising factor of the HUSID diagram in the form

$$\mathrm{AI} = \int_{0}^{s} a^2 \, dt \tag{7. 2. 8}$$

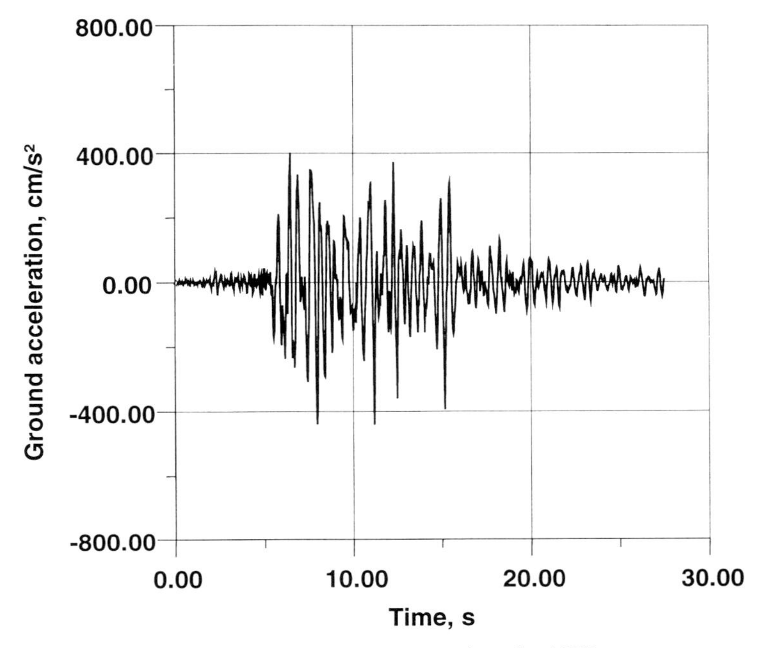

Fig. 7.2-6: "Oliva" record in Petrovac, Montenegro earthquake 1979

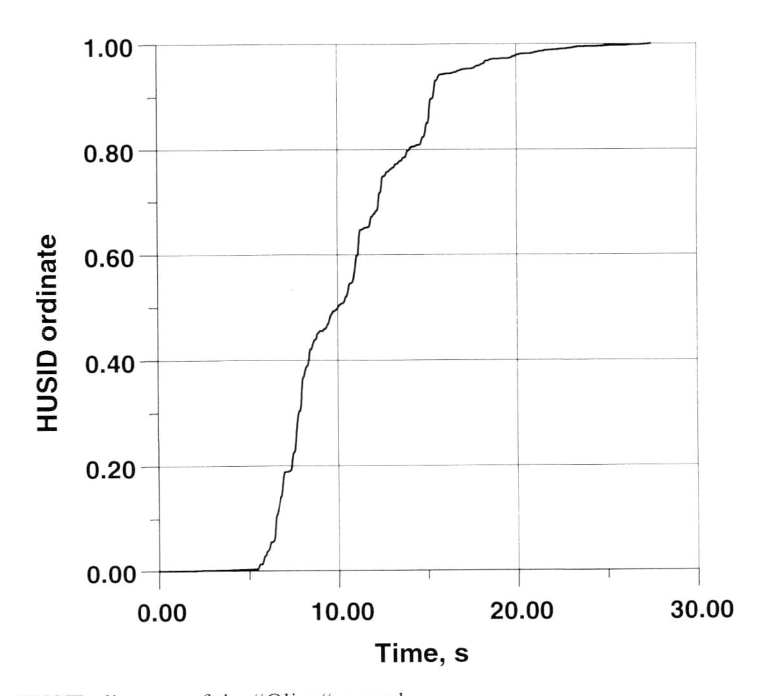

Fig. 7.2-7: HUSID diagram of the "Oliva" record

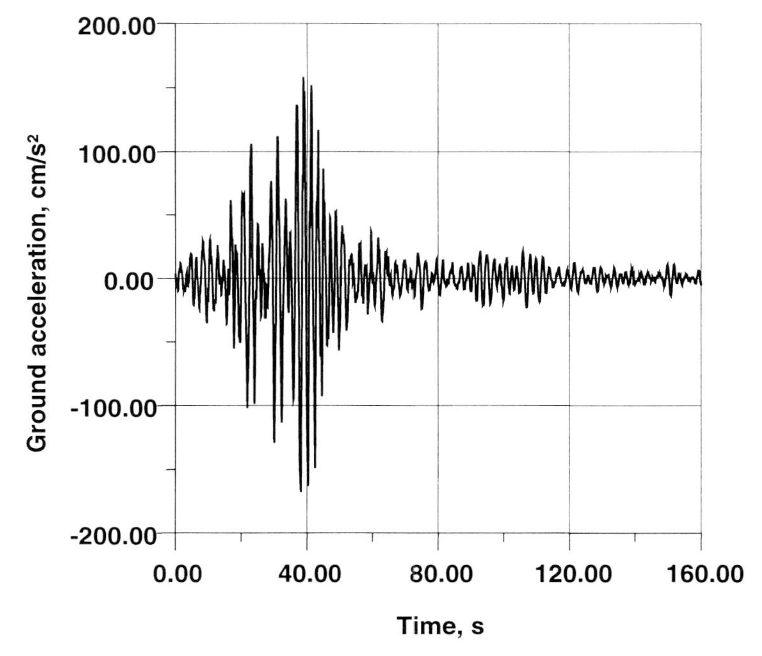

Fig. 7.2-8: SCT record, Mexico City earthquake 1985

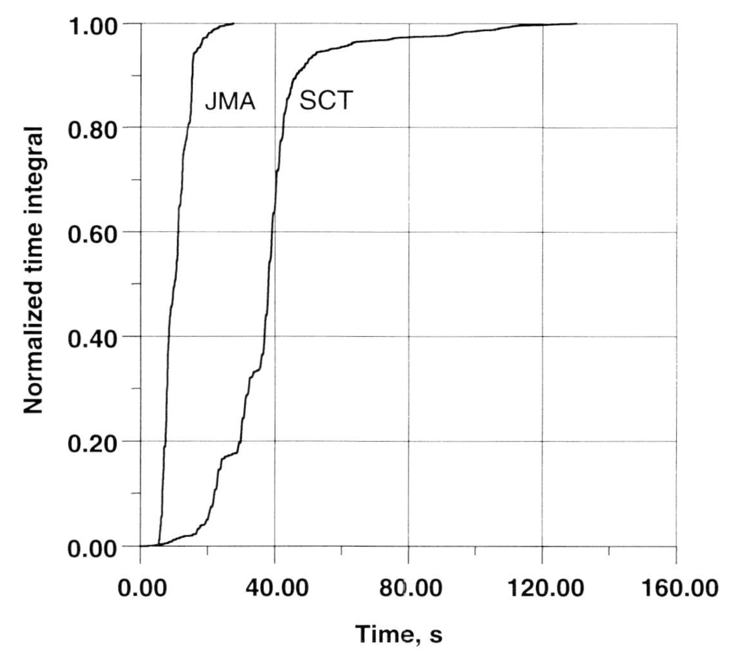

Fig. 7.2-9: HUSID diagrams for the JMA and SCT records

Its original form, as proposed by ARIAS [7.6], is

$$AI = \frac{\pi}{2g} \int_0^s a^2 \, dt \tag{7.2.9}$$

Similarly to the RMS value, the ARIAS intensity is also a measure for the energy contained in the accelerogram. For the "Oliva" record, the ARIAS intensity after Eq. (7.2.9) equals 4.46 m/s, for the JMA record 8.43 m/s, 0.68 m/s for the "Albatros" record and 2.43 m/s for the SCT record.

5 Strong motion duration

One definition of the "effective" duration of shaking based on a given accelerogram according to BOLT [7.7] is the time segment during which the acceleration peaks stay above a given threshold (e.g. 0.05g). Other definitions are based on the HUSID diagram and set the strong motion duration equal to the time between its 5% and 95% ordinates (TRIFUNAC and BRADY, [7.8], NOVIKOVA and TRIFUNAC [7.9]) or between 0 and 90%. A long strong motion duration may lead to many loading cycles in the structure with corresponding high damage levels. For the "Oliva" record, the 5%-95% strong motion duration amounted to 10.6 s, for "Albatros" it was equal to 9.7 s, for JMA 8.1 s and finally, for the long Mexico City SCT record, it reached 36.5 s.

In order to give an idea of their value range, a table summarising the values of the time-domain parameters discussed in this section for the four records mentioned is given below.

Record	PGA	AI	Duration 5%-95%
	m/s^2	m/s	s
„Albatros"	2.29	0.68	9.7
„Oliva"	4.43	4.46	10.6
SCT	1.68	2.43	36.5
JMA	8.20	8.43	8.1

7.2.2 Frequency domain parameters

1 Response spectra

Fig. 7.2-10 schematically depicts the general case of a structure excited by the motion of its single support, showing the two limiting cases of a very soft structure (idealised as a slender grass stem) and a rigid body (shown as a block welded to the ground). In the former case, the natural frequency of the system tends to zero, while the rigid block's natural frequency is infinite and the natural period equal to zero. In the case of the soft structure ($\omega_1 \approx 0$), the mass remains stationary due to its inertia while the ground is experiencing displacements according to the excitation. It follows that the acceleration of and the inertia force acting on the mass are equal to zero, while the relative displacement Δ between the mass and the support equals the displacement obtained by a double time integration of the ground acceleration. In the case of the rigid structure, the relative displacement between the block and the ground is zero and the block experiences the full ground acceleration.

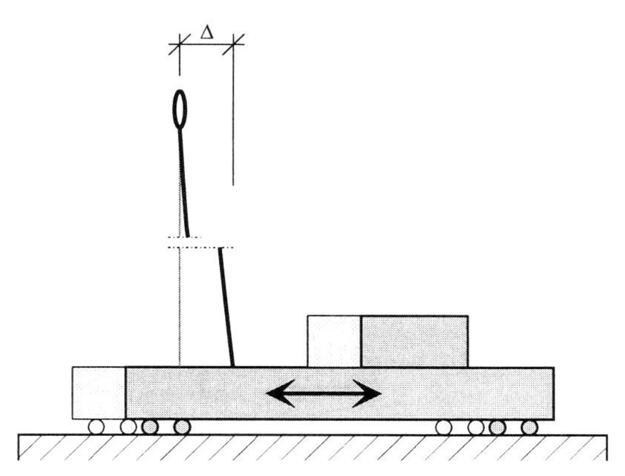

Fig. 7.2-10: A soft and a stiff system subject to support excitation

Seismic response spectra are functions which quantitatively describe the behaviour of simple systems (single-degree-of-freedom oscillators) to given ground motions and at the same time express the distribution of the energy of the accelerogram over the interesting frequency range. For standard structural dynamics applications, the latter usually includes frequencies between 0.1 and 10 Hz (corresponding to periods between 0.1 and 10 s). Linear seismic response spectra for a given accelerogram $\ddot{u}_g(t)$ are evaluated by subjecting a series of viscously damped single-degree-of-freedom systems (with variable natural frequencies ω_1 but a common damping ratio D) to a support excitation equal to $\ddot{u}_g(t)$ and determining their maximum displacement response. The pertinent differential equation of motion is given by

$$\ddot{u} + 2 \cdot D \cdot \omega_1 \, \dot{u} + \omega_1^2 \cdot u = -\ddot{u}_g(t) \qquad (7.2.10)$$

Its solution (computed for example by means of the DUHAMEL integral) yields the absolute maximum displacement $u_{max} = S_d$ of the mass relative to the support. The minus sign of the right-hand-side of Eq. (7.2.10) is usually suppressed, since we are only interested in the absolute value of the maximum relative displacement. The computed S_d values as functions of the damping ratio D and the natural frequencies $\omega_{1,i}$ are the ordinates of the so-called displacement spectrum. Pseudo relative velocity ordinates S_v and pseudo absolute acceleration ordinates S_a are defined as functions of the computed S_d values according to

$$S_d = (1/\omega_1) \cdot S_v = (1/\omega_1^2) \cdot S_a \qquad (7.2.11)$$

The spectral values S_v and S_a do not necessarily represent maximum values for the velocity and acceleration of the single-degree-of freedom system as would be the case if its response to the accelerogram were purely harmonic, which explains the prefix "Pseudo".

For a given base excitation, a computer program can easily determine the maximum relative displacement S_d for a fixed D and a series of ω_1 values, and by using Eq. (7.2.11) furnish the (relative) displacement, (pseudo relative) velocity and (pseudo absolute) acceleration spectrum. It is recommended to plot the results in a tripartite logarithmic diagram as shown schematically in Fig. 7.2-11, with the pseudo relative velocity spectral ordinates S_v plotted over period or frequency.

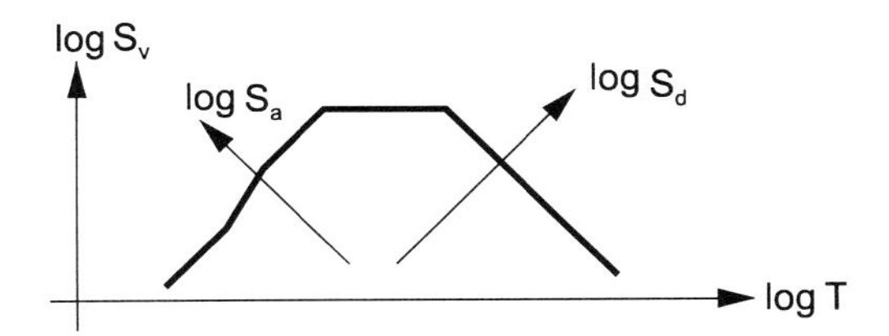

Fig. 7.2-11: Logarithmic plot of a seismic response spectrum (schematically)

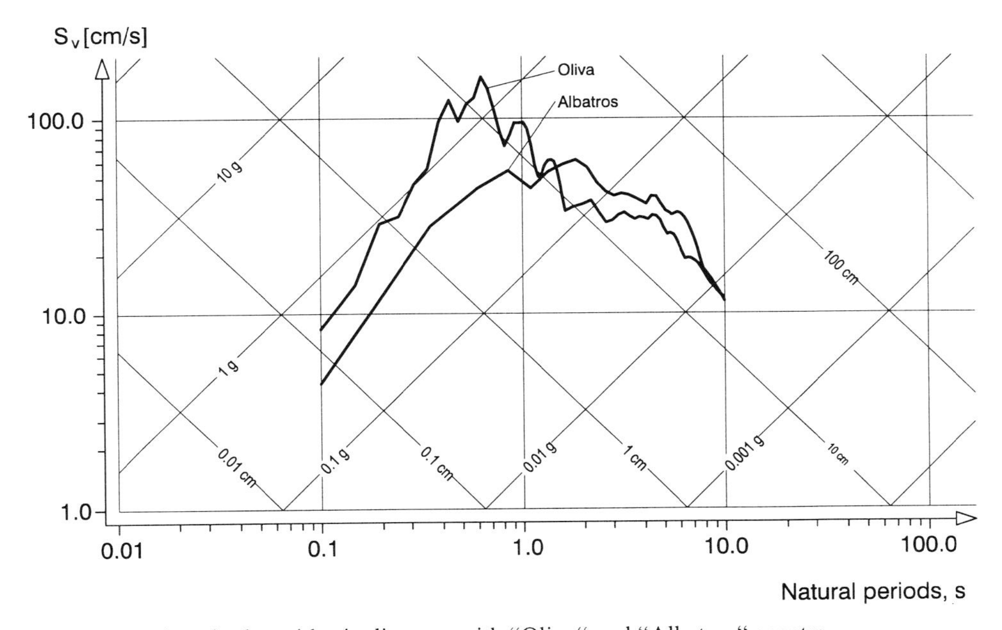

Fig. 7.2-12: Tripartite logarithmic diagram with "Oliva" and "Albatros" spectra

From such a tripartite logarithmic diagram one can also directly obtain the S_a and S_d ordinates along orthogonal axes rotated by 45° with respect to the (T, S_v) co-ordinate system, as shown in Fig. 7.2-12. For very small periods (rigid systems), the spectral acceleration is constant and equal to the PGA, for very large periods (very flexible systems) we obtain a constant spectral displacement value equal to the maximum ground displacement.

Fig. 7.2-13 shows the acceleration response spectra of the "Oliva" and "Albatros" records for a damping ratio D = 5% and Fig. 7.2-12 shows the same spectra in tripartite logarithmic plots, as computed by the program SPECTR. This program requires as input the accelerogram (input file ACC) and it produces the output file SPECTR which contains in columns 2, 3 and 4 the spectral ordinates for displacement (in cm), pseudo relative velocity (in cm/s) and absolute acceleration (in g), while column 1 contains the period values (in s). Columns 5 and 6 of file SPECTR contain the ordinates of the absolute and relative energy spectra, which are discussed in the next section.

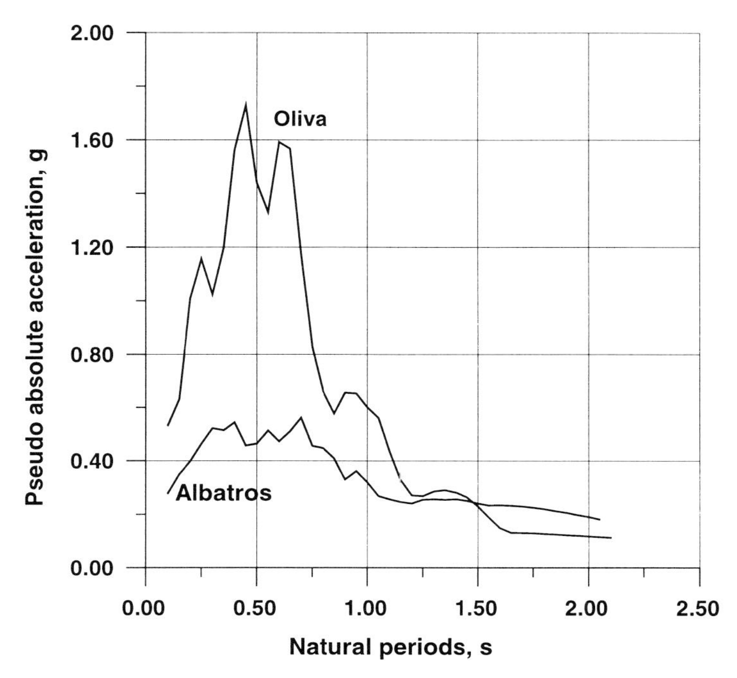

Fig. 7.2-13: Acceleration spectra, Montenegro 1979 earthquake

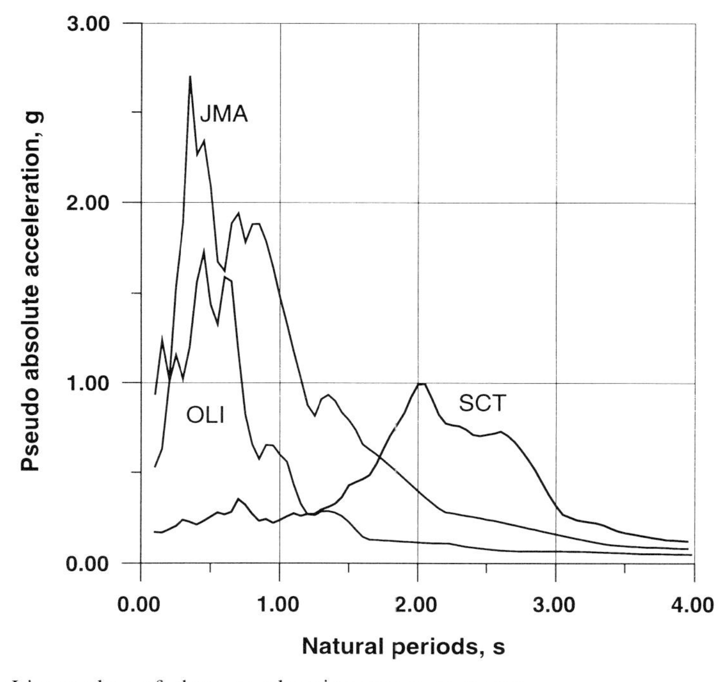

Fig. 7.2-14: Linear plots of three acceleration response spectra

Fig. 7.2-14 compares the acceleration spectrum of the "Oliva" record to those of the Mexico City (SCT) and Kobe (JMA) records, illustrating the variability range involved.

A response spectrum can also define an effective acceleration value by calculating the mean of the spectral acceleration over certain period ranges. For example, an Applied Technology Council (ATC) proposal defines an effective acceleration value of $a_{eff} = \overline{S}_a / 2{,}5$, $\overline{S}_a$ being the mean spectral acceleration in the period range from 0.1 to 0.5 s.

2 Energy response spectra

The differential equation of motion for a single-degree-of-freedom system subject to the support excitation $\ddot{u}_g$ is given by

$$m(\ddot{u} + \ddot{u}_g) + c\dot{u} + k u = 0 \tag{7. 2. 12}$$

Here, u is the displacement of the mass in relation to its support. Integrating this equation over the entire duration of the earthquake yields the expression:

$$\int (c\dot{u})du + \int (k u)du + \frac{1}{2}m(\dot{u} + \dot{u}_g)^2 = \int m(\ddot{u} + \ddot{u}_g)du_g \tag{7. 2.13}$$

or

$$\int (c\dot{u})\dot{u}\,dt + \int (k u)\dot{u}\,dt + \frac{1}{2}m(\dot{u} + \dot{u}_g)^2 = \int m(\ddot{u} + \ddot{u}_g)\dot{u}_g\,dt \tag{7. 2. 14}$$

The single terms are, from left to right, the energy dissipated by damping, the elastically stored work, the kinetic energy, and, as sum of these three terms, the "absolute" input energy, the latter corresponding to the work done by the inertia force $m(\ddot{u}+\ddot{u}_g)$ along the displacement of the system's support. If the single-degree-of-freedom system with the differential equation

$$m\ddot{u} + c\dot{u} + k u = -m\ddot{u}_g \tag{7. 2. 15}$$

is considered as a fixed system (no support displacement possible) with loading equal to $-m\ddot{u}_g$, the following energy balance equation results:

$$\int (c\dot{u})\dot{u}\,dt + \int (k u)\dot{u}\,dt + \frac{1}{2}m\dot{u}^2 = -\int m\ddot{u}_g\,\dot{u}\,dt \tag{7. 2. 16}$$

The terms expressing the work dissipated by the damping force and the restoring force stay the same, while the third term on the left-hand side expresses the „relative“ kinetic energy and the right-hand side the "relative" input work, which can be visualised as the work done by the force $-m\ddot{u}_g$ along the displacement u relative to the support. The "absolute" and the "relative" input energy expressions are accordingly given by

$$E_{abs} = \int_{t=0}^{t=s} m(\ddot{u} + \ddot{u}_g)\dot{u}_g\,dt \tag{7. 2. 17}$$

and

$$E_{rel} = (-)\int_{t=0}^{t=s} m\ddot{u}_g\,\dot{u}\,dt \tag{7. 2. 18}$$

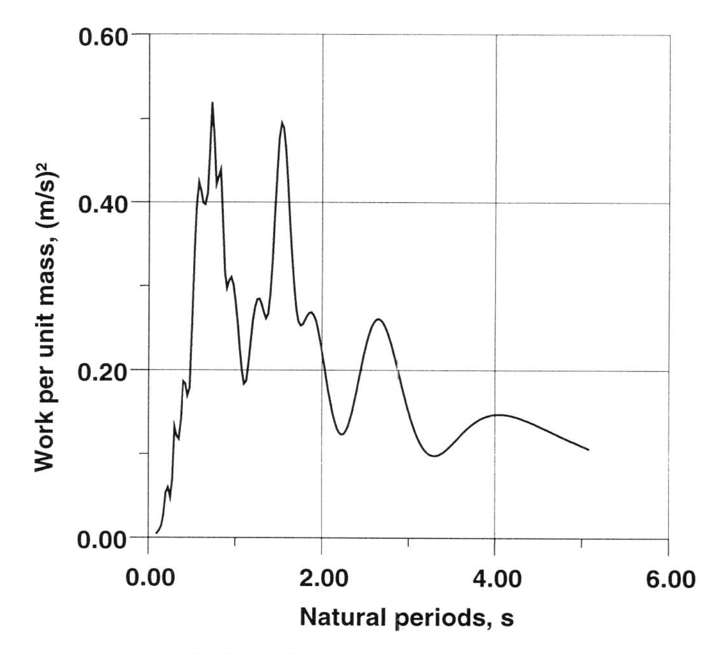

Fig. 7.2-15: Energy spectrum, "Albatros" record

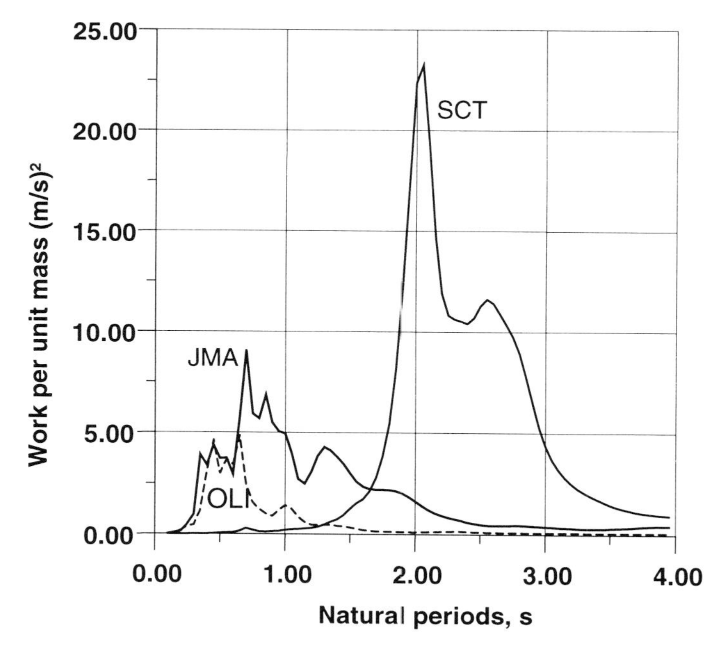

Fig. 7.2-16: Energy spectra of three records

These expressions can be evaluated for a series of single-degree-of-freedom systems with constant damping and variable natural frequency, yielding corresponding energy spectra. The differences between the maximum values of E_{abs} and E_{rel} usually are not very pronounced [7.9], [7.10]. Fig. 7.2-15 shows the energy spectrum of the "Albatros" record, Fig. 7.2-16 the energy spectra for the Mexico City (SCT), "Oliva" (OLI) and Kobe (JMA) records; the high damage potential of the SCT record is much more obvious here than in Fig. 7.2-14. These energy spectra have been computed by the SPECTR program; they can be found in columns 5 (for the „absolute") and 6 (for the „relative" energy spectrum) of the output file SPECTR.

3 Inelastic response spectra

If non-linear single-degree-of-freedom systems according to section 3.7 are subject to support excitations, corresponding "inelastic response spectra" can be computed, the ordinates of which are generally smaller than for the linear elastic spectra just discussed. The most popular non-linear displacement-restoring force law is the elastic-perfectly plastic law shown in Fig. 3.7-1. The ratio of the maximum elasto-plastic displacement u_{max} to its elastic limit u_{el} is defined as maximum ductility μ:

$$\mu = \frac{u_{max}}{u_{el}} \qquad (7.2.19)$$

This ductility value is the additional parameter needed to characterise the inelastic response spectrum. Its ordinates refer to the maximum elastic displacement u_{el}, with $S_d = u_{el}$, $S_v = u_{el} \cdot \omega$ and $S_a = u_{el} \cdot \omega^2$. Inelastic response spectra of given accelerograms for arbitrary target ductilities μ can be computed by means of the program NLSPEC. Its input file is again ACC, containing the accelerogram, and the output file NLSPK contains the natural periods (in s), the spectral ordinates $S_d = u_{el}$ for the displacement (in cm), S_v for the pseudo relative velocity (in cm/s) and S_a for the pseudo absolute acceleration (in g) in the first four columns. In column 5, the maximum ductility μ that was actually reached is given, which does not always coincide exactly with the target ductility that has been specified. The program's CPU time requirements are quite noticeable, especially for long records, due to the iterative process used for the determination of the correct spectral ordinates. It is also possible that no satisfactory solution can be found for a combination of target ductility, natural period and accelerogram, since not every ductility value is physically attainable for a certain system and excitation. In this case, after having reached the internally determined maximum iteration cycle number, the program simply uses the best approximation found so far and proceeds to the next period value.

Fig. 7.2-17 depicts an inelastic response spectrum for a target ductility $\mu = 2.5$, computed with the program NLSPEC. The underlying accelerogram is a synthetically generated record, the elastic response spectrum of which is shown in Fig. 7.2-17 together with its elastic target spectrum.

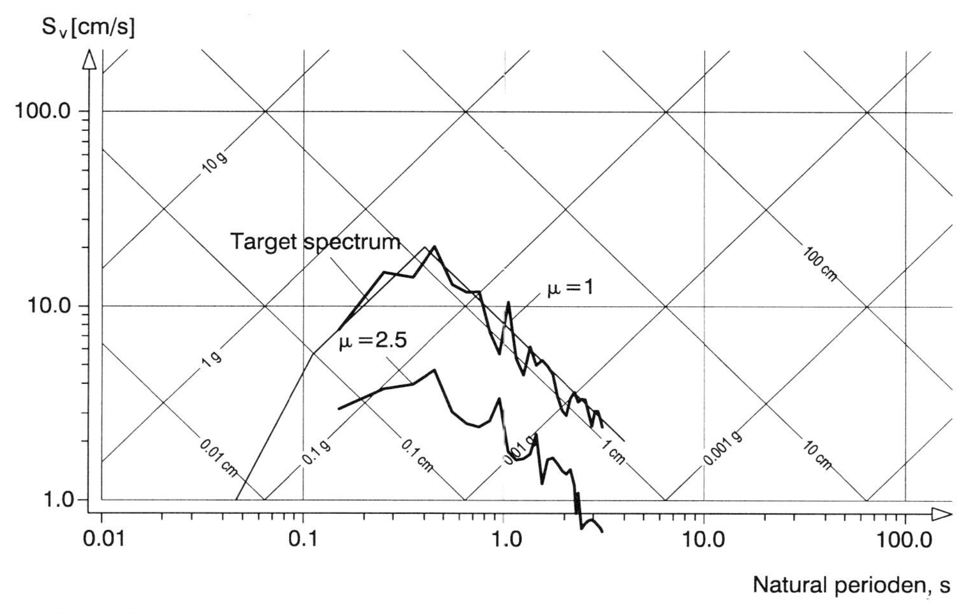

Fig. 7.2-17: Elastic and inelastic response spectrum

4 Spectral intensity

The area under the pseudo relative velocity spectrum in the period range between 0.1 and 2.5s has been introduced by HOUSNER as spectral intensity SI:

$$SI(D) = \int_{T=0,1}^{2,5} S_v(T, D)\, dT \tag{7. 2. 20}$$

The value of SI is also computed by the program SPECTR.

The programs introduced in section 7.2 are summarised in the following table with their input and output files:

Program name	Input file	Output file(s)
BASKOR	ACC	VEL.UNK DIS.UNK ACC.KOR VEL.KOR DIS.KOR
INTEG	ACC	VEL DIS
HUSID	ACC	HUS
SPECTR	ACC	SPECTR
NLSPEC	ACC	NLSPK

7.3 Site-dependent elastic response spectra

A building, a bridge or some other civil engineering structure cannot be designed on the basis of the response spectrum of a single measured record alone, since it is most unlikely that the structure will be subjected to exactly the same ground acceleration during its life span. A straightforward approach for developing elastic response spectra that describe the seismic hazard of a region, which, however, requires that a sufficient number of measured records is available, is as follows:

- The measured accelerograms are divided into several groups according to the underlying soil type (e.g. rock, alluvium, soft soil).

- In every group, the records are scaled to the same peak or effective ground acceleration, ARIAS intensity, spectral intensity or some other parameter.

- The response spectra of the scaled records are determined. The resulting spectrum is the curve corresponding to the mean plus a factor multiplied by the standard deviation σ with some smoothing done in order to eliminate the typical "valleys" due to local conditions.

In a more general approach, the probabilistic seismic hazard analysis for a certain site must consider all possible earthquake foci (fault zones) in the region with their respective magnitude-recurrence relations and attenuation laws in order to finally derive uniform hazard spectra, the ordinates of which have the same probability of being exceeded in a given period of time.

The relationships between the maximum ground displacement $d_g = d_{max}$, the maximum ground velocity $v_g = v_{max}$ and the peak ground acceleration $a_g = a_{max}$ on one hand and the maximum spectral values S_a, S_v and S_d on the other hand have been investigated (e.g. in [7.11]) for different spectral damping values and soil conditions. For stiff soil and D=2%, NEWMARK [7.11] suggests the following expressions:

$$\max S_a = \beta_a \cdot a_{max} \approx 4 \cdot a_{max} \tag{7. 3. 1}$$

$$\max S_v = \beta_v \cdot v_{max} \approx 3 \cdot v_{max} \tag{7. 3. 2}$$

$$\max S_d = \beta_d \cdot d_{max} \approx 2 \cdot d_{max} \tag{7. 3. 3}$$

For D=5%, the factors β_a, β_v and β_d are roughly equal to 2.5, 2.0 and 1.8. If the ground motion maxima a_{max}, v_{max} and d_{max} are known, they can be plotted as three straight lines in the tripartite logarithmic diagram shown in Fig. 7.2-12. This yields a site-dependent "basic spectrum", which can be scaled with the damping-dependent factors β_a, β_v and β_d and further modified in order to yield a site-dependent elastic spectrum. Normally, reliable values for all three parameters a_{max}, v_{max} and d_{max} are not available, but a_{max} at least can be determined for example from the probable local seismic intensity using an equation similar to (7.1.8). This peak ground acceleration can then be used for scaling soil-dependent standard spectra, as given, for example, by NEWMARK and HALL [7.11]. There, a spectrum characterised by $a_{max} = 1 \cdot g$, v_{max} = 122 cm/s and d_{max} = 91 cm is recommended for stiff soil, and suitable scaling factors for different damping values are given. For $a_{max} = 0.33 \cdot g$ and D=5%, this yields the spectrum shown in Fig. 7.3-1.

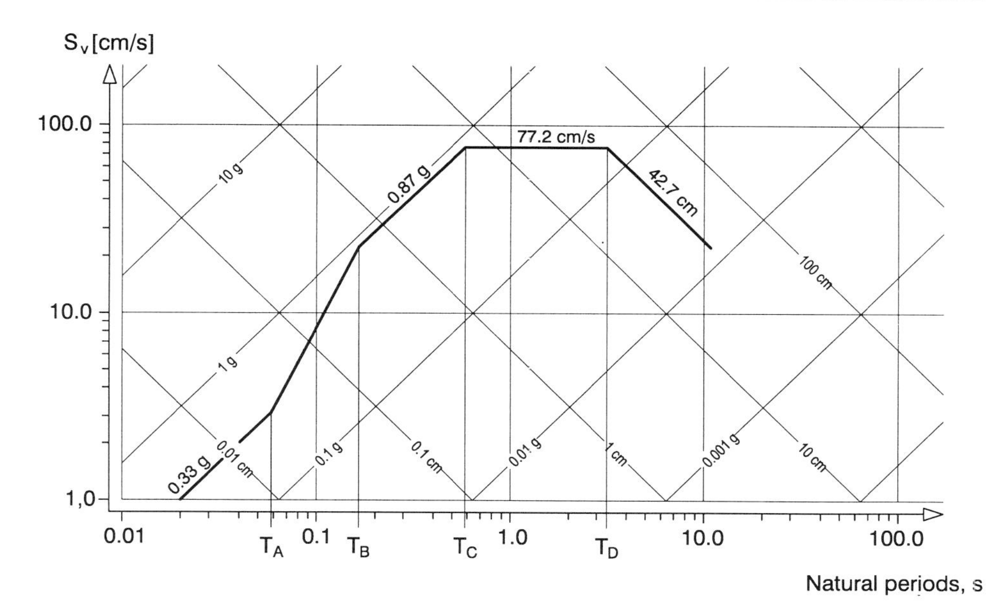

Fig. 7.3-1: NEWMARK-HALL standard response spectrum

Generally, a response spectrum can be divided into four regions (Fig. 7.3-1):

$T \leq T_A$: Constant spectral acceleration $S_a = a_{max}$; T_A may also be zero (as in EC 8).
$T_A \leq T \leq T_B$: Transition zone, increasing spectral acceleration.
$T_B \leq T \leq T_C$: Constant spectral acceleration, $\max S_a = f_a \cdot a_{max}$
$T_C \leq T \leq T_D$: Constant spectral velocity, $\max S_v = f_v \cdot v_{max}$
$T_D \leq T$: Constant spectral displacement, $\max S_d = f_d \cdot d_{max}$

Eurocode 8 (EC 8) distinguishes between three soil types A, B and C, with the shear wave velocity v_s according to Eq. (7.1.1) as main criterion:

Soil class	Properties
A	Rock with v_s not less than 800 m/s, Stiff soil with $v_s \geq 400$ m/s in 10 m depth
B	Sand and gravel or cohesive soils with $v_s \geq 200$ m/s in 10 m depth
C	$v_s < 200$ m/s in the top 20 m

In EC 8, the elastic acceleration response spectra are given by the following expressions:

$$0 \leq T \leq T_B: \quad S_e(T) = a_g \cdot S \cdot \left[1 + \frac{T}{T_B}(\eta \cdot \beta_0 - 1)\right] \tag{7.3.4}$$

$$T_B \leq T \leq T_C: \quad S_e(T) = a_g \cdot S \cdot \eta \cdot \beta_0 \tag{7.3.5}$$

$$T_C \leq T \leq T_D: \quad S_e(T) = a_g \cdot S \cdot \eta \cdot \beta_0 \left[\frac{T_C}{T}\right]^{k_1} \tag{7.3.6}$$

$$T_D \leq T: \quad S_e(T) = a_g \cdot S \cdot \eta \cdot \beta_0 \left[\frac{T_C}{T_D}\right]^{k_1} \left[\frac{T_D}{T}\right]^{k_2} \tag{7.3.7}$$

Here, S_e is the spectral acceleration ordinate, η a correction factor to be applied if the damping is not equal to 5%, S a soil-dependent parameter, k_1 and k_2 soil-dependent exponents and β_0 the spectral acceleration amplification factor for D=5%. Recommended values of these parameters are summarised in the following table for all three soil classes:

Soil	S	β_0	k_1	k_2	T_B in s	T_C in s	T_D in s
A	1.0	2.5	1.0	2.0	0.10	0.40	3.0
B	1.0	2.5	1.0	2.0	0.15	0.60	3.0
C	0.9	2.5	1.0	2.0	0.20	0.80	3.0

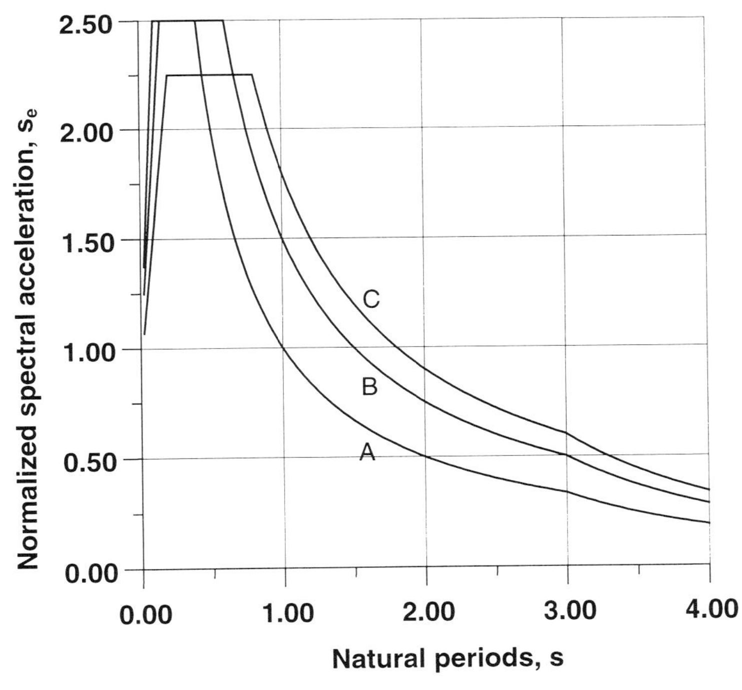

Fig. 7.3-2: Elastic acceleration response spectra according to EC 8, linear plot

Fig. 7.3-2 shows linear plots of these spectra for a_g = 1.0 m/s^2 and $\eta = 1$. For damping ratios $D \neq 5\%$ the correction factor η according to EC 8 is given by:

$$\eta = \sqrt{\frac{7}{2+D}} \geq 0{,}7 \tag{7.3.8}$$

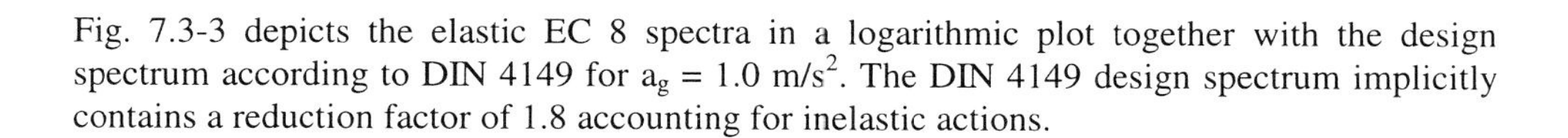

Fig. 7.3-3 depicts the elastic EC 8 spectra in a logarithmic plot together with the design spectrum according to DIN 4149 for $a_g = 1.0$ m/s^2. The DIN 4149 design spectrum implicitly contains a reduction factor of 1.8 accounting for inelastic actions.

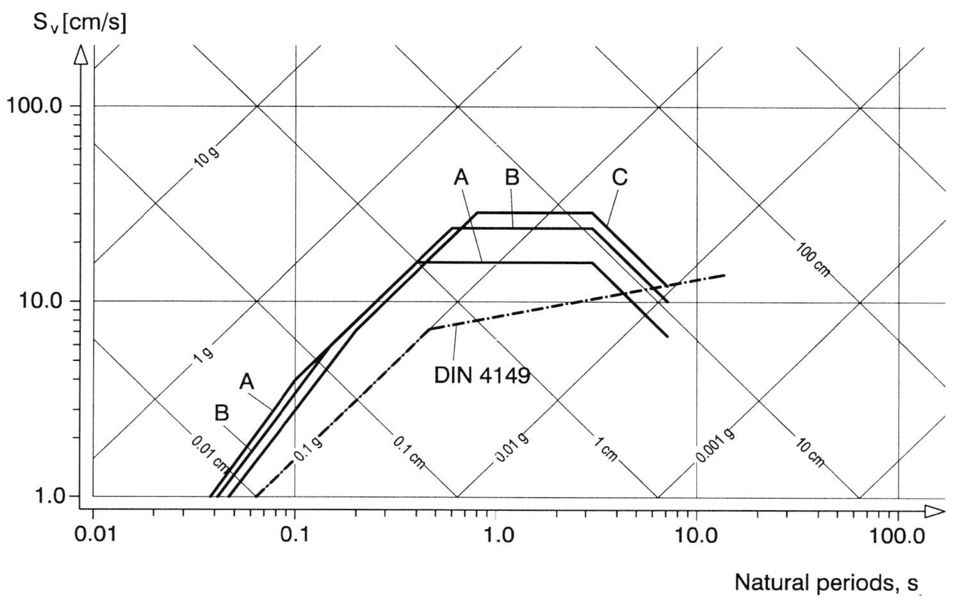

Fig. 7.3-3: EC 8 elastic spectra and DIN 4149 design spectrum (logarithmic)

The expression for the design spectrum contained in DIN 4149 is given by Eq. (7.3.9) for natural periods $T \geq 0.45$ s :

$$S_e = 0.528 \cdot T^{-0,8} \tag{7. 3. 9}$$

Here, no limits are imposed on the spectral displacements for large periods. In EC 8, an expression is given for the value d_g of the maximum ground displacement:

$$d_g = 0.05 \cdot a_g \cdot S \cdot T_C \cdot T_D \tag{7. 3. 10}$$

As an example, for soil class A and $a_g = 1$ m/s^2 , this leads to $d_g = 0.06$ m.

The linear representation of response spectra which are described by piecewise linear curves in the logarithmic (T, S_v) diagram, may be carried out by the program LINLOG. Its input file ESLOG contains (format free) the (T, S_v) value pairs, in units of s and cm/s, defining the polygonal spectrum; the total number of these points is entered interactively. The output file ASLIN contains four columns, giving the natural periods in s in the first and the spectral ordinates of the relative displacement (in cm), pseudo relative velocity (in cm/s) and pseudo absolute acceleration (in g) in the last three columns.

Program name	Input file	Output file
LINLOG	ESLOG	ASLIN

As an example, we consider the design acceleration spectrum of DIN 4149 shown in Fig. 7.3-3. It can be defined by the three (T,S_v) pairs of the following table:

Period T [s]	Pseudo relative velocity [cm/s]
0.01	0.156
0.45	7.030
3.00	10.46

Fig. 7.3-4 shows the linear plot of this acceleration spectrum, as computed by the program LINLOG.

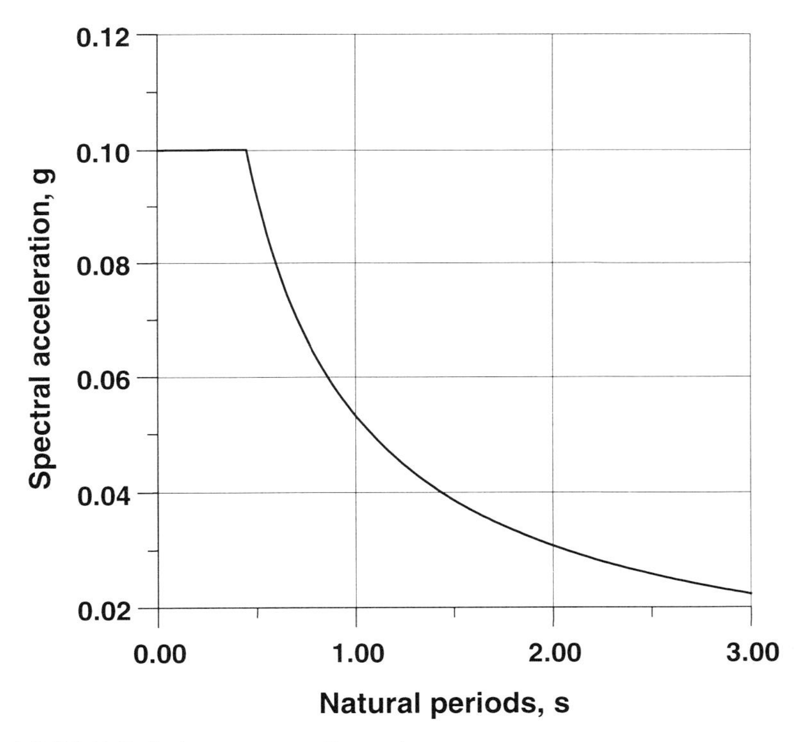

Fig. 7.3-4: DIN 4149 design spectrum, linear plot

7.4 The generation of artificial accelerograms

The checking of the non-linear behaviour of seismically excited structures is usually carried out by means of time-domain direct integration methods, which furnish the system's response to explicit ground motion records. It is not always possible to obtain suitable accelerograms measured near the site in question to be used directly in such an investigation, in which case numerically generated records must be computed, taking into account all available site information (e.g. soil type). The simplest approach lies in utilising an elastic response spectrum for the site and generating an acceleration record, the response spectrum of which closely matches the original "target spectrum" in the period range of interest. A suitable value

for the duration of the record must be chosen additionally, since this information is not contained in the target spectrum.
It is possible to modify an existing (usually measured) accelerogram, so that its spectrum fits some given target spectrum, by judiciously scaling this record's frequency content in the period ranges of interest through selective filtering. It is much simpler though, to construct a synthetic record as the sum of sine waves with random phase angles (uniformly distributed between zero and 2π) and iteratively adjust it so it fits the target spectrum; a deterministic intensity function is chosen for modulating the stationary signal resulting from adding the single harmonic components. Fig. 7.4-1 shows the trapezoidal function used in the program SYNTH, the application of which is explained below.

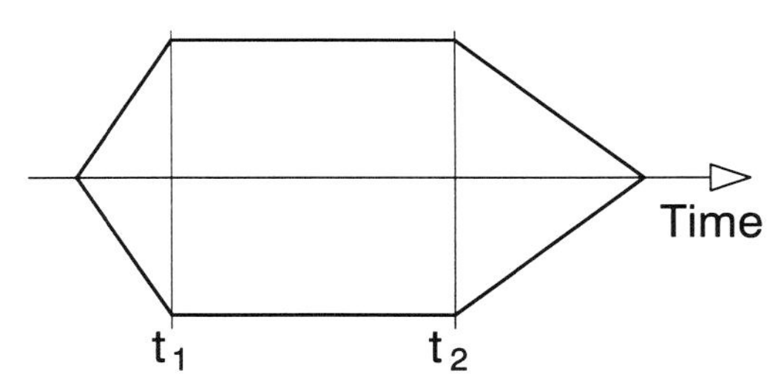

Fig. 7.4-1: Trapezoidal intensity function

As an example, we consider the spectrum B in Fig. 7.3-3 as target spectrum and aim to generate a suitable artificial record with 10s duration. The target spectrum is described by NK pairs of (T, S_v) values, and it is assumed that it is piecewise linear in a tripartite logarithmic plot such as Fig. 7.3-3. The period range of interest, where the target spectrum is to be approximated by the spectrum of the generated record, is defined by entering the two periods TANF and TEND (TANF < TEND) bracketing this range. The interval TEND-TANF should lie within the spectral polygon defined by the NK (T, S_v) values and it should not be overly broad; standard values are TANF=0.1s and TEND=2.5s. Since the existence of a record that fits the target spectrum optimally over a wide period range cannot be guaranteed, it might become necessary to call the program more than once with different integer starting values IY, IY being between 1 and 1024, in order to produce different sets of random phase angles. The input file ESYN of the program SYNTH contains the following data (in free format, with each new entry in a new row):

1. Arbitrary integer number IY between 1 and 1024.
2. Number NK of the pairs (T, S_v) to be input for describing the target spectrum.
3. Number N of the points of the artificial record to be generated, with a constant time step of 0.01 s.
4. Number of the time step where the ramp at the start of the trapezoidal intensity function of Fig. 7.4-1 ends.
5. Number of the time step where the constant part of the trapezoidal intensity function of Fig. 7.4-1 ends.
6. Number of the iteration cycles to be performed, normally 5 to 15.
7. Periods TANF and TEND bracketing the spectral region to be approximated.
8. Damping of the spectrum (e.g. 0.05 for D=5%).
9. NK pairs of (T, S_v) values for the description of the target spectrum, T in s and S_v in cm/s; each number pair in a new row.

The input file ESYN for approximating the spectrum B in Fig. 7.3-3 (with D=5%) in the interval between 0.1 and 2.5 s is as follows:

```
13
5
1000
150
800
16
0.1
2.5
0.05
0.025,  0.49736
0.15,   5.968
0.60,   23.873
3.0,    23.873
4.0,    17.905
```

A record consisting of 1000 points (10 s duration with the fixed time step of 0.01 s) is to be generated, with the initial ramp of the trapezoidal intensity function ending at 1.5 s and the descending ramp starting at 8 s. The input data are written in the file KONTRL for control purposes, while the generated record is written in the file ASYN (2E14.7 format, with time points in the first and acceleration ordinates in m/s^2 units in the second column). It is recommended to check the time histories of the corresponding velocity and displacement time histories (e.g. by means of the program INTEG), and, if necessary, carry out a baseline correction using the program BASKOR. Fig. 7.4-2 shows the computed record which has also been subjected to a baseline correction, Fig. 7.4-3 the ground displacement time histories before and after this baseline correction. In Fig. 7.4-4 the spectrum of the corrected record is finally compared to the target spectrum.

A disadvantage of such spectrum compatible records is their relatively high energy content; their advantage, on the other hand, lies in the ease of their generation with a minimum of required input data.

It is also possible to construct inelastic response spectra from such spectrum compatible motions. To that effect, five, ten or more records are generated for an elastic target spectrum, their (elastic and inelastic) spectra are computed and mean spectra are evaluated. As an example, Fig. 7.4-5 shows the mean spectra of five records generated to a (logarithmically) bilinear target spectrum. The mean inelastic spectra shown correspond to maximum target ductilities of μ= 2.5, 3.0 and 3.5. Comparing these mean inelastic spectra with the inelastic spectrum shown in Fig. 7.2-13, which belonged to a single record, illustrates the smoothing effect due to forming the mean of several spectra.

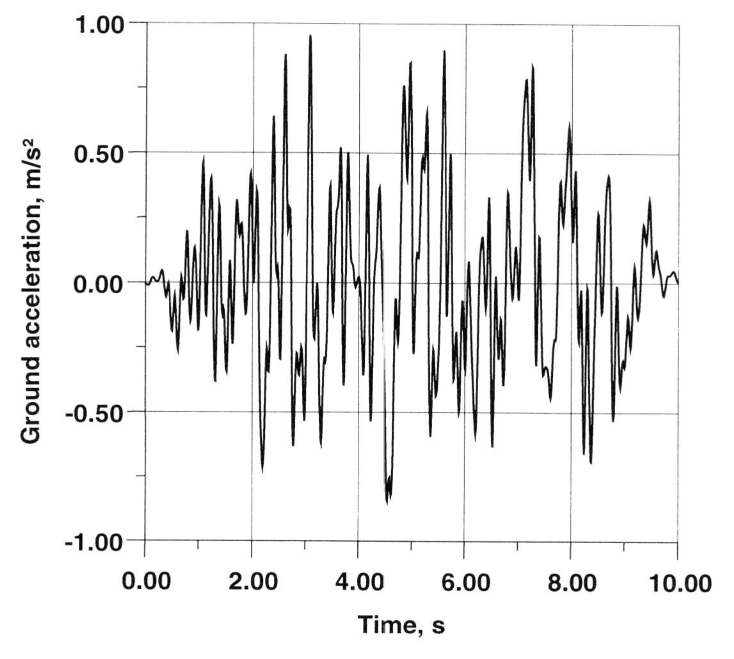

Fig. 7.4-2: Artificial record, baseline corrected

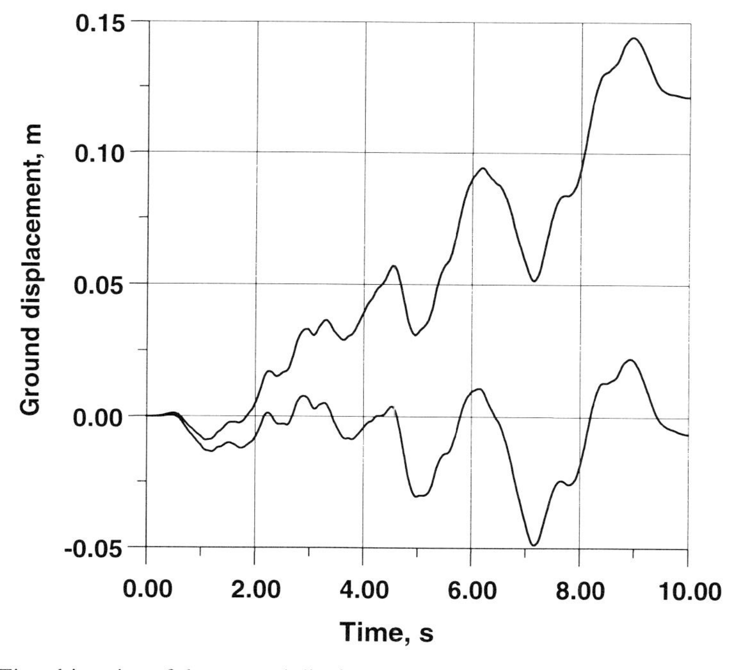

Fig. 7.4-3: Time histories of the ground displacement with/without baseline correction

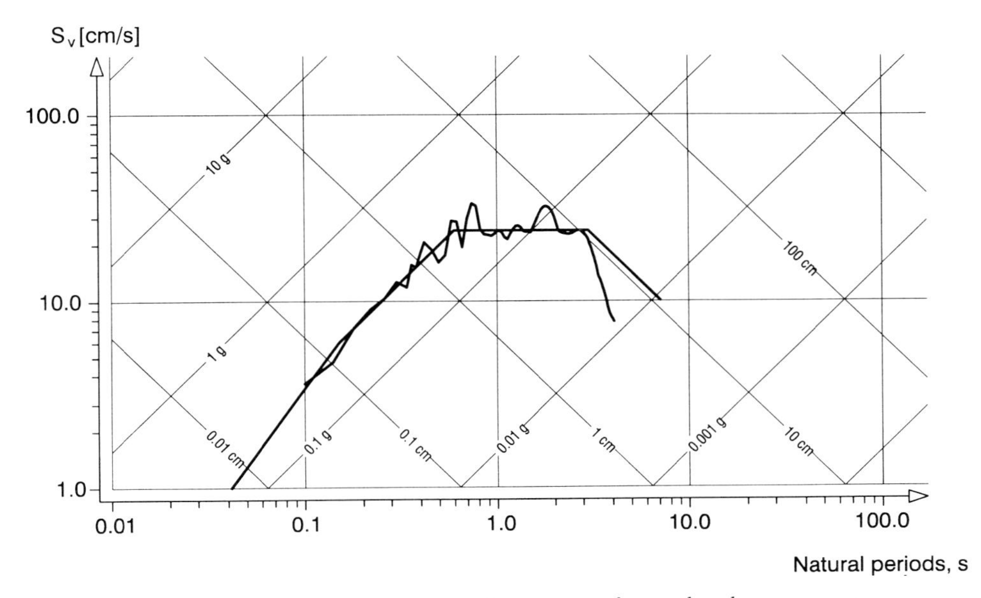

Fig. 7.4-4: Spectrum of the generated and baseline corrected record and target spectrum

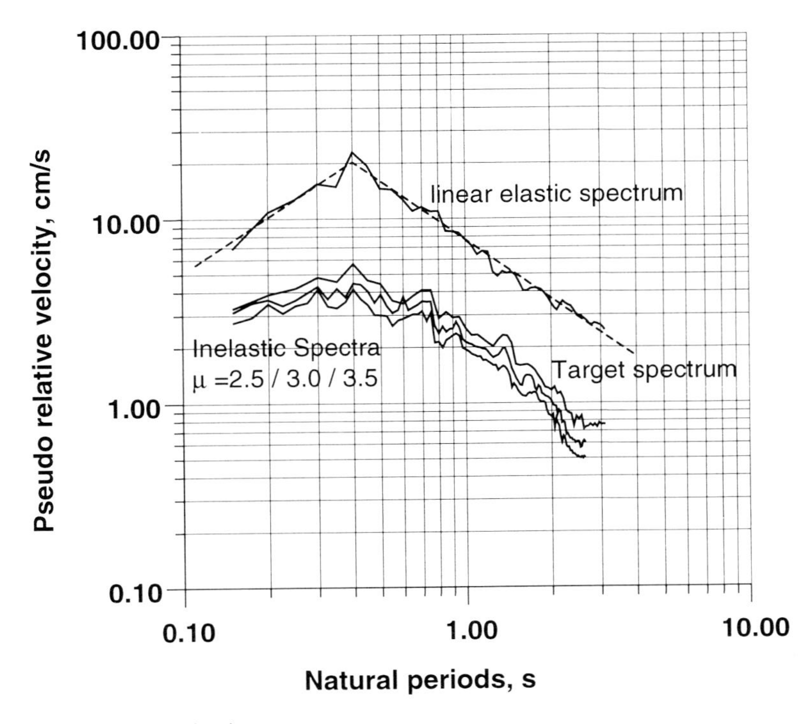

Fig. 7.4-5: Elastic and inelastic spectra, mean curves

The input and output files of the program SYNTH are repeated below:

Program name	Input file	Output files
SYNTH	ESYN	KONTRL ASYN

Another approach for generating artificial acceleration records uses filtered and modulated white noise as the basic ingredient; local site natural frequencies can be considered directly in the filtering process. The single steps are described below:

- A meaningful value for the (double sided) power spectral density S_0 of the white noise process is assumed, e.g. $S_0 = 0.001\ m^2/s^3$. Corresponding stationary time signals with a constant time step Δt may be generated as a series of random numbers with zero mean and unit standard deviation with the single values multiplied by a factor c equal to

$$c = \sqrt{\frac{2\pi \cdot S_0}{\Delta t}} \tag{7. 4. 1}$$

- This time series is then selectively filtered (in the frequency domain) by high-pass, low-pass and band-pass filters as discussed in section 3.5. The most important step is the filtering by the KANAI-TAJIMI-low pass filter according to Eq. (3.5.19) which is repeated here for convenience

$$H(\omega) = \frac{1 + \frac{\omega^2}{\omega_0^2}(4\xi_0^2 - 1) - i2\xi_0 \frac{\omega^3}{\omega_0^3}}{\left[1 - \left(\frac{\omega^2}{\omega_0^2}\right)^2\right]^2 + 4\xi_0^2 \left(\frac{\omega^2}{\omega_0^2}\right)^2} \tag{7. 4. 2}$$

  The coefficients ω_0 and ξ_0 can be regarded as the natural circular frequency and the damping of the soil, respectively. Their values usually lie about 20 rad/s for ω_0 and 0.50 for ξ_0. The high-pass filter according to Eq. (3.5.18), repeated below, can additionally be used for eliminating unwanted long-period components; a meaningful choice is $\omega_H = 1\ \mathrm{rad/s}$.

$$H(\omega) = \frac{\omega^2 + i\omega\omega_H}{\omega_H{}^2 + \omega^2} \tag{7. 4. 3}$$

- Finally, after transforming the stationary signal back to the time-domain, it is modulated by a suitable intensity function, such as those shown in Figs. 7.4-1 and 7.4-6.

A much more reliable method for generating site-dependent accelerograms uses measured records from the neighbourhood of the site in question and manipulates them in order to incorporate all additionally available information, especially concerning the geologic situation. As an example for this approach, the determination of accelerograms for the seismic investigation of Cologne cathedral is presented in the following, which has been carried out by Dr. K. HINZEN, head of the Bensberg Seismological Station of Cologne University.

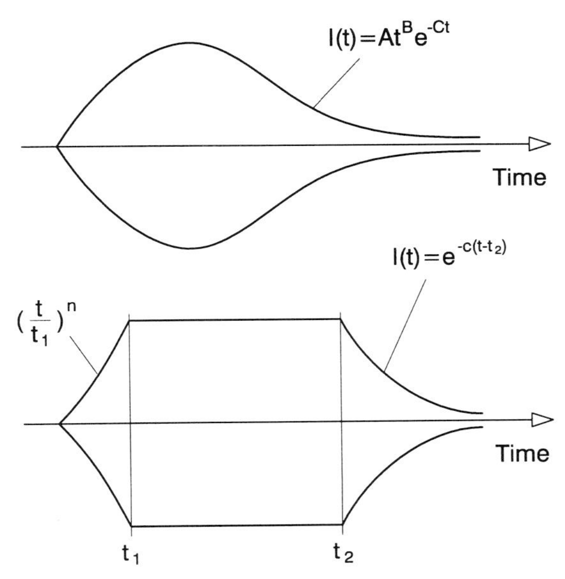

Fig. 7.4-6: Some more time-domain intensity functions

The three components of the Roermond earthquake of 1992 with a moment magnitude of 5.2 as measured in Bergheim (approx. 30 km from the Cologne cathedral) were the accelerograms initially available; Fig. 7.4-7 shows the NS component of this event. Bedrock lies at a depth of 750m below the Bensberg station and 240m below the Cologne cathedral; the measured Bergheim records were de-convoluted by a one-dimensional model (by means of the computer program SHAKE91, [7.12]) down to the bedrock, the result corrected for the 30 km distance to the cathedral and the corresponding free-field accelerogram for the surface was again calculated with the SHAKE91 program for the soil profile at the cathedral site. The single steps of this procedure are summarised as follows:

- Recording all three components of the Roermond earthquake at Bergheim.
- Computing the bedrock acceleration at Bergheim with SHAKE91.
- Transforming this result for the distance to Cologne.
- Computing the accelerations at the surface in Cologne, also with SHAKE91.

The seismic safety was, however, evaluated for a stronger earthquake than the Roermond event. A historical earthquake with an assumed moment magnitude of 6.0 which took place in the year 1878 at Tollhausen, at about 30 km west of the cathedral, was chosen as most unfavourable excitation. Of course, no accelerograms of this event are available, so empirical spectra were developed, spectrum compatible motions were generated by the program SYNTH and these were fitted to the existing soil profiles by means of the program SHAKE91. The single steps are summarised below:

- Computation of an empirical response spectrum for the bedrock,
- Generation of spectrum compatible motions by SYNTH using a suitable deterministic intensity function for near-field events,
- Taking into account of the soil profile at the cathedral site by means of the SHAKE91 program.

Fig. 7.4-8 shows the resulting record, designated with TOL60 for comparison purposes . Fig. 7.4-9 depicts its response spectrum, together with the response spectra of the SCT Mexico City record and spectrum of the NS component of the Kalamata/Greece event of 1986, measured at the City hall.

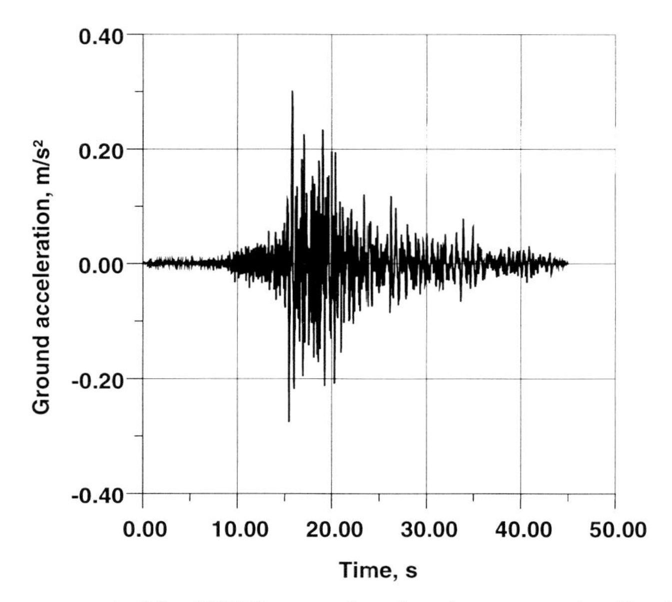

Fig. 7.4-7: NS component of the 1992 Roermond earthquake, measured at Bergheim

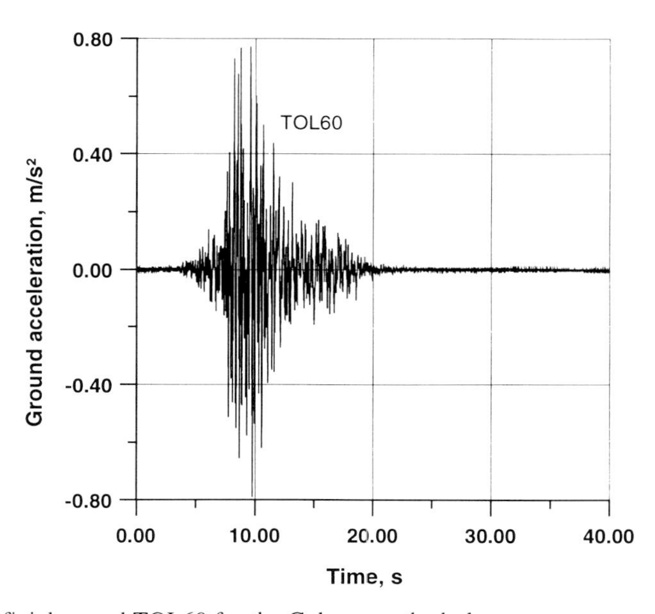

Fig. 7.4-8: Artificial record TOL60 for the Cologne cathedral

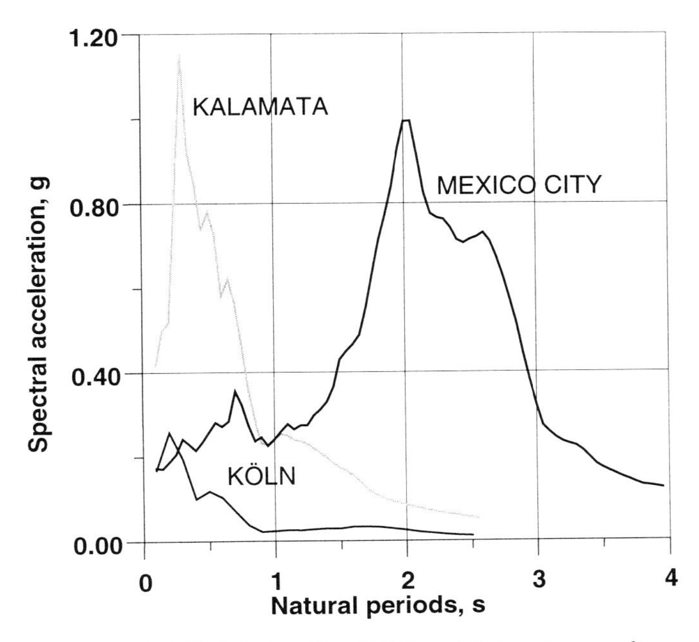

Fig. 7.4-9: Spectra of the TOL60, Mexico City (SCT) and Kalamata records

The site-dependent frequency distribution of the earthquake's energy is of prime importance for the safety of civil engineering structures, as has been drastically demonstrated by the catastrophic Mexico City earthquake of 1985, where the amplification of frequency components about 0.5 Hz due to local geological conditions led to severe damage and also collapse of buildings, the periods of which lay in this range. The aspects of local earthquake site response are being studied world-wide; an important part deals with the experimental determination of the local ground natural frequencies and amplification factors. A simple method for determining local natural frequencies has been proposed by NAKAMURA [7.13]. It is based on measuring both the horizontal and the vertical ambient vibration components at a site; under the assumption that the vertical component is not markedly modified on its way from the bedrock to the surface and that both components originally were the same in the bedrock, the ratio of the spectrum of the horizontal component to the one of the vertical component serves as an indicator of the influence of the local soil profile. The accuracy of the simple one-dimensional NAKAMURA approach has often been questioned, but it seems to be able to furnish dependable values at least for the natural frequencies (COUTEL and MORA, [7.14]).

7.5 Determination of seismic structural response

7.5.1 Introductory remarks

The currently valid seismic design philosophy, which is implicitly adopted in most of the existing building codes of practice, may be summarised as follows:

- The structure must be able to withstand frequent, low-level seismic loading without any damage.

- Medium intensity earthquakes, as might be expected to occur once during the lifetime of the structure, are permitted to cause a certain amount of damage. The actual performance of the structure must not, however, be impaired, which calls for a verification of the serviceability limit state.

- For very strong earthquakes, the intensity of which approaches the maximum intensity registered or thought possible at the site, widespread structural damage is permitted, as long as total collapse is avoided. Here, the collapse limit state must be verified in order to guard against injuries or loss of life.

The validity of the design philosophy just sketched is, however, being increasingly questioned in view of large casualty numbers and high monetary losses experienced during recent earthquakes such as the Northridge event in 1994 (61 casualties, damages 30 billion US-\$) or the Kobe earthquake in 1995 (over 5000 casualties, damages approx. 150 billion US-\$). The central question is what to do to provide a higher degree of in-built seismic safety in an attempt to avoid such horrendous casualty and loss figures. It must be remembered that the seismic risk increases proportionally to the monetary value of the civil engineering structures forming our urban environment, as long as their seismic vulnerability remains constant. Generally, the seismic damage potential is proportional to the monetary value of a structure times its vulnerability and the pertinent seismic hazard at the site in question. The seismic strengthening of existing structures and also the earthquake resistant design of new structures require an (implicit or explicit) assessment of their inelastic behaviour prior to failure, as described in section 8.1.

As a first step, we consider the N story building shown in Fig. 7.5-1 subject to the support point motion u_g as dynamic excitation. The N horizontal story displacements relative to the support are the master degrees of freedom; they are contained in the vector $\underline{V}^T = (V_1, V_2, ..., V_N)$ and they are associated with the N story masses of the building.

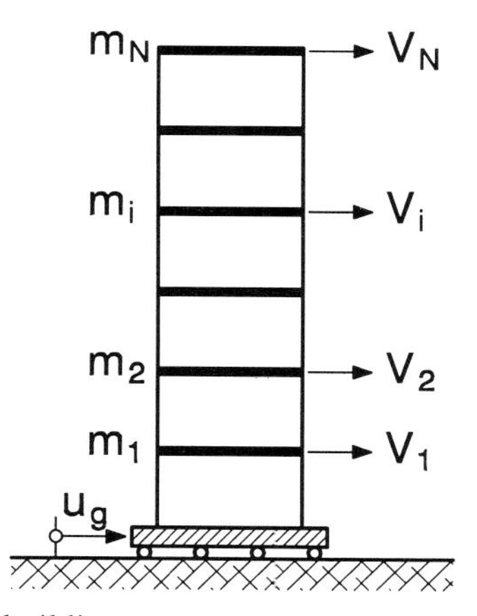

Fig. 7.5-1: Seismically excited building

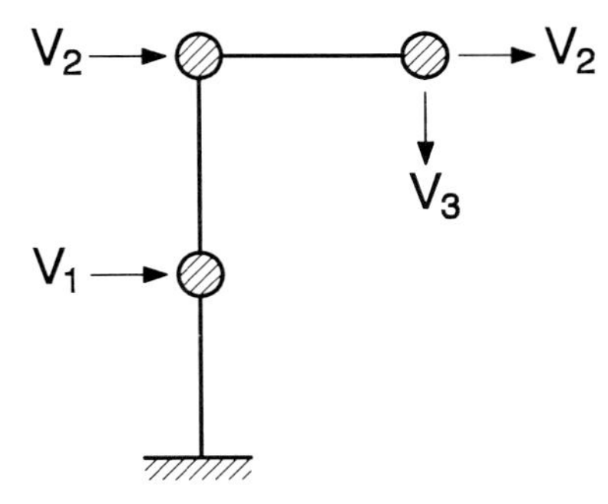

Fig. 7.5-2: Another discrete system with degrees of freedom

The system of differential equations of motion of such multi-degree-of-freedom systems is generally given by

$$\underline{M}\left(\ddot{\underline{V}}+\underline{r}\,\ddot{u}_g\right)+\underline{C}\,\dot{\underline{V}}+\underline{K}\,\underline{V}=0 \qquad (7.5.1)$$

or

$$\underline{M}\,\ddot{\underline{V}}+\underline{C}\,\dot{\underline{V}}+\underline{K}\,\underline{V}=-\underline{M}\,\underline{r}\,\ddot{u}_g \qquad (7.5.2)$$

In these expressions, $\underline{K}$ is the condensed (N,N) stiffness matrix in the N master degrees of freedom, $\underline{M}$ the diagonal mass matrix and $\underline{C}$ the viscous damping matrix (which is explicitly needed only if Eq. (7.5.2) is to be solved by Direct Integration methods). The column vector $\underline{r}$ contains the displacements in the N master degrees of freedom due to a unit displacement of the support in the direction of the seismic action. In the standard case of the building shown in Fig. 7.5-1 subject to horizontal seismic excitation, it is equal to

$$\underline{r}^T=(1,1,1,\ldots,1) \qquad (7.5.3)$$

In the case of the system shown in Fig. 7.5-2, we obtain

$$\underline{r}^T=(1,1,0) \qquad (7.5.4)$$

for a horizontal seismic component and

$$\underline{r}^T=(0,0,1) \qquad (7.5.5)$$

for a vertical seismic component.

Most of this section deals with the determination of the displacements and internal forces of seismically excited two-dimensional frame structures. There are various possible approaches to this task, as outlined below:

1. Application of equivalent static loads without dynamic analysis.
2. Modal analysis using response spectra for describing the seismic input.
3. Frequency domain analysis based on the power spectral density function of the excitation.
4. Time history analysis in the time domain with accelerograms as input.

The first two approaches are the most important for practical purposes, mainly because of the simple loading models they require. While direct integration methods need suitable accelerograms and frequency domain methods corresponding power spectral density functions, which must usually be determined separately, the response spectra which serve as input to modal analytic investigations can usually be extracted directly from the relevant Codes of Practice. The overall effort for the modal analysis-response spectrum procedure can be reduced further by considering only a single modal contribution (normally the one corresponding to the first natural period).

Three-dimensional models are discussed in section 7.6, which deals with lateral load-carrying systems. It is important to note that from the approaches mentioned above, only direct integration procedures in the time domain are in a position to quantitatively describe the non-linear structural behaviour to be expected under heavy seismic loading.

7.5.2 Modal analysis-response spectrum approach

Starting point is the differential equation (7.5.1) of a discrete multi-degree-of-freedom system

$$\underline{M}\,\ddot{\underline{V}}_{abs} + \underline{C}\,\dot{\underline{V}} + \underline{K}\,\underline{V} = 0 \qquad (7.5.6)$$

which can be also expressed in the form of Eq. 7.5.2:

$$\underline{M}\,\ddot{\underline{V}} + \underline{C}\,\dot{\underline{V}} + \underline{K}\,\underline{V} = -\underline{M}\,\underline{r}\,\ddot{u}_g = \underline{P} \qquad (7.5.7)$$

The modal decomposition of Eq. (7.5.7) with assumed proportional damping ($\underline{\Phi}^T\,\underline{C}\,\underline{\Phi} = \mathrm{diag}[2\,D_i\,\omega_i]$) yields N uncoupled equations in the modal co-ordinates η_i, $i = 1,2,\ldots N$:

$$\ddot{\eta}_i + 2D_i\,\omega_i\,\dot{\eta}_i + \omega_i^2\,\eta_i = \underline{\Phi}_i^T\,\underline{P} \qquad (7.5.8)$$

Here, $\underline{\Phi}_i$ is the ith eigenvector with the circular natural frequency ω_i in rad/s; the modal damping ratio D_i must be chosen to fit the situation. For each modal contribution i, a participation factor β_i is introduced according to

$$\beta_i = (-)\frac{\underline{\Phi}_i^T\,\underline{M}\,\underline{r}}{\underline{\Phi}_i^T\,\underline{M}\,\underline{\Phi}_i} = \frac{L_i}{M_i} = \frac{L_i}{1} = \underline{\Phi}_i^T\,\underline{M}\,\underline{r} \qquad (7.5.9)$$

Due to the scaling procedure for the eigenvectors utilised in the program JACOBI for the eigenproblem solution, the modal mass $M_i = \underline{\Phi}_i^T\,\underline{M}\,\underline{\Phi}_i$ is unity. The negative sign in the second term of Eq. (7.5.9) is usually suppressed, and Eq. (7.5.8) becomes

$$\ddot{\eta}_i + 2D_i\omega_i\dot{\eta}_i + \omega_i^2\eta_i = \beta_i\ddot{u}_g(t) \qquad (7.5.10)$$

The solution of this ordinary differential equation is given by the DUHAMEL integral as

$$\eta_i(t) = \beta_i \cdot \bar{S}_{d,i} \qquad (7.5.11)$$

with

$$\bar{S}_{d,i}(t) = \frac{1}{\omega_{Di}}\int_0^t \ddot{u}_g e^{-D_i\omega_i(t-\tau)}\sin\omega_{Di}(t-\tau)d\tau \qquad (7.5.12)$$

as already introduced in section 3.3. The absolute maximum value of $\bar{S}_{d,i}$ is set equal to the ordinate S_d of the relative displacement spectrum according to Eq. (7.2.11):

$$\max\left|\bar{S}_d(t)\right| = S_d = (1/\omega_1)\cdot S_v = (1/\omega_1^2)\cdot S_a \qquad (7.5.13)$$

The absolute maximum value max η_i of the modal co-ordinate η_i is given in Eq. (7.5.14) as a function of the ordinate $S_{d,i}$ of the displacement spectrum, $S_{v,i}$ of the pseudo relative velocity spectrum or $S_{a,i}$ of the pseudo absolute acceleration spectrum:

$$\max \eta_i = \beta_i \cdot S_{d,i} = \beta_i \cdot \frac{S_{v,i}}{\omega_i} = \beta_i \cdot \frac{S_{a,i}}{\omega_i^2} \tag{7. 5. 14}$$

The spectral ordinates themselves are functions of the circular natural frequency ω_i and damping D_i of the ith mode. The modal displacement maximum values are finally given by

$$\max \underline{V}_i = \max \eta_i \cdot \underline{\Phi}_i = \beta_i \cdot S_{d,i} \cdot \underline{\Phi}_i = \beta_i \cdot \frac{S_{v,i}}{\omega_i} \cdot \underline{\Phi}_i = \beta_i \cdot \frac{S_{a,i}}{\omega_i^2} \cdot \underline{\Phi}_i \tag{7. 5. 15}$$

The internal forces in the ith mode can be computed from the corresponding modal displacements by means of the well-known methods of matrix structural analysis. Alternatively, we can compute modal equivalent static loads to be applied to the structure as external loads. The maximum elastic restoring forces of the ith mode are given by

$$\max \left(\underline{K} \cdot \underline{V}\right)_i = \underline{K} \beta_i S_{d,i} \underline{\Phi}_i \tag{7. 5. 16}$$

They are equal to the modal inertia forces, as can be seen from the following derivation:

$$\begin{aligned} \max \left(\underline{K} \cdot \underline{V}\right)_i &= \underline{K} \beta_i S_{d,i} \underline{\Phi}_i = \omega_i^2 \underline{M} \beta_i S_{d,i} \underline{\Phi}_i = \\ &= \underline{M} \beta_i S_{d,i} \omega_i^2 \underline{\Phi}_i = \underline{M} \beta_i S_{a,i} \underline{\Phi}_i = \max \left(\underline{M} \cdot \ddot{\underline{V}}_{abs}\right)_i \end{aligned} \tag{7. 5. 17}$$

This leads to the following expression for the equivalent static load H_E acting in the ith mode in the direction of the degree of freedom k

$$H_{E,k,i} = \beta_i \; S_{a,i} \, m_k \, \Phi_{i,k} \tag{7. 5. 18}$$

Here, m_k is the mass associated with the degree of freedom k and $\Phi_{i,k}$ is the corresponding ordinate in the ith mode shape.

A question of great practical importance concerns the number of modal contributions that are needed in order to achieve a certain accuracy of the results. In this context, the so-called „effective modal mass" $M_{i,eff}$ associated with the ith mode shape is introduced as follows:

$$M_{i,eff} = \beta_i^2 M_i = \beta_i^2 (\underline{\Phi}_i^T \underline{M} \underline{\Phi}_i) \tag{7. 5. 19}$$

It is shown in the following that the sum of all N effective modal masses is equal to the effective total mass $M_{Tot,eff}$, which is also given by

$$M_{Tot,eff} = \underline{r}^T \underline{M} \underline{r} \tag{7. 5. 20}$$

To that effect, we introduce

$$\underline{r} = \underline{\Phi} \underline{\beta} \tag{7. 5. 21}$$

where $\underline{\Phi}$ is the modal matrix with the N mode shapes as columns and $\underline{\beta}$ a column vector containing the N participation factors. Multiplying both sides of Eq. (7.5.21) with $\underline{\Phi}_i^T \underline{M}$ leads to

$$\underline{\Phi}_i^T \underline{M} \underline{r} = \underline{\Phi}_i^T \underline{M} \underline{\Phi} \underline{\beta} \tag{7. 5. 22}$$

or

$$L_i = M_i \, \beta_i ; \; \beta_i = \frac{L_i}{M_i} \qquad (7.5.23)$$

as already given by Eq. (7.5.9). Introducing Eq. (7.5.21) in the expression (7.5.20) for the effective total mass yields

$$M_{Tot,eff} = \underline{\beta}^T \underline{\Phi}^T \underline{M} \underline{\Phi} \underline{\beta} = \underline{\beta}^T \underline{M}_i \underline{\beta} = \sum_{i=1}^{N} \beta_i^2 M_i = \sum_{i=1}^{N} M_{i,eff} \qquad (7.5.24)$$

A useful rule states that in order to achieve a satisfactory accuracy one should consider enough modal contributions so that the sum of their effective modal masses (this sum being equal to the sum of the squares of the participation factors with unit modal masses, $M_i = 1$) is at least equal to 90% of the effective total mass. The latter is equal to the sum of the story masses for building models like the one shown in Fig. 7.5-1, where all components of the vector $\underline{r}$ are equal to 1. Careful attention is required if rotations are also present among the master degrees of freedom, with the associated mass moments of inertia as entries on the diagonal of the mass matrix: The corresponding inertia forces are activated only for rotational support excitations, in which case the vector $\underline{r}$ contains the displacements of the translational degrees of freedom and the rotations of the rotary degrees of freedom due to a unit support rotation.

The product of the effective modal mass and the spectral acceleration S_a equals the total seismic base shear force F_i in the ith mode:

$$F_i = M_{i,eff} \cdot S_{a,i} \qquad (7.5.25)$$

Analogously, a "total base shear" can be expressed as the product of the effective total mass and the spectral acceleration for the first natural period of the structure. It is denoted by F_b in Eurocode 8 (EC 8) and it is given by

$$F_b = \frac{W}{g} \cdot S_a(T_1) \qquad (7.5.26)$$

where W is the total weight of the structure. This is exploited further in section 7.5.3 in the context of the simplified response spectrum method.

The computer program MDA2DE computes modal displacements and equivalent modal static loads for plane frame structures, the natural periods and mode shapes of which have been computed by the program JACOBI; the spectral ordinates corresponding to the single natural periods must be input interactively. MDA2DE requires the input files MDIAG, OMEG, PHI and RVEKT; the latter contains the displacements in all (master) degrees of freedom due to a unit support displacement in the direction of the seismic excitation.

The three-story frame shown in Fig. 4.4-1 subject to a horizontal base excitation is considered as an example. The horizontal displacements of the three stories are the master degrees of freedom, and the vector $\underline{r}$ contains accordingly three ones. The program MDA2DE yields the following results for the participation factors and effective modal masses:

Natural periods	Participation factors β_i	$M_{i,eff}$ in t
$T_1 = 0.755$ s	8.15	66.5 (= 98% of $M_{tot,eff}$)
$T_2 = 0.182$ s	1.18	1.39
$T_3 = 0.106$ s	0.31	0.10

These results show that the contribution of first mode is the most important, which is generally true for cantilever-type building structures like the one shown in Fig. 7.5-1. For spectral ordinates $S_a = 1$ m/s^2 in all three modes, the MDA2DE program yields the following modal displacements and corresponding equivalent static forces:

1st mode	Story 1	Story 2	Story 3
Displacement, m	0.0118	0.0157	0.0170
Equiv. static load, kN	24.477	32.594	9.432

2nd mode	Story 1	Story 2	Story 3
Displacement, m	$0.1374 \cdot 10^{-3}$	$-0.4170 \cdot 10^{-4}$	$-0.2125 \cdot 10^{-3}$
Equiv. static load, kN	4.928	-1.496	-2.032

3rd mode	Story 1	Story 2	Story 3
Displacement, m	$0.5595 \cdot 10^{-5}$	$-0.1033 \cdot 10^{-4}$	$0.2117 \cdot 10^{-4}$
Equiv. static load, kN	0.595	-1.098	0.600

The computation of the complete displacement and internal force distribution for each mode can be carried out by the program RAHMEN as explained in section 4.2. The equivalent static loads given above correspond to the degrees of freedom No. 4, 8 and 12 of the original discretisation shown in Fig. 4.4-2. For the 1st mode, the load vector is for example given by:

```
0.0, 0.0,    0.0,     24.477, 0.0
0.0, 0.0,    32.594,  0.0,    0.0
0.0, 9.432,  0.0,     0.0,    0.0
```

Figs. 7.5-3 and 7.5-4 show the computed bending moment diagrams for the first two modes. As we can see, the dominant influence of the first mode is also present for the bending moments, although here it is not as strong as for the displacements.

The modal analysis-response spectrum method that has just been outlined is very popular due to its simplicity and it is often found in codes of practice for the earthquake resistant design of buildings, such as DIN 4149. In contrast to approximate methods on a strictly "statical" basis, it also considers the dynamic properties of the structure. Its main drawback lies in the difficulty of combining the computed modal maximum values of internal forces and displacements, which naturally do not occur at the same time. It is customary to combine all p computed modal contributions for internal forces or displacements E_i to obtain a result E_E by means of the so-called SRSS combination rule, which simply evaluates the square root of the sum of the squares of all modal contributions:

$$E_E = \sqrt{E_1^{\,2} + E_2^{\,2} + \ldots + E_p^{\,2}} \tag{7. 5. 27}$$

If, however, modes exist, the natural periods of which coincide or lie at a distance of less than 10% to each other, that is

$$\frac{T_i}{T_a} \geq 0.90;\ T_i < T_j \tag{7. 5. 28}$$

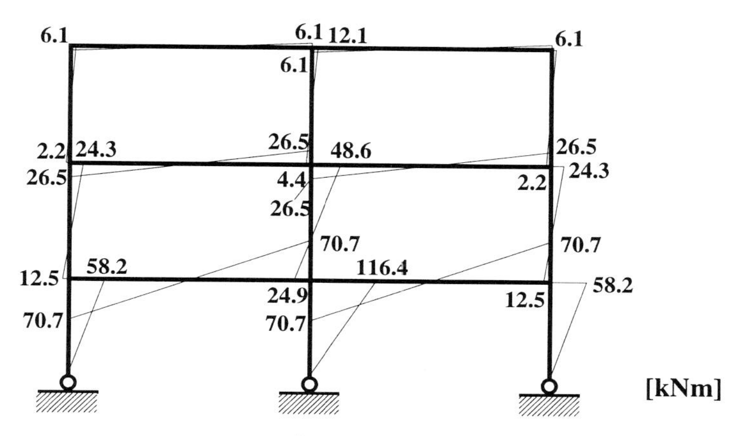

Fig. 7.5-3: Bending moment diagram, 1st mode

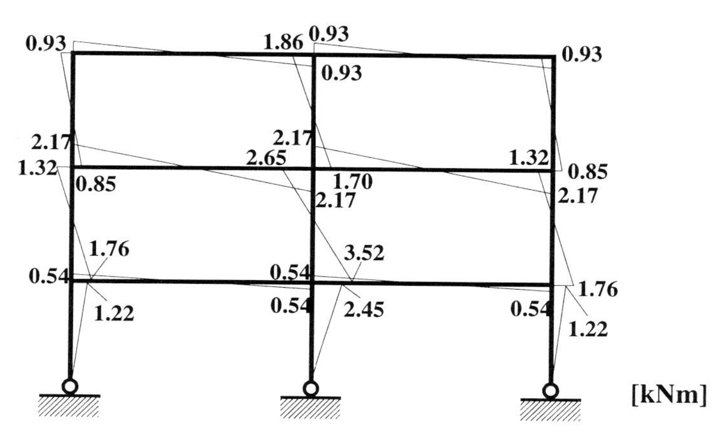

Fig. 7.5-4: Bending moment diagram, 2nd mode

the SRSS rule is not accurate enough and the so-called Complete Quadratic Combination or CQC rule must be employed. Such closely spaced modes might appear, for example, in the case of three-dimensional models with combined torsional and translational mode shapes. For a vector $\underline{S}^T$ consisting of p modal contributions, $\underline{S}^T = (E_1, E_2, ..., E_p)$, the CQC rule [7.15] yields:

$$\underline{S}_E = \sqrt{\sum_{i=1}^{p} \sum_{j=1}^{p} E_i E_j \, \varepsilon_{ij} \, \alpha_{ij}} \tag{7. 5. 29}$$

with

$$\varepsilon_{ij} = \frac{8\sqrt{D_i D_j}(D_i + rD_j)\, r^{1.5}}{\left(1-r^2\right)^2 + 4D_i D_j r\left(1+r^2\right) + 4\left(D_i^{\,2} + D_j^{\,2}\right) r^2}\,; \quad r = \frac{\omega_j}{\omega_i} \le 1 \tag{7. 5. 30}$$

and

$$\alpha_{ij} = \frac{\beta_i \, \beta_j}{\left| \beta_i \, \beta_j \right|} \tag{7.5.31}$$

being equal to +1 or −1 according to the signs of the participation factors. If all modal damping ratios are equal, Eq. (7.5.30) simplifies to

$$\varepsilon_{ij} = \frac{8D^2(1+r)r^{1.5}}{\left(1-r^2\right)^2 + 4D^2 r(1+r)^2} \tag{7.5.32}$$

The correlation coefficients ε_{ij} are equal to 1 for r = 1; if Eq. (7.5.28) holds, all ε_{ij} with $i \neq j$ can be neglected and the usual SRSS combination rule results.

The problem of the correct combination of the different modal contributions especially in the case of complex three-dimensional structures is even more difficult, because for every probable maximum of a single internal force component (e.g. a bending moment), the corresponding (and not the maximum) values of axial and shear forces are needed. In the simplest case, when a single internal force component controls the design (e.g. the bending moment when no axial forces are present), the determination of the probable maximum of this internal force by the SRSS or CQC rule is sufficient. If two or more internal force components are simultaneously needed (e.g. bending moments, axial force and shear forces for a beam-column), it is customary to assume a concurrent appearance of the single maximum values. More detailed expressions for the combination of multi-component internal forces have been given by GUPTA / SINGH [7.16], ANASTASSIADIS [7.17] and ROSENBLUETH / CONTRERAS [7.18], among others. It is an important advantage of the direct integration approach for given accelerograms as discussed in section 7.5.4 that it automatically yields the maximum values of single internal force components complete with all other simultaneously appearing variables and thus dispenses with the need for combination rules. The problem of combining the effects from the horizontal and the vertical seismic components will be discussed in section 7.6 in the context of the discretisation of three-dimensional structures.

7.5.3 Equivalent static loads, simplified response spectrum approach

A weak point of the standard modal analysis-response spectrum approach is the uncertainty associated with the combination of modal contributions considered. The resulting internal force diagrams, which have been determined e.g. by the SRSS rule, do not satisfy the equilibrium conditions, since the signs have been lost, and the determination of critical combinations of the single internal force components (e.g. maximum bending moment and corresponding shear and axial forces for a beam column) is also by no means straightforward. A simpler approach, which yields quite acceptable results at least for regular structures, starts with the determination of the base shear as product of the structure's weight times a coefficient depending primarily on the site (earthquake intensity, soil class) and the geometrical and mechanical properties of the structure expressed for example by its natural period. If necessary, the first natural period is computed by some approximate expression without having to set up a complete mathematical model of the structure and solve the eigenvalue problem. The determined base shear is then distributed to the single stories of the building according to some simple law and the subsequent static computation of displacements and internal forces yields the information required for designing or checking. An added advantage is that here the earthquake excitation is formally considered as yet another loading case, with the computed internal forces satisfying the equilibrium equations.

As an example, Eq. (7.5.33) shows the distribution law given in EC 8 for the base shear F_b:

$$H_i = F_b \frac{W_i h_i}{\sum_{i=1}^{n} W_i h_i} \qquad (7.5.33)$$

H_i is the equivalent static force in the ith story, the weights of the single stories are denoted by W_i and h_i are the heights of the stories in respect to the base of the building. Additional details are given in the context of the relevant codes of practice (DIN 4149 and EC 8) in sections 7.7 and 7.8.

7.5.4 Direct integration solution

Direct integration may be employed to directly solve the coupled differential equation system Eq. (7.5.1) as well as for solving one or more of the uncoupled modal equations (7.5.10) in the time domain. The latter approach is recommended for buildings and other cantilever-type structures subject to seismic loading, because considering just a few modes usually yields sufficiently accurate results. The comparison of the sum of the effective modal masses of the modes considered to the effective total mass serves as an indicator of whether enough modal contributions have been considered (90% being a useful estimate). Advantages of the Direct Integration approach are the determination of maximum internal forces which satisfy the equilibrium conditions together with the simultaneously appearing further internal force components and also the possibility to employ various accelerograms with different characteristics in order to gain some insight on the response variability. According to EC 8, at least 5 different records should be used, conforming to certain requirements, for example with respect to their duration. A drawback of the direct integration of uncoupled modal equations is that it cannot consider non-linear effects in the response of the system in contrast to the direct integration of the coupled differential equation system.

The computer program MODBEN carries out such a direct integration solution of the uncoupled modal differential equations of plane frame structures with subsequent addition of the results. It input files are the output files OMEG and PHI of the program JACOBI, the file MDIAG with the NDU masses associated with the master degrees of freedom, AMAT as output file of the program KONDEN, the file ACC containing the accelerogram, RVEKT containing the vector $\underline{r}$ with the displacements in the NDU degrees of freedom due to a unit support displacement and, finally, in the files V0 and VP0 the initial (t=0) displacements and velocities in the NDU master degrees of freedom (which are usually zero). It produces three output files: THIS.MOD contains the time histories of the displacements in the master degrees of freedom (with the time points in the first column and the displacements in the remaining NDU columns), THISDG contains the displacements in all (NDU + NDPHI) degrees of freedom (without the time points), and, finally, THISDU contains the displacements in the master degrees of freedom without the associated time points. This file, THISDU, serves as input for the program INTFOR which computes all internal forces and displacements of the frame structure, their maximum values, and, optionally, their time histories. Apart from THISDU, INTFOR requires the input file EKOND of the program KONDEN and the output file AMAT of KONDEN as input. The results of INTFOR are the maximum and minimum values of the internal forces, complete with the time points of their appearance and the values of the remaining internal forces at that time (file MAXMIN), the complete distribution of internal forces and displacements at a certain time point to be entered interactively (file FORSTA), and, finally, the time history of a certain internal force or displacement component (optionally, file THHVM).

The solution of the coupled differential equation system is carried out by the program NEWBEN, which is based on the NEWMARK integration scheme. Input files are again the accelerogram (file ACC), the $\underline{r}$ vector (file RVEKT), the stiffness matrix in the master degrees of freedom (file KMATR), the diagonal of the mass matrix (file MDIAG), the initial conditions (files V0 and VP0) and, finally, the damping matrix (file CMATR) which can be produced, for example by the programs CRAY or CMOD. The output of NEWBEN consists of the time histories of (alternatively) displacements, velocities or accelerations in the NDU master degrees of freedom (file THIS.NEW) and the output file THISDU (as with MODBEN), which serves as input for the INTFOR computer program.

The input and output files of the programs introduced in this section are summarised below:

Program name	Input files	Output files
MDA2DE	MDIAG OMEG PHI RVEKT	STERS2
MODBEN	MDIAG OMEG PHI V0 VP0 ACC RVEKT AMAT	THIS.MOD THISDG THISDU
NEWBEN	MDIAG KMATR CMATR V0 VP0 ACC RVEKT	KONTRL THIS.NEW THISDU

As an example we consider the three-story frame that has already been investigated by modal analysis in section 7.5.2. It is subjected to the "Albatros" accelerogram scaled by a factor of 0.238, so that the spectral acceleration for the first natural period $T_1 = 0.755$ s of the structure becomes equal to 1.0 m/s^2, as assumed in section 7.5.2. A damping of 5% was assumed for all three modes. Fig. 7.5-5 shows the time history of the horizontal displacement of the roof and of the first story, the latter being also the "first story drift". Story drifts, defined as relative displacements of the adjacent stories, are quite useful damage indicators because they are directly related to the damage of both structural and non-structural elements in this story. The maximum displacement is reached at t = 9.36 s, as computed by the program MODBEN and output to the monitor; the computation of the complete displacement and internal force distribution by the program INTFOR should therefore be carried out for this instant (the results to be written in the file FORSTA). The bending moment diagram at t=9.36s is shown in Fig. 7.5-6; the agreement between the results obtained by direct integration and by modal analysis is very good.

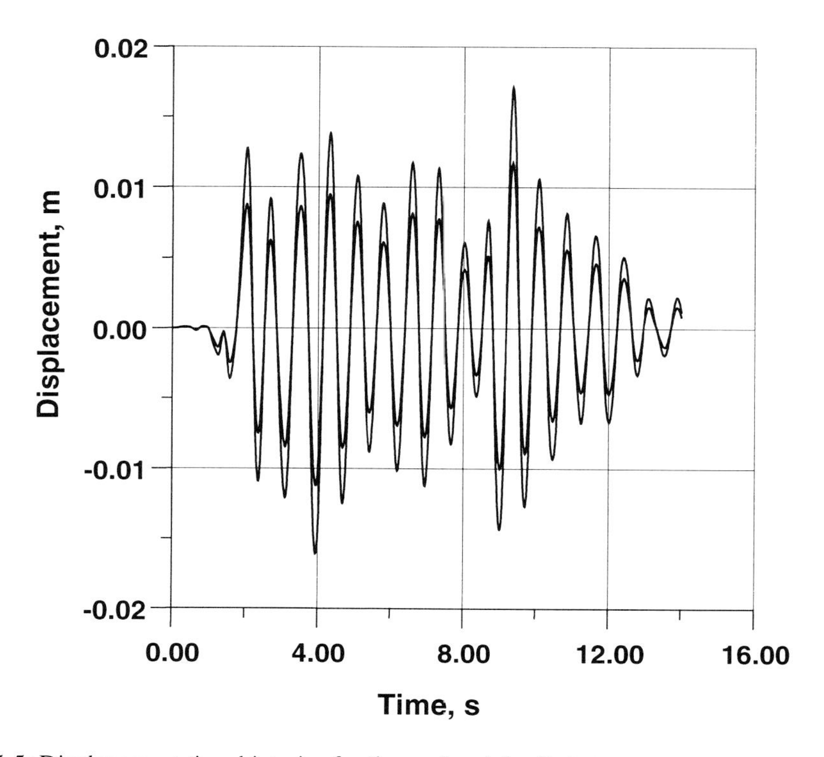

Fig. 7.5-5: Displacement time histories for the roof and the first story

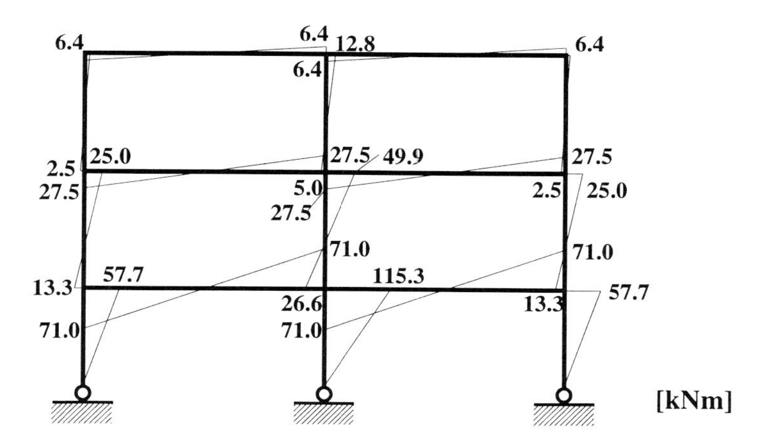

Fig. 7.5-6: Bending moment diagram at time t = 9.36 s

The program INTFOR further determines maximum values of an arbitrary internal force component at every node and outputs these values together with the times at which they occurred and the values of the remaining internal force components at this instant (output in the file MAXMIN). As an example, the results for the bending moment maximum values of the ground floor central column (element no. 7) are reproduced below. The bending moment maximum values are found in column 4, the associated horizontal and vertical forces in columns 2 and 3, and column 1 contains the time points at which the maximum bending moment values were reached.

```
Element No. 7
Max. pos., element end 1: 3.9400   .3173E+02   .1943E-14   .2003E-04
Max. neg., element end 1: 9.3600  -.3295E+02  -.2017E-14  -.2115E-04
Max. pos., element end 2: 9.3600   .3295E+02   .2017E-14   .1153E+03
Max. neg., element end 2: 3.9500  -.3175E+02  -.1944E-14  -.1111E+03
```

Fig. 7.5-7 depicts the time histories of the bending moment at the top of the central ground floor column and also of the shear force present in this element. Fig. 7.5-8 compares the time histories of the horizontal roof displacement computed by the programs MODBEN and NEWBEN; the damping matrix used with NEWBEN has been determined by the program CMOD assuming 5% modal damping for all three modes. The agreement is almost total.

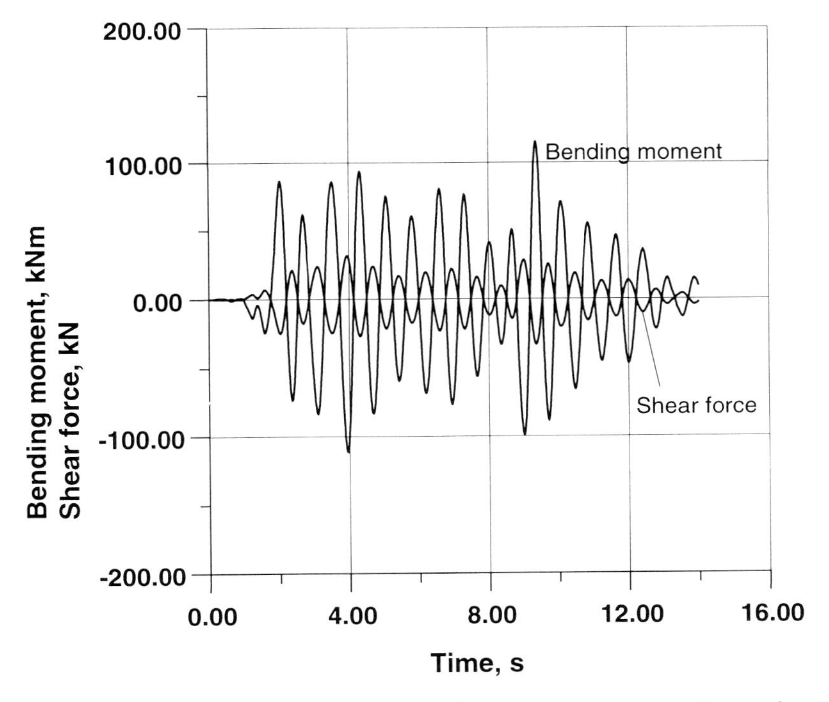

Fig. 7.5-7: Bending moment and shear force time histories, central ground floor column

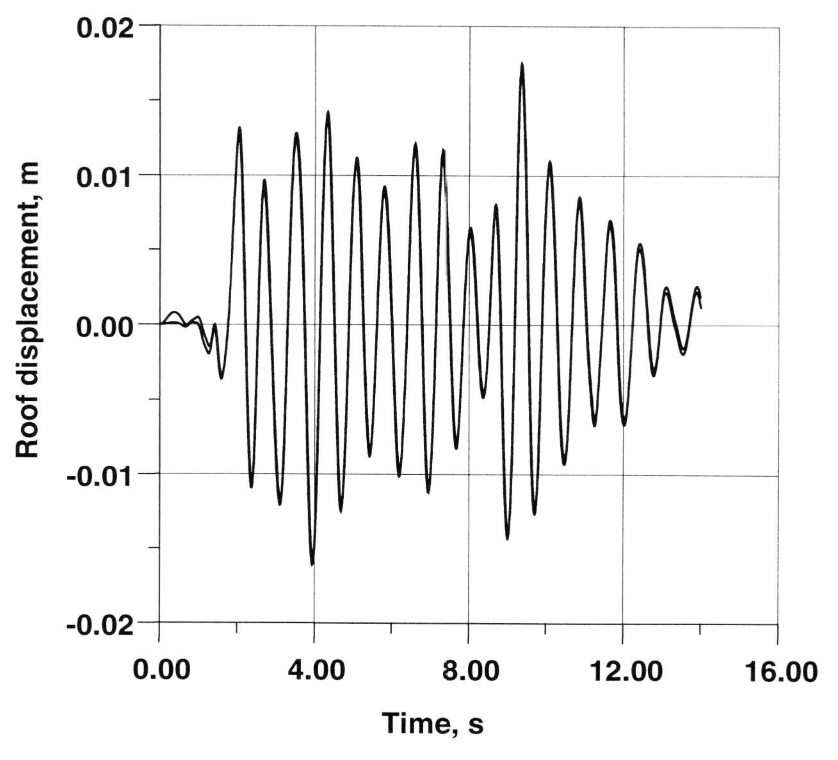

Fig. 7.5-8: Comparison between MODBEN and NEWBEN, roof displacement time history

7.6 Three-dimensional building models for seismic analyses

7.6.1 Introduction

This section deals with the practical modelling of standard lateral load-carrying systems for buildings and the properties, advantages and disadvantages of their usual idealisations.

The systems shown in Fig. 7.6-1 are encountered quite often in practice. They consist of

1. Moment-resisting frames or frame/truss systems.
2. Frames with infill masonry or diagonal struts.
3. Shear walls with or without openings.
4. Coupled frame-and-shear-wall systems.
5. Open cores with U-, L- or T-sections
6. Closed cores with openings, e.g. U-sections with strong coupling beams at the story levels.

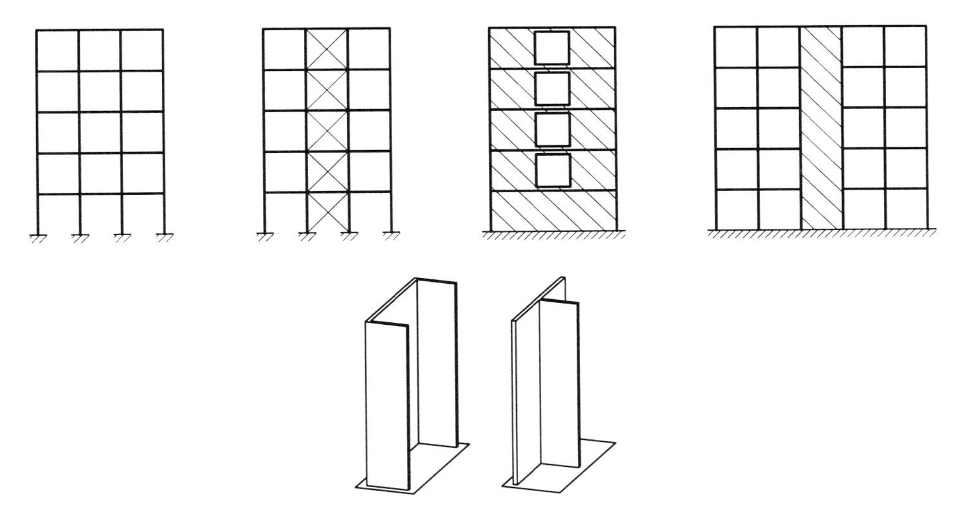

Fig. 7.6-1: Some lateral load-carrying systems

The lateral load-carrying mechanisms of these systems are quite heterogeneous. “Open” systems such as moment-resisting frames or isolated shear walls transmit the lateral loads to the foundation via bending and shear, while cores can also transmit loads by torsion. In open cores, warping action is an essential part of their load-carrying behaviour, while closed cores usually act in ST. VENANT (warp-free) torsion.

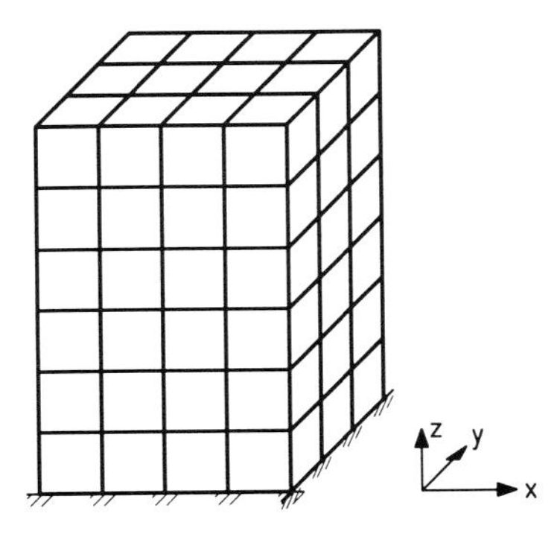

Fig. 7.6-2: Three-dimensional frame building

This section deals only with the “open” systems of the types 1 to 4 of the overview given above.

Fig. 7.6-2 shows a building stiffened by moment-resisting frames as an example for illustrating the relative complexity of possible numerical models. This six-story building (n=6) is laterally stiffened by four frames in its longitudinal and five frames in the transverse direction, thus having $4 \cdot 5 = 20$ nodes per story (m=20).

- A general three-dimensional frame model would need six degrees of freedom per node (three translational and three rotational), therefore requiring a total of $6 \cdot 6 \cdot 20 = 720$ kinematic degrees of freedom.
- A first simplifying assumption that can be used in order to reduce the overall number of degrees of freedom lies in considering all floors as rigid in their own plane. In this case, two horizontal translations and one rotation about the vertical axis suffice for describing the position of every floor with all its nodes, so that only three unknown further deformation components per node remain. The model requires $3 \cdot 6 \cdot 20$ degrees of freedom for the nodes plus three degrees of freedom for each of the six floors, that is 378 degrees of freedom in all. This is a reduction to 52.5% of the full three-dimensional model.
- Additionally, the degrees of freedom corresponding to the axial deformations of the columns may be neglected (effectively introducing an infinite column axial stiffness), thus saving further $6 \cdot 20$ degrees of freedom. There remain 258 degrees of freedom, or 35,8% of the original number.
- Assuming further that all girders are torsionally flexible, so that the single moment-resisting frames are active only in their own plane, we end up with just the three degrees of freedom per floor mentioned above (two horizontal translations and the rotation about the vertical axis). These $3 \cdot 6 = 18$ degrees of freedom are the absolute minimum needed for discretising the structure, being just 2.5% of the original 720 degrees of freedom.

The variant last mentioned is the "pseudo three-dimensional" model for buildings that is often used in practice because of its simplicity. In it, the contribution of each shear wall or frame reaching from the foundation to the roof of the n story building is characterised by its (n,n) lateral stiffness matrix, this being simply the condensed stiffness matrix with the n horizontal story displacements in the plane of the shear wall or moment-resisting frame as master degrees of freedom. The derivation of these lateral stiffness matrices for different systems will be discussed in section 7.6.2; the remainder of this section deals with the derivation of the general stiffness matrix of the three-dimensional building as function of the lateral stiffness matrices of the single shear walls and frames.

Fig. 7.6-3 depicts schematically the floor plan of a building which is stiffened for lateral loading through "open" elements such as shear walls, frames or mixed systems. It is further assumed for the sake of simplicity that all these elements are continuous from foundation to roof level, without any change of their geometrical or mechanical properties. As already mentioned, the floor diaphragms are assumed to be rigid in their own plane, which means that the position of all points of a floor is known once the horizontal displacements u_x, u_y of the mass centre C_M of the floor and the floor rotation ϑ about the vertical axis are known. The translations in x and y direction and the rotation about the z axis are then the master degrees of freedom. It is further assumed that all shear walls or frames carry loads only in their own plane. A further simplification lies in not considering the vertical degrees of freedom, which means that the compatibility of vertical deformations at columns that belong to more than one frame is not enforced. This may lead to serious errors for the internal forces in such common columns; in order to take this vertical compatibility into account, the corresponding vertical degrees of freedom can be retained until after the spatial stiffness matrix of the structure has been set up and then eliminated by the usual condensation procedure. This approach is not, however, adopted in the following.

The mass centre of a floor is denoted by C_M, and its stiffness centre by C_S. The latter can be determined as the centre of gravity of the lateral stiffness values of the single wall or frame elements, which are assumed to extend from the foundation to the top story with constant dimensions and stiffness. It is quite easy to compute the position of C_S by hand or by means of a short program. By definition, lateral loads through C_S in the direction of the main axes of inertia cause purely translational displacements without rotary components.

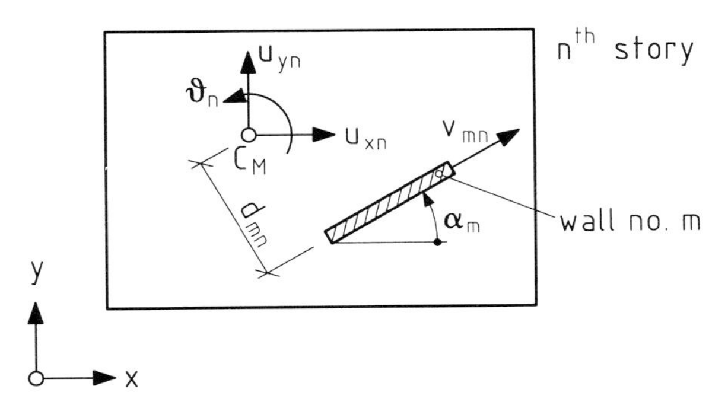

Fig. 7.6-3: A floor plan with an arbitrary situated shear wall

We consider the shear wall m in Fig. 7.6-3, which shows the plan of floor n. Its distance to the mass centre C_M is d_{mn} and its angle relative to the global x-axis is α_m (positive counter-clockwise). The horizontal displacement v_{mn} in its own plane can be expressed by the displacements u_x and u_y of C_M and the rotation ϑ_n of the (rigid) floor n as

$$v_{mn} = (\cos\alpha_m \quad \sin\alpha_m \quad d_{mn}) \begin{pmatrix} u_{xn} \\ u_{yn} \\ \vartheta_n \end{pmatrix} \qquad (7.6.1)$$

The horizontal displacements of the shear wall m at the n floor levels are given by the column vector $\underline{V}_m$ in the form

$$\underline{V}_m = \underline{A}_m \underline{U} \qquad (7.6.2)$$

with

$$\underline{V}_m^T = (v_{m1}, v_{m2}, \ldots, v_{mn}),$$
$$\underline{U}^T = (u_{x1}, u_{x2}, \ldots, u_{xn}, u_{y1}, u_{y2}, \ldots, u_{yn}, \vartheta_1, \vartheta_2, \ldots, \vartheta_n) \qquad (7.6.3)$$
$$\underline{A}_m = \left[\cos\alpha_m \underline{I} \quad \sin\alpha_m \underline{I} \quad \mathrm{diag}\left[d_{mn}\right]\right]$$

Here, I is the (n,n) unit matrix and $\mathrm{diag}\left[d_{mn}\right]$ the (n,n) diagonal matrix containing the distance values from the mass centres C_M in the single stories to the shear wall m.

The (3n,3n) stiffness matrix $\underline{\tilde{K}}_m$ of the mth shear wall in the global co-ordinate system is given by

$$\underline{\tilde{K}}_m = \underline{A}_m^T \, \underline{K}_m \, \underline{A}_m \qquad (7.6.4)$$

with $\underline{K}_m$ as the lateral (n,n) stiffness matrix of the mth wall in the n horizontal displacements at the floor levels as master degrees of freedom. The (3n,3n) total stiffness matrix $\underline{\tilde{K}}_{tot}$ for the lateral load-carrying system of the building is obtained by summing up the stiffness matrices $\underline{\tilde{K}}_m$ of all walls present:

$$\underline{\tilde{K}}_{tot} = \sum \underline{\tilde{K}}_m \tag{7. 6. 5}$$

The associated load vector $\underline{P}$ has 3n components which express, in this sequence:

- The force components in x direction through C_M of all n floors.
- The force components in y direction through C_M of all n floors.
- The torsion moments in all n floors (force times distance from C_M).

For a given load vector $\underline{P}$, the system $\underline{P} = \underline{\tilde{K}}_{ges}\ \underline{U}$ can be solved for the displacements and rotations contained in $\underline{U}$. The (n,n) displacement vector $\underline{V}$ of the an arbitrary wall is obtained by multiplying the $3 \cdot n$ vector $\underline{U}$ by the transformation matrix $\underline{A}$ according to Eq. (7.6.2). Next, the part of the floor shear corresponding to each wall can be determined by multiplying the lateral stiffness matrix by the vector $\underline{V}$ containing the n horizontal displacements in the wall's own plane. These computations can be carried out sequentially as follows:

- At the beginning, the (n,n) lateral stiffness matrices $\underline{K}_m$ of all shear walls and frames of the n-story building must be evaluated by means of the programs described in section 7.6.2.

- The computer program TRA3D transforms each of these (n,n) lateral stiffness matrices to a (3n,3n) stiffness matrix $\underline{\tilde{K}}_m$ relative to the global (x, y, ϑ) co-ordinate system. The input data for each wall are its lateral stiffness matrix and, to be entered interactively, the distance of the wall from the mass centre and the angle α (positive counter-clockwise) between the x axis and the wall longitudinal axis.

- Once all (3n,3n) global stiffness matrices have been computed, they are summed by the program MATSUM to form the stiffness matrix KMATR of the entire lateral load-carrying system.

- In preparing for modal analysis, the eigenvalue problem is solved by calling the program JACOBI. The input file MDIAG containing the diagonal of the mass matrix is set up by entering the n story masses in x direction, followed by the (same) n masses in y direction, and, finally, the n mass moments of inertia of the floors about the vertical axis through the centre of mass.

- The computer program MDA3DE (being analogous to the program MDA2DE for the two-dimensional case) is then evoked to compute the 3n equivalent modal static loads for all selected modes. The direction of the (horizontal) excitation is given by its angle relative to the global x axis. Only the modes which contribute significantly to the system's response are considered; the user decides whether to consider a mode or not during the program run using its participation factors as guide. For each mode considered, the 3n load components (forces in x direction, forces in y direction and torsional moments about the vertical axis through C_M) are output in a file the name of which must also be specified by the user.

- The computer program DISP3D computes the 3n displacement values of the building for a given set of equivalent modal static loads. These displacements consist of the n story displacements of C_M in the x direction, the n story displacements of C_M in the y direction and finally the n story rotations about a vertical axis. They are written to separate files for each mode, the names of which must be specified by the user.

- Finally, the computer program FORWND computes the n story shears in the plane of each shear wall. It requires the lateral stiffness matrix of this wall and its position relative to C_M (given by the angle α and the distance d).

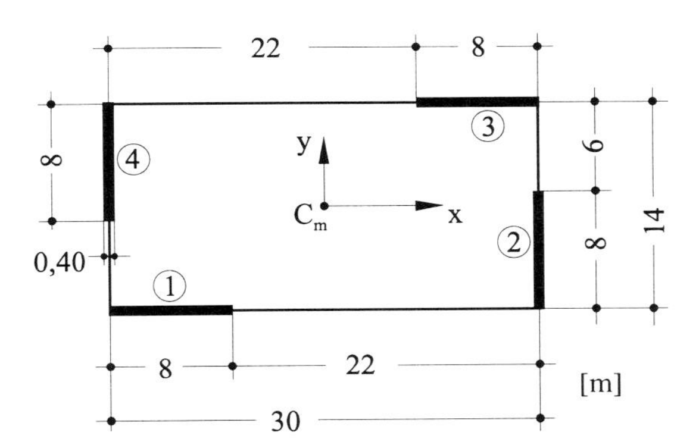

Fig. 7.6-4: Floor plan with shear walls

The procedure will be illustrated by an example. Fig. 7.6-4 shows the floor plan of a 10 story building with a constant story height of 3.50 m, the lateral load-carrying system of which consists solely of the walls shown. All shear walls are 8 m long and 0.40 m thick, the concrete's YOUNG's modulus equals $2.9 \cdot 10^7$ kN/m^2 and its POISSON ratio 0.20. In order to take the decreased stiffness due to concrete cracking under service loads into account, the value of the YOUNG's modulus is decreased by 40% (PENELIS/KAPPOS [7.19]). In this case, the shear walls are quite slender (8 m long and 35 m high), so that they may be considered as simple cantilever flexural beams, the (10,10) lateral stiffness matrix of which can be determined by the program KONDEN presented in section 4.3. Next, the program TRA3D is invoked four times in order to supply the four (30,30) matrices M1, M2, M3 and M4 for the global stiffness matrices of the four shear walls. In all four cases, the lateral stiffness matrix used as input is the same; the additional data for describing the position of each shear wall are given below:

Wall no., name of output file	Angle α	Distance d
M1	0°	7.0
M2	90°	15.0
M3	180°	7.0
M4	270°	15.0

The MATSUM program then sums the four (30,30) matrices to obtain the overall system stiffness matrix KMATR. The diagonal of the mass matrix (file MDIAG) is given by

```
546., 546., 546., 546., 546., 546., 546., 546., 546., 462.
546., 546., 546., 546., 546., 546., 546., 546., 546., 462.
49868., 49868., 49868., 49868., 49868.,
49868., 49868., 49868., 49868., 42196.,
```

With KMATR and MDIAG as input files, the program JACOBI computes all 30 eigenvalues and mode shapes and outputs this information in the files OMEG and PHI. The first five natural periods are:

```
T1 = 1.201873s
T2 = 1.201873s
T3 = 0.693902s
T4 = 0.192202s
T5 = 0.192202s
```

The existence of multiple eigenvalues is obvious.

Next, the program MDA3DE is invoked which yields the equivalent static loads in the single modes. For a seismic excitation along the y axis ($\alpha = 90^\circ$), the modes no. 2, 5, 8, 11 and 13 were chosen, because of their relatively high participation factors which together account for over 90% of the effective total seismic mass. The modal loading vectors were output to files named LV1, LV2, ... , LV5 during the program run; the general output file STERS3 which contains additional explanatory text as shown below presents all five contributions.

```
Mode shape No.   2

Part of the effective mass considered so far:   64.37%
Spectral acceleration as input in g  =  .102
 1         .000        13.402        .000
 2         .000        51.125        .000
 3         .000       109.462        .000
 4         .000       184.764        .000
 5         .000       273.505        .000
 6         .000       372.359        .000
 7         .000       478.297        .000
 8         .000       588.692        .000
 9         .000       701.434        .000
10         .000       689.652        .000
Modal loads (x, y, torsion):
       .000    3462.692        .000
--------------------------------------------------

Mode shape no.   5

Part of the effective mass considered so far:   84.17%
Spectral acceleration as input in g  =  .102
 1         .000        41.146        .000
 2         .000       134.966        .000
 3         .000       238.986        .000
 4         .000       316.493        .000
 5         .000       340.652        .000
 6         .000       297.587        .000
 7         .000       187.299        .000
 8         .000        21.850        .000
 9         .000      -179.193        .000
10         .000      -334.733        .000
Modal loads (x, y, torsion):
       .000    1065.052        .000
--------------------------------------------------

Mode shape no.   8
```

```
Part of the effective mass considered so far:  90.98%
 Spectral acceleration as input in g  =  .102
  1         .000        59.734         .000
  2         .000       161.726         .000
  3         .000       210.541         .000
  4         .000       161.144         .000
  5         .000        32.525         .000
  6         .000      -107.548         .000
  7         .000      -181.211         .000
  8         .000      -142.412         .000
  9         .000         1.332         .000
 10         .000       170.343         .000
 Modal loads (x, y, torsion):
        .000      366.175         .000
 -------------------------------------------------

 Mode shape no.  11

 Part of the effective mass considered so far:  94.45%
 Spectral acceleration as input in g  =  .102
  1         .000        72.727         .000
  2         .000       148.591         .000
  3         .000        99.108         .000
  4         .000       -42.670         .000
  5         .000      -137.783         .000
  6         .000       -91.405         .000
  7         .000        46.828         .000
  8         .000       131.020         .000
  9         .000        63.121         .000
 10         .000      -102.596         .000
 Modal loads (x, y, torsion):
        .000      186.943         .000
 -------------------------------------------------

 Mode shape no.  13

 Part of the effective mass considered so far:  96.54%
 Spectral acceleration as input in g  =  .102
  1         .000        79.780         .000
  2         .000       106.446         .000
  3         .000       -17.411         .000
  4         .000      -108.716         .000
  5         .000       -25.713         .000
  6         .000        98.711         .000
  7         .000        66.825         .000
  8         .000       -68.608         .000
  9         .000       -83.944         .000
 10         .000        65.107         .000
 Modal loads (x, y, torsion):
        .000      112.479         .000


  Effective total seismic mass:     5376.000
  Part of this mass considered:     5190.122
  Sum of the loads (x, y, torsion):
        .000     5193.341         .000
```

The next program, DISP3D, computes the displacements of the entire system due to the modal load vectors LV1, LV2, ..., LV5. The results are output in the files DIS1, DIS2, ..., DIS5 which are subsequently used by the program FORWND to compute the story shears of the single walls and do not include any explanatory text. This is the case for the output file DISPL of the program DISP3D, which presents both the input loads and the computed story displacements. As an example, the results written in DISPL for the loading vector LV1 are as follows:

```
STORY         X FORCE          Y FORCE          TORSIONAL MOM.
    1      .0000000E+00     .1340200E+02      .0000000E+00
    2      .0000000E+00     .5112486E+02      .0000000E+00
    3      .0000000E+00     .1094615E+03      .0000000E+00
    4      .0000000E+00     .1847639E+03      .0000000E+00
    5      .0000000E+00     .2735048E+03      .0000000E+00
    6      .0000000E+00     .3723588E+03      .0000000E+00
    7      .0000000E+00     .4782970E+03      .0000000E+00
    8      .0000000E+00     .5886925E+03      .0000000E+00
    9      .0000000E+00     .7014339E+03      .0000000E+00
   10      .0000000E+00     .6896523E+03      .0000000E+00
 STORY     X DISPL.         Y DISPL.          ROTATION, RAD
    1      .7327424E-20     .8981204E-03     -.1264754E-18
    2      .2795206E-19     .3426076E-02     -.4825930E-18
    3      .5984710E-19     .7335443E-02     -.1005034E-17
    4      .1010180E-18     .1238175E-01     -.1627625E-17
    5      .1495363E-18     .1832862E-01     -.2309106E-17
    6      .2035838E-18     .2495322E-01     -.3029236E-17
    7      .2615046E-18     .3205254E-01     -.3771721E-17
    8      .3218623E-18     .3945057E-01     -.4521461E-17
    9      .3835026E-18     .4700581E-01     -.5270179E-17
   10      .4456177E-18     .5461925E-01     -.6018387E-17
```

The program FORWND computes the modal story shears of an arbitrary wall characterised by its lateral stiffness matrix and its position relative to C_M, the latter being given by the angle α and the distance d. For wall no. 2 in Fig. 7.6-4, the results for the first mode are:

```
  6.701001074968837     25.562431240822150     54.730759328374520
 92.381940628272690    136.752416117879100    186.179420838239800
239.148498405041900    294.346227730211200    350.716928243364600
344.826142243447900
```

The influence of accidental eccentricities in plan must be routinely taken into account, even for symmetric lateral load-carrying systems. The corresponding eccentricity value in this case is given by $\pm 0.05 \cdot L$ according to EC 8, L being the dimension of the building perpendicular to the seismic excitation. The resulting torsion due to the story shear times eccentricity must be considered as an additional loading case. The combination of the single modal contributions can be carried out by means of the SRSS or the CQC rule.

The sequence of computer programs just described is repeated below with the input and output files required. The number of stories of the building is denoted by n in the following table.

In determining the internal forces of the single walls, both horizontal components of the seismic excitation (or their geometric sum) must be considered as acting simultaneously (e.g. according to EC 8, section 3.3.5). Another combination rule for two orthogonal horizontal components which is also mentioned in the EC 8 code is given by

$$\begin{aligned} &E_{Edx} \text{ "+" } 0.30 \cdot E_{Edy} \\ &0.30 \cdot E_{Edx} \text{ "+" } E_{Edy} \end{aligned} \qquad (7.6.6)$$

Here, "+" stands for "to be combined with", E_{Edx} are internal forces due to seismic excitation in the x direction and E_{Edy} internal forces due to seismic excitation in the y direction. For substructures which are also sensitive to the vertical component of the seismic excitation, such as long horizontal cantilevers, Eq. (7.6.6) can be analogously expanded to include the vertical term E_{Edz}:

$$\begin{aligned} &E_{Edx} \text{ "+" } 0.30 \cdot E_{Edy} \text{ "+" } 0.30 \cdot E_{Edz} \\ &0.30 \cdot E_{Edx} \text{ "+" } E_{Edy} \text{ "+" } 0.30 \cdot E_{Edz} \\ &0.30 \cdot E_{Edx} \text{ "+" } 0.30 \cdot E_{Edy} \text{ "+" } E_{Edz} \end{aligned} \quad (7.6.7)$$

Program name	Input files	Output files
TRA3D	Freely chosen names for the (n,n) lateral stiffness matrices of the single walls, e.g. L1, L2, ...	Freely chosen names for the (3n,3n) global stiffness matrices of the walls, e.g. M1, M2, ...
MATSUM	The (3n,3n) global stiffness matrices computed by TRA3D, e.g. M1, M2, ..	KMAT3D as sum of M1, M2, ...
MDA3DE	MDIAG OMEG PHI	The output is summarised in the file STERS3. Freely chosen names for the (3n,1) modal load vectors, e.g. LV1, LV2, ..
DISP3D	KMAT3D The (3n,1) vectors generated by MDA3DE, e.g. LV1, LV2,...	Freely chosen names for the (3n,3n) modal displacement vectors, e.g. DIS1, DIS2, ... Input and output are summarised in the file DISPL.
FORWND	The lateral (n,n) stiffness matrix of the wall, e.g. L1 The (3n,3n) displacement vector from DISP3D, e.g. DIS1	Freely chosen name for the file in which the n story shear forces should be output.

Another topic is the consideration of the torsional action for both horizontal components of the seismic excitation. In addition to the accidental eccentricity which is also present for symmetric buildings and to the actual eccentricity as distance between the mass and the stiffness centres, an additional eccentricity which takes the influence of coupled torsional/translational modes into account is normally introduced. The relevant procedures according to DIN 4149 and EC 8 are described in section 7.7.

7.6.2 Lateral stiffness matrices for different wall types

The following types of “open” lateral stiffening wall types are considered in this section:

1. Moment-resisting frames or frame/truss systems.
2. Reinforced concrete shear walls.
3. Coupled frame-and-shear-wall systems.

ad 1): For arbitrary plane frames with or without diagonal struts, the computer program KONDEN may be used for determining the lateral stiffness matrices as condensed stiffness matrices in the story horizontal displacements as master degrees of freedom. If the column sections are large, it might be useful to take into account the rigid end regions of the girders as discussed under case 3.

ad 2): Like a cantilever beam, a slender reinforced concrete shear wall acts predominantly in flexure, in spite of the "shear" attribute in its name. Walls with a height to width ratio of 3 or greater can be modelled by beam elements, which can also consider shear deformations; squat walls with or without openings can be discretised by means of standard two-dimensional finite elements. Fig. 7.6-5 shows, as an example, a standard rectangular element with 8 degrees of freedom, these being the displacements u_1, v_1, ... u_4, v_4 of its nodes.

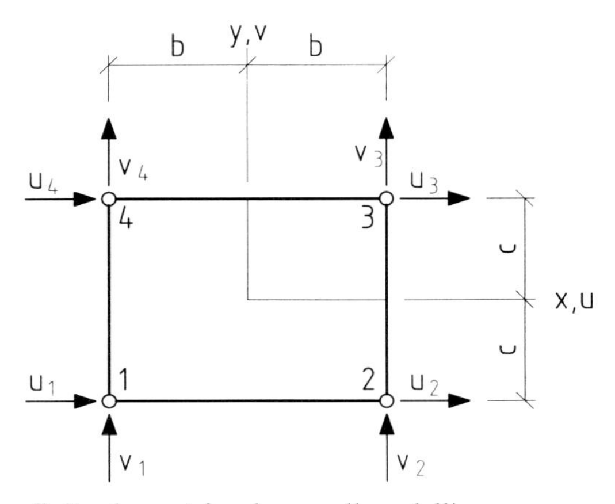

Fig. 7.6-5: Rectangular finite element for shear wall modelling

The (8,8) stiffness matrix for this element with bilinear interpolation functions is given by NILSSON/SAMUELSSON (see [7.20]) as follows:

$$\underline{K} = \frac{Qt}{12\left(1-m^2\right)} \begin{pmatrix} A_1 & C_1 & A_2 & -C_2 & A_4 & -C_1 & A_3 & C_2 \\ & B_1 & C_2 & B_3 & -C_1 & B_4 & -C_2 & B_2 \\ & & A_1 & -C_1 & A_3 & -C_2 & A_4 & C_1 \\ & & & B_1 & C_2 & B_2 & C_1 & B_4 \\ & & & & A_1 & C_1 & A_2 & -C_2 \\ \text{Symm.} & & & & & B_1 & C_2 & B_3 \\ & & & & & & A_1 & -C_1 \\ & & & & & & & B_1 \end{pmatrix} \qquad (7.6.8)$$

For plane stress, $Q = E$, $m = \nu$ whereas for plane strain $Q = \dfrac{E}{1-\nu^2}$, $m = \dfrac{\nu}{1-\nu}$.

The single coefficients are given by:

$$\begin{aligned} A_1 &= \left(4-m^2\right)c/b + 1.5\,(1-m)\,b/c \\ A_2 &= -\left(4-m^2\right)c/b - 1.5\,(1-m)\,b/c \\ A_3 &= \left(2+m^2\right)c/b - 1.5\,(1-m)\,b/c \\ A_4 &= -\left(2+m^2\right)c/b - 1.5\,(1-m)\,b/c \end{aligned} \qquad (7.6.9)$$

$$\begin{aligned} C_1 &= 1.5(1+m) \\ C_2 &= 1.5(1-3m) \end{aligned} \qquad (7.6.10)$$

The coefficients $B_1, \ldots, B_4$ are equal to the coefficients $A_1, \ldots, A_4$ with b and c interchanged. It is obvious that the coefficients of the stiffness matrix are functions of the side ratio c/b of the rectangular element with constant thickness t. The corresponding displacement vector is given by

$$\underline{V}^T = (u_1,\ v_1,\ u_2,\ v_2,\ u_3,\ v_3,\ u_4,\ v_4) \qquad (7.6.11)$$

While for (not overly large) plane frames it is quite easy to introduce and number the active kinematic degrees of freedom by hand, in the case of walls to be modelled by rectangular finite elements this is no longer a viable option because of the (normally) large number of degrees of freedom present. The computer program LATWND generates the (n,n) condensed lateral stiffness matrix for the special case of an n-story solid shear wall (no openings) with constant thickness and material properties. All n story heights are assumed to be equal; story-high rectangular finite elements are employed and the number of elements per story can be chosen by the user. It is recommended to choose their number in such a way that the side ratio is close to unity. Further data to be input interactively include the story number, the wall thickness, the wall's YOUNG's modulus and its POISSON ratio.

Program name	Input file	Output file
LATWND	None required	LATWD contains the (n,n) lateral stiffness matrix of the wall in the n horizontal story displacements.

As an example, the program LATWND is employed to compute the (10,10) stiffness matrix of the 8 m wide, 0.4 m thick and 35 m high reinforced concrete wall that has been idealised as a cantilever beam in the example of Fig. 7.6-4. Three elements per story (h = 3.5 m) were used, with corresponding widths of 2.667 m, $\nu = 0.20$ and E = $1.74\ 10^7$ kNm^2. For comparison purposes, Fig. 7.6-6 shows the deformed shapes for both idealisations due to a horizontal load of 100 kN acting at a height of 35 m; the agreement is close enough for practical purposes.

ad 3): As already mentioned, slender shear walls exhibit a beam-like behaviour under lateral loading, with convex deformed shapes when seen in the direction of the loading. On the other hand, moment-resisting frames with girders that are, typically, much stiffer than the columns, behave more like „shear beams", with concave deformed shapes (seen in the direction of the loading). Coupled shear walls and moment-resisting frames can be modelled by the computer programs already mentioned; a substructuring approach is usually best. As an example (Fig. 7.6-7), we consider the combination of a five story moment-resisting frame (girder bending stiffness equal to $4.5 \cdot 10^5$ kNm^2, column bending stiffness equal to $0.63 \cdot 10^5$ kNm^2) with a shear wall, the bending stiffness of which is equal to EI = $7.031 \cdot 10^6$ kNm^2. The loading consists of the horizontal forces shown in the figure; the two systems are connected by inextensional bars at floor levels 2 to 5, which, however, offer no flexural resistance. Fig. 7.6-8 depicts a possible discretisation of the entire coupled system; the computer program RAHMEN yields the deformed shape shown in Fig. 7.6-9.

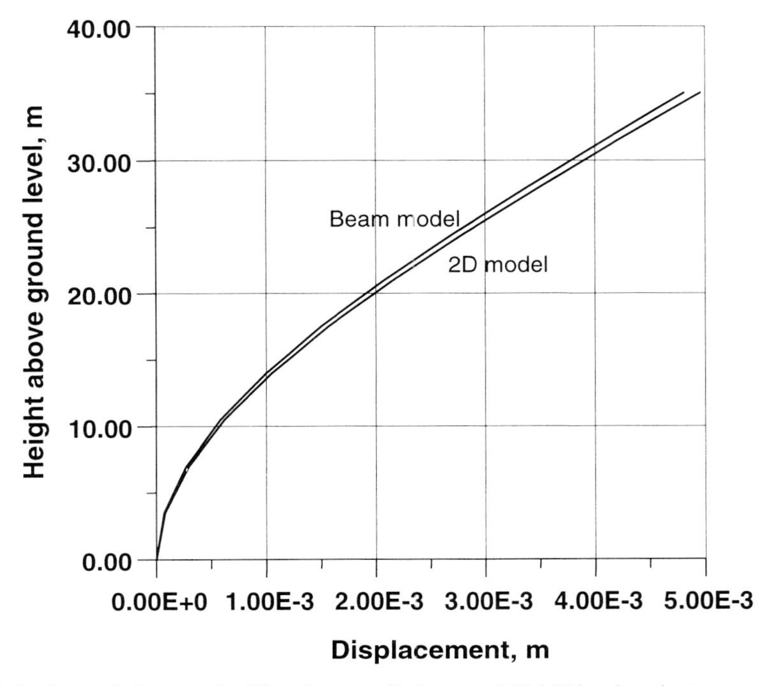

Fig. 7.6-6: Deformed shapes of a 35m shear wall due to a 100 kN load at the top

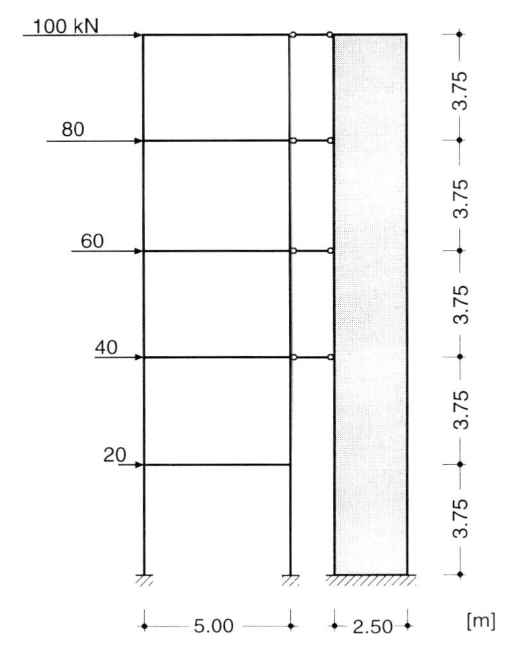

Fig. 7.6-7: Coupled wall-frame system

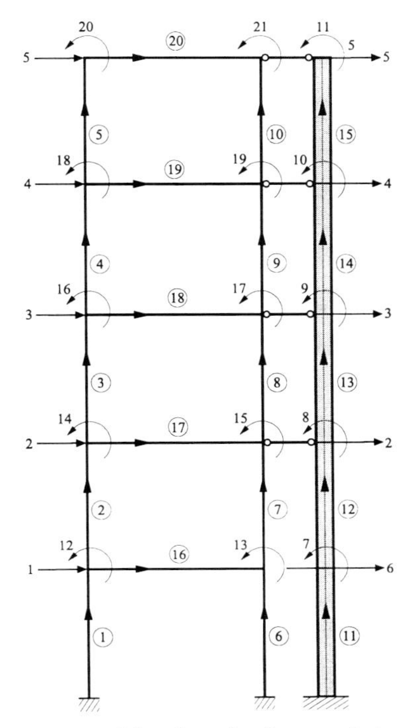

Fig. 7.6-8: Active kinematic degrees of freedom for the coupled system

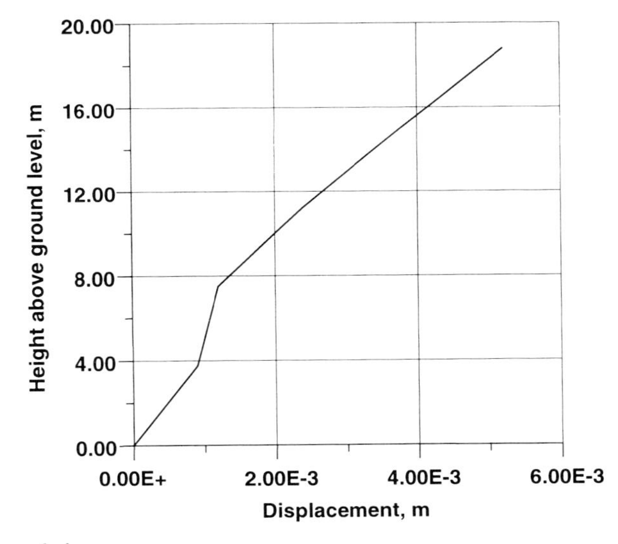

Fig. 7.6-9: Deformed shape

The same results are obtained when the substructuring approach is used. Here, the shear wall shown at the left-hand side of Fig. 7.6-10 can be expressed by the (4,4) condensed stiffness matrix in the degrees of freedom 3, 5, 7 and 9 which is generated by the program KONDEN. Its output file KMATR can be renamed to FEDMAT and used as the input spring stiffness matrix while analysing (by the program RAHMEN) the moment-resisting frame shown at the right in Fig. 7.6-10. The contents of the input file INZFED of the program RAHMEN are the numbers 2, 3, 4 and 5 of the degrees of freedom connected to each other and the ground by the spring matrix in FEDMAT.

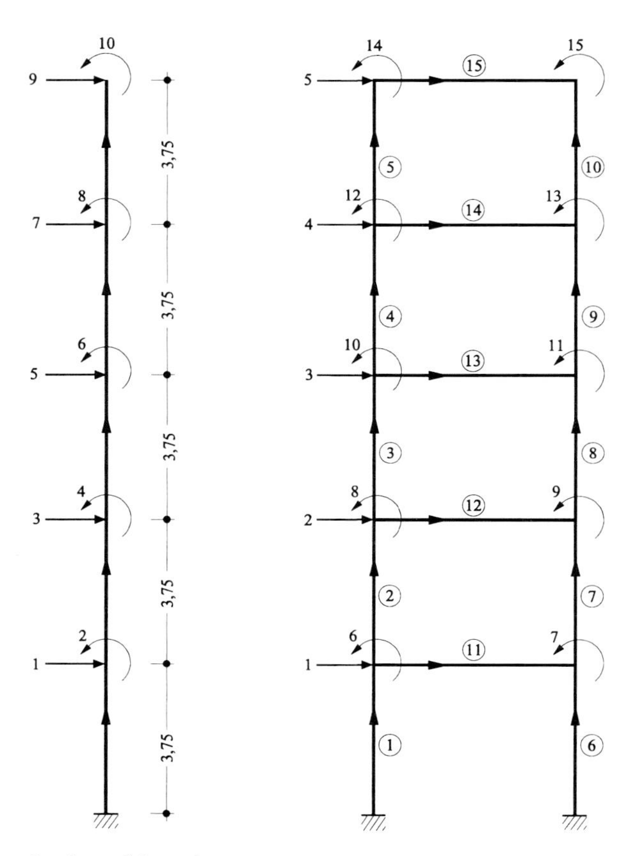

Fig. 7.6-10: Discretisation of the substructures

Beam elements with rigid end zones as shown in Fig. 7.6-11 are generally used in cases where frames are connected to shear walls by moment-resisting girders or two shear walls are interconnected by lintel beams. The end portions $\alpha\ell$, $\beta\ell$ have sections with infinite inertia so that they remain straight and the additional degrees of freedom can be eliminated by kinematic condensation. This is done in the computer program STAKON for the calculation of the condensed stiffness matrix of a moment-resisting frame containing such members. The input file ESTAK of the program STAKON contains the following data:

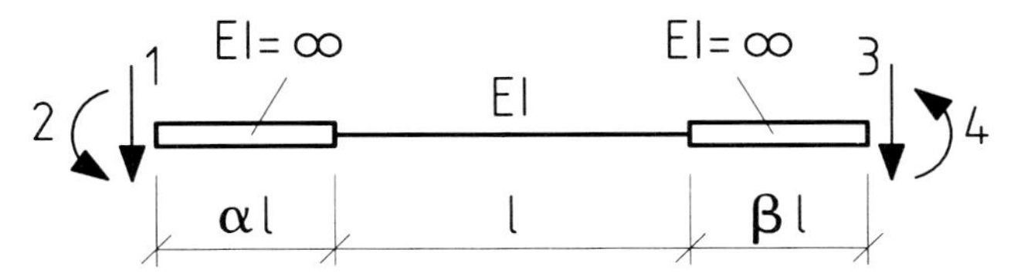

Fig. 7.6-11: Beam element with rigid end zones

The first NELRIE rows (NELRIE being the number of members with rigid end zones) contain the data EI (in kNm^2), ℓ (length of the flexible portion of the member in m), EA (axial stiffness in kN), α (angle between the global x-axis and the member axis in degrees, positive counter-clockwise), AL (length of the rigid zone at the left-hand member end in m) and BL (length of the rigid zone at the right-hand member end in m) of each member. The next NELRIE rows contain the incidence vectors of these members, with six integer numbers per row. In the following NELSTU rows (NELSTU being the number of members without rigid end sections) the data EI, ℓ, EA and α are given, followed by the corresponding incidence vectors in further NELSTU rows (again with six integer numbers per row). Finally, the numbers of the NDU master degrees of freedom (generally the horizontal floor displacements) are given.
The output file ASTAK contains the (NDU,NDU) condensed stiffness matrix, and the output file AMAT, the transformation matrix for determining the displacements in the slave degrees of freedom from known deformations in the master degrees of freedom. Spring supports as single springs or multi-dimensional coupled spring matrices may also be considered, as in the program RAHMEN, by supplying the spring matrices in the input file FEDMAT and the corresponding incidence vectors in the file INZFED.
The computer program STARAH may be used for the evaluation of deformations and internal forces due to static loads. Its input file ESTAR is like the input file ESTAK of the program STAKON, the difference being that instead of the numbers of the NDU master degrees of freedom at the end of the file, ESTAR contains the NDOF static load components corresponding to the NDOF system degrees of freedom. The output file ASTAR presents the deformations and internal forces of all members.

Program name	Input files	Output files
STAKON	ESTAK FEDMAT INZFED	ASTAK AMAT
STARAH	ESTAR FEDMAT INZFED	ASTAR

As an example, the ten story frame shown in Fig. 7.6-12 will be investigated. The bending stiffness values are given by:

$$EI_{Girder} = 5.285\ 10^5\ kNm^2$$
$$EI_{Wall} = 3.263\ 10^7\ kNm^2$$
$$EI_{Column} = 1.208\ 10^5\ kNm^2$$

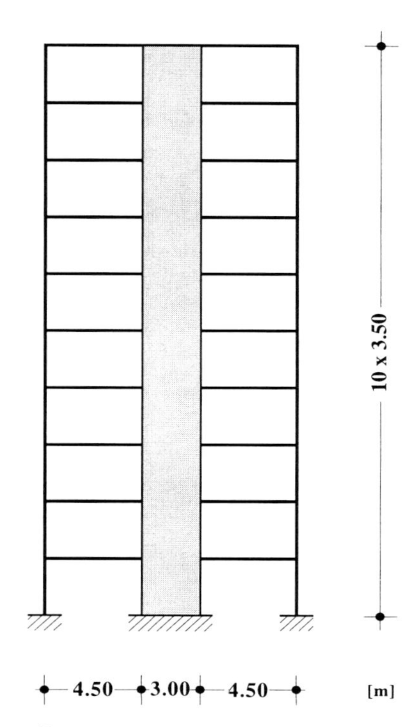

Fig. 7.6-12: 10-story frame-wall system

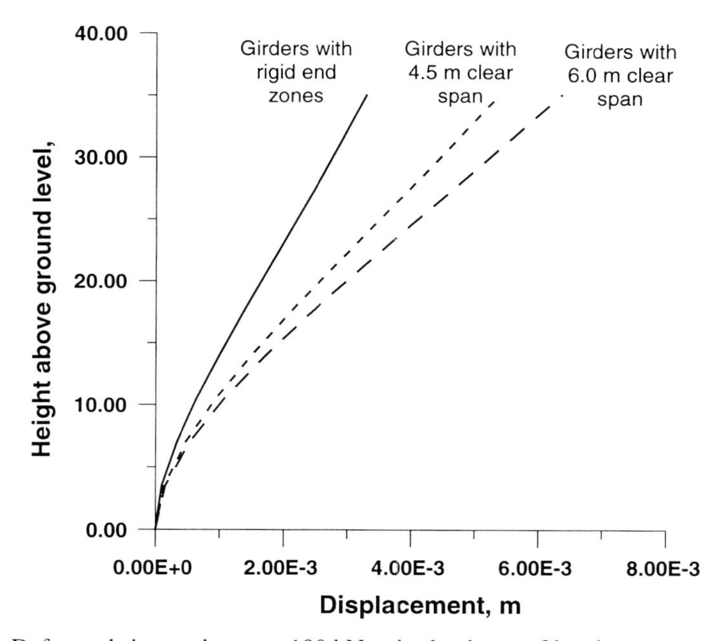

Fig. 7.6-13: Deformed shapes due to a 100 kN point load at roof level

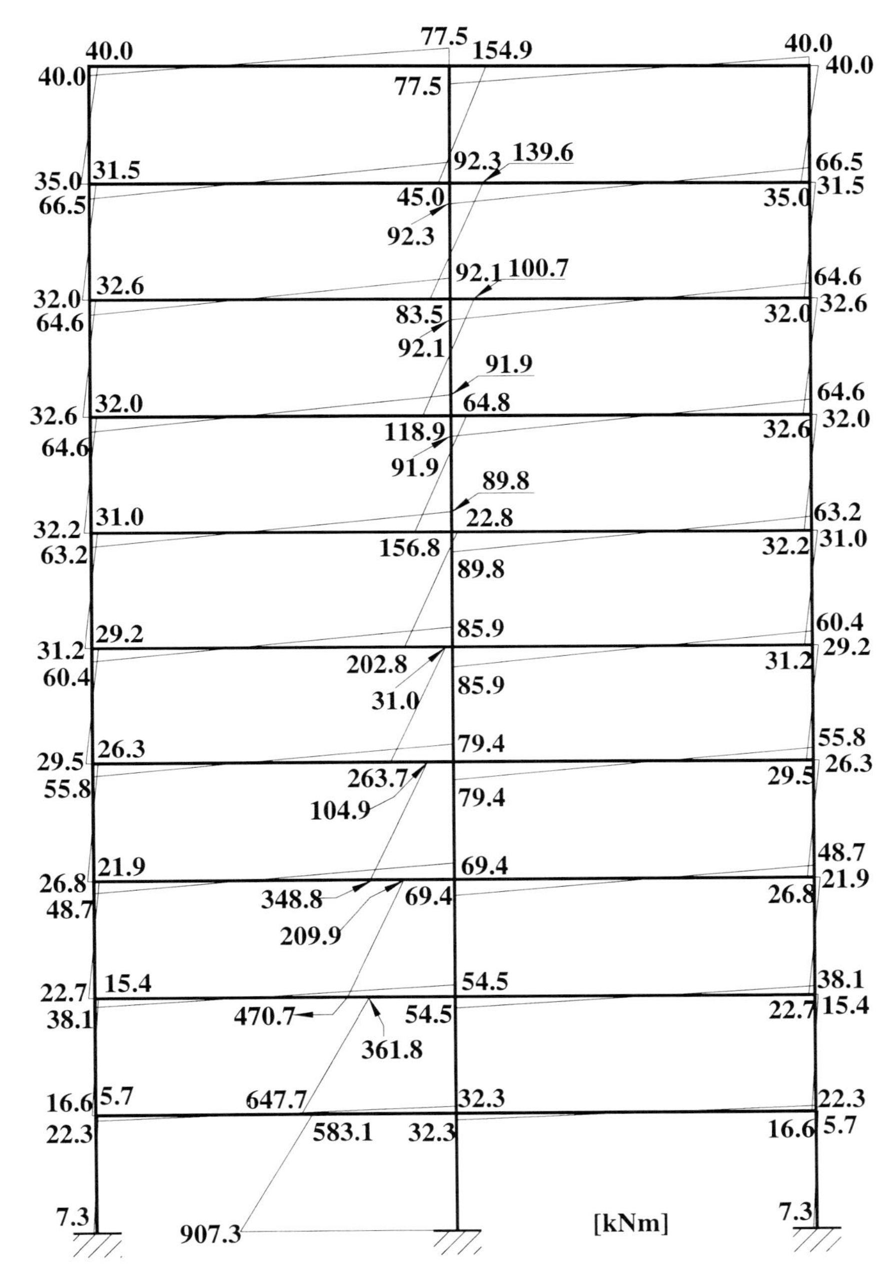

Fig. 7.6-14: Bending moment diagram without consideration of the rigid girder portions

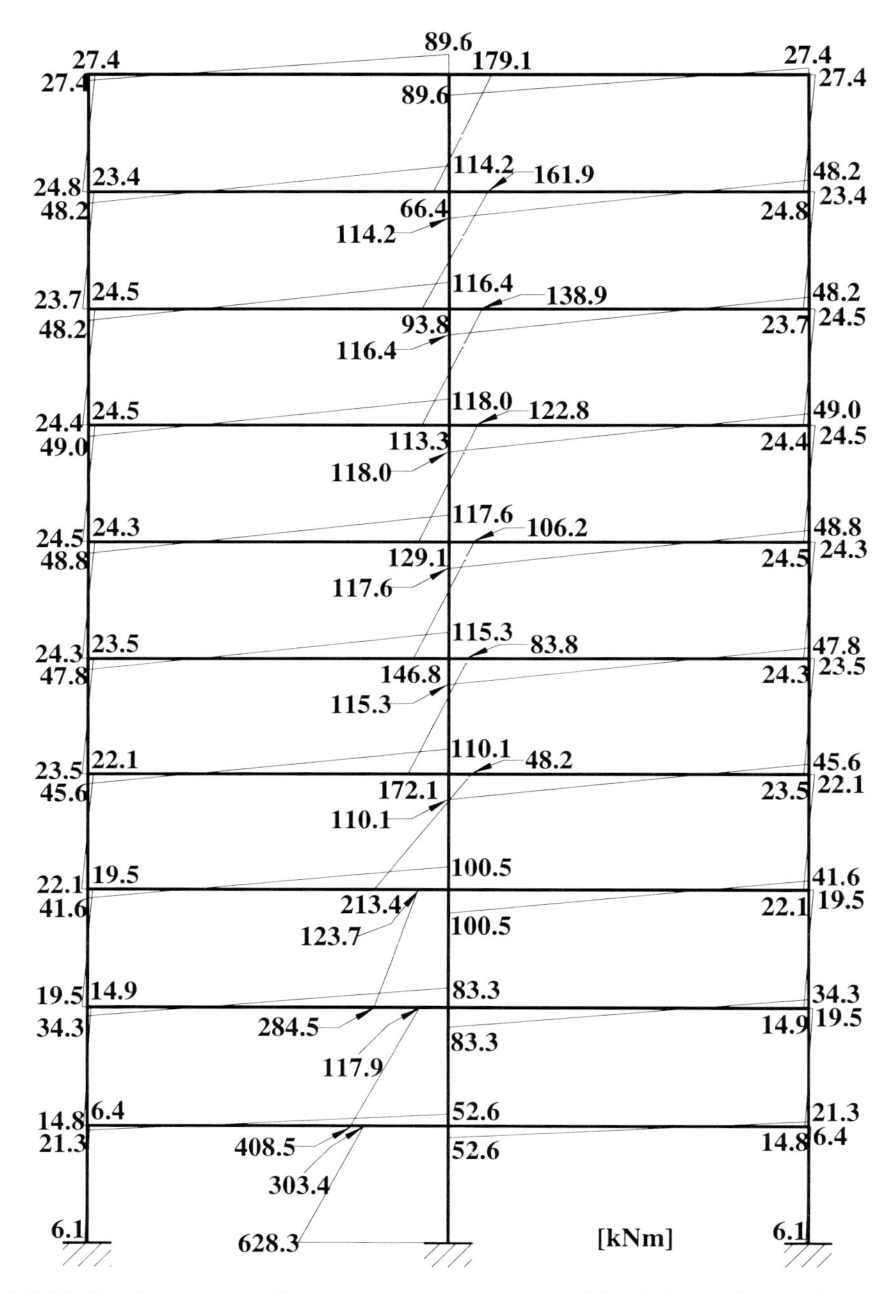

Fig. 7.6-15: Bending moment diagram under consideration of the rigid member portions

The determination of the condensed stiffness matrix has been carried out both with the program KONDEN (assuming girder spans of 6.00 m and, alternatively, 4.50 m), and with the program STAKON. The deformed shapes for a horizontal point load of 100 kN acting at roof level have been determined for all three cases; the results are shown in Fig. 7.6-13, with the corresponding bending moment diagrams depicted in Figs. 7.6-14 and 7.6-15. Fig. 7.6-14 shows the bending moment diagram computed by the program RAHMEN without consideration of the rigid girder zones (assuming a girder span of 6.00m), while Fig. 7.6-15 shows the corresponding diagram computed by the program STARAH, which takes the rigid girder zones into account.

7.7 Seismic design according to DIN 4149

This section deals with the rules laid down in the German DIN 4149 for the earthquake resistant design of buildings; it does not intend to reproduce and explain all single steps contained in the code but rather to outline the underlying concepts and assumptions. The corresponding rules of the Eurocode 8 are dealt with in Section 7.8 and illustrated by a practical example.

The German DIN 4149, entitled „Buildings in German Seismic Areas", stems from April 1981. It does not apply to structures which merit special safety checks, such as nuclear power plants, but to ordinary building structures, which are classified into three categories according to their importance. Buildings of special importance, such as hospitals or fire stations, belong to Category 3, while one-story sheds, adequately stiffened residential or office buildings up to a height of 22m and other simple structures fall into Category 1. Germany has been divided into five seismic zones according to different design ground acceleration values; for structures in zones 1 to 4 the structural stability checks must include the seismic load case. The basic design ground acceleration values a_0 for the different zones are as follows:

Zone 1: $a_0 = 0.25\ m/s^2$
Zone 2: $a_0 = 0.40\ m/s^2$
Zone 3: $a_0 = 0.65\ m/s^2$
Zone 4: $a_0 = 1.00\ m/s^2$

The actual value of the horizontal ground acceleration to be considered in the calculations is denoted by cal a and can be computed as

$$\text{cal a} = a_0\,\kappa\alpha \tag{7. 7. 1}$$

Here, α $(0.5 \le \alpha \le 1)$ is a reduction factor depending on the seismic zone and the structural importance category and κ (generally $1.0 \le \kappa \le 1.4$) is a function of the quality of the subsoil, with $\kappa = 1.0$ for hard rock and $\kappa = 1.4$ for soft to medium soil deposits. For very loose soils, κ can also exceed 1.4. The influence of the vertical acceleration component is ordinarily not taken into account.

Apart from the self-weight of the structure, the additional loads contributing to the inertia forces are considered with values between 0.5 kN/m^2 for residential buildings and 2.0 kN/m^2 for assembly halls. The seismically induced stresses and deformations are computed by modal response spectrum analysis, with the equivalent static loads for the ith mode given by

$$H_{E,j,i} = m_j \cdot \beta \cdot \gamma_{j,i} \cdot \text{cal a} \tag{7. 7. 2}$$

The underlying structural model is shown in Fig. 7.7-2; m_j is the mass of the jth story, β the spectral ordinate of the normalised response spectrum given by Eq. (7.7.3), $\gamma_{j,i}$ a coefficient depending on the ordinate $\psi_{j,i}$ of the jth mass in the ith mode (see Eq. 7.7.4), and cal a the nominal value of the horizontal ground acceleration after Eq. (7.7.1). The design response spectrum $\beta=\beta(T)$ is defined by (Fig. 7.7-1):

$$T \leq 0.45s : \beta = 1$$
$$T > 0.45s : \beta = 0.528T^{-0.8} \quad (7.7.3)$$

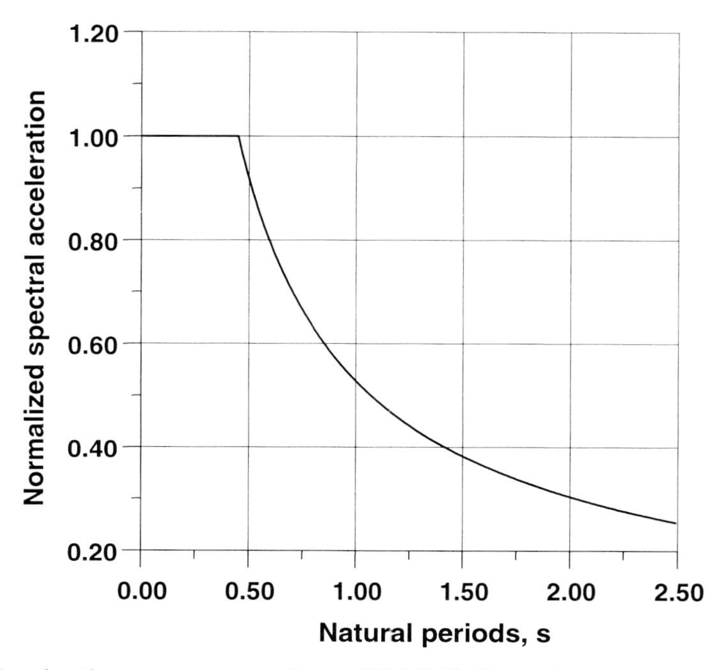

Fig. 7.7-1: Acceleration spectrum according to DIN 4149, linear plot

It should be mentioned that in this design spectrum a reduction factor of 1.8 has been incorporated, in order to account for the ductile behaviour of the structure.

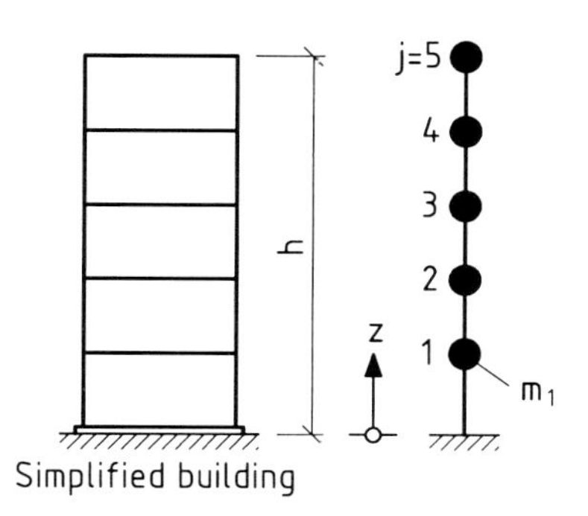

Fig. 7.7-2: Nomenclature according to DIN 4149

The coefficient $\gamma_{j,i}$ is given by

$$\gamma_{j,i} = \psi_{j,i} \frac{\sum_{j=1}^{n} m_j \, \psi_{j,i}}{\sum_{j=1}^{n} m_j \, \psi_{j,i}^2} \tag{7.7.4}$$

with $\psi_{j,i}$ as the ordinate of the ith modal shape corresponding to the jth discrete story mass. The combination of the modal stresses and deformations that have been computed for the sets of modal equivalent static forces is carried out by the SRSS rule. For regular structures with a fundamental period $T_1 \leq 1$ s, DIN 4149 presents an approximate method, in which the horizontal equivalent static loads are given by

$$H_{E,j} = 1.5 \cdot m_j \cdot \beta(T_1) \cdot \frac{z_j}{h} \cdot \text{cal a} \tag{7.7.5}$$

The fundamental period T_1 of the structure may be computed approximately by means of an expression given in DIN 4149.

Torsional effects are considered by assuming the equivalent static loads H_E to act at a distance "e" from the stiffness centre. For each lateral stiffening element (e.g. a shear wall), the most unfavourable of the following two values must be assumed (see Fig. 7.7-3):

$$\max e = e_0 + e_1 + e_2 \tag{7.7.6}$$

or

$$\min e = e_0 - e_2 \tag{7.7.7}$$

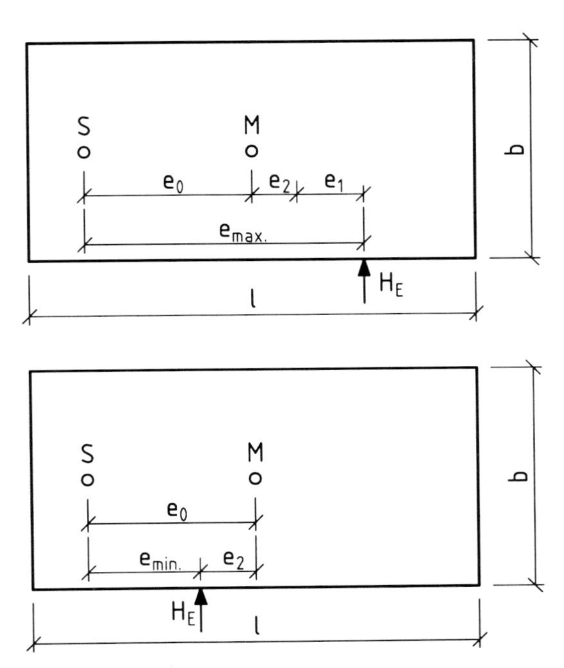

Fig. 7.7-3: Accidental and actual eccentricities

Here, e_2 is an accidental eccentricity equal to 5% of the building's dimension perpendicular to the direction of the seismic action, e_0 is the distance between the mass and stiffness centres and e_1 is given by the following expression for H_E parallel to side "b" of the rectangular plan:

$$e_1 = 0.1(\ell + b)\sqrt{\frac{10\,e_0}{\ell}} \leq 0.1(\ell + b) \qquad (7.7.8)$$

For more details the reader is referred to the actual Code.

Reinforced concrete buildings must be designed for the computed internal forces divided by 1.75. A series of restrictions are posed in order to guarantee sufficient ductility; for example, the maximum permissible tensile reinforcement ratio is limited and the introduction of compressive reinforcement as well as an adequate confinement of the concrete's compression zone are recommended. The normalised compressive axial force in frame columns given by

$$n = \frac{N}{A_b\,\beta_R} \qquad (7.7.9)$$

with the concrete strength β_R, the axial force N (negative in compression) and the concrete section area A_b should stay above -0.50; for $n < -0.23$ special measures must be taken (e.g. additional stirrups).

7.8 Seismic design according to Eurocode 8

7.8.1 Theoretical background

Eurocode 8 (EC 8 for short) deals with the seismic design of structures in Europe [7.21]. A purely linear-elastic design, which does not permit plastic zones to appear anywhere in the structure, even in the case of quite strong earthquakes, is in principle feasible, but economically unattractive. Usually, a certain amount of seismic energy is allowed to be dissipated through plastic action, as long as excessive deformations or even collapse are avoided, so that the design philosophy outlined at the beginning of section 7.5.1 is generally adopted.

Two specific properties of a building structure are of prime importance when dealing with its seismic behaviour. They are:

- Its strength and stiffness in resisting lateral loads, and
- its deformation capacity (or ductility in a broad sense).

Strength and ductility of a structure are by no means independent of each other. It is the engineer's task to design structures featuring a favourable combination of both properties while respecting the provisions laid down in the pertinent codes of practice. Generally, the following remarks apply:

- Structures with relatively low strength (lateral load-carrying capacity), require a higher in-built deformation capacity (ductility supply), if the safety is not to be compromised.
- On the other hand, structures featuring high strength against lateral loads do not normally require large ductility reserves.

Modern seismic codes such as the EC 8 include provisions which aim to guarantee that both sufficient strength and ductility are available in well-designed structures. Providing for sufficient ductility reserves in the structure is done by adhering to the rules of the so-called capacity design method. In this approach, critical zones of the structure are defined, where plastic actions are encouraged to occur, the regions in question having been especially endowed with sufficiently high ductility reserves. The remainder of the structure is guaranteed to stay in the elastic region and can be accordingly designed with strength rather than ductility in mind.

The earthquake hazard of a certain location is usually described by the elastic response spectrum for the horizontal ground acceleration; vertical accelerations are normally not considered with the exception of special cases such as horizontal cantilevers and beams with spans exceeding 20 m.

While in the design spectrum given in DIN 4149 the beneficial effect of the inelastic action is considered by an implicit reduction factor for the spectral ordinates of 1.8, in EC 8 elastic site-dependent response spectra are given first, which in a second step are transformed into design spectra by explicitly introducing reduction factors depending on the structure's ductility. The pertinent factor is the so-called behaviour factor q which quantifies the structure's ductile properties; the larger the ductility, and with it the factor q, the smaller the design spectrum ordinates and the design seismic loads to be considered. In EC 8, three ductility classes are regarded, namely class H (for high ductility), class M (for medium ductility) and class L (for low ductility). In this context, a simple definition of (maximum) ductility is the ratio between the maximum elasto-plastic value of some displacement to the corresponding elastic limit displacement:

$$\mu_\Delta = \frac{\Delta_{max}}{\Delta_{el}} \tag{7.8.1}$$

Thus, the behaviour factor q is also equal to the ratio of the restoring force F_{el} that would have occurred for linear system behaviour throughout to the maximum restoring force F_R occurring in an elastic-perfectly plastic system. Two possible idealisations may be used for establishing a relationship between the maximum ductility value μ_Δ and the behaviour factor q: The first one assumes that both the elastic and the elastic-perfectly plastic system suffer the same maximum displacement, as seen in Fig. 7.8-1. This is typically the case for long-period systems, with T larger than, say, 1.5 s. In this case, we have simply

$$\mu_\Delta = q \tag{7.8.2}$$

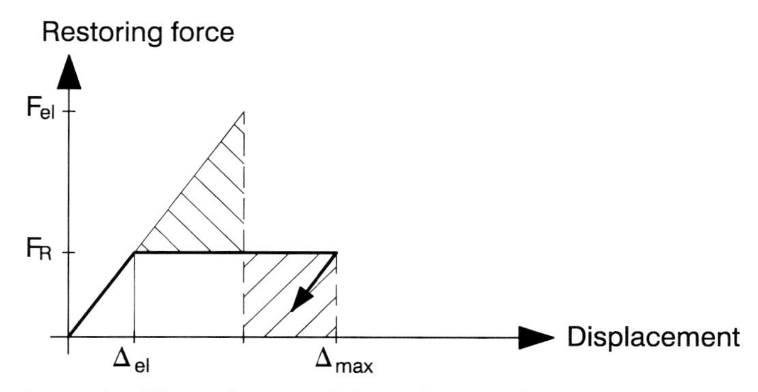

Fig. 7.8-1: Maximum ductility and system deformation, equal displacement case

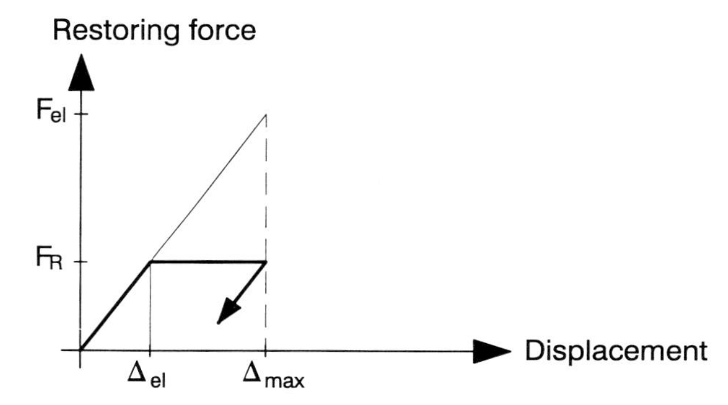

Fig. 7.8-2: Maximum ductility and system displacement, equal work case

The second idealisation assumes that the work depicted by the areas shown in Fig. 7.8-2 is the same for the elastic and the elastic-perfectly plastic system. This leads to the relationship

$$\mu_\Delta = \frac{q^2+1}{2}, \quad q = \sqrt{2\mu_\Delta - 1} \tag{7. 8. 3}$$

This expressions yields good results for stiff systems with short natural periods, up to approx. 0.1 s. For very stiff systems with still smaller natural periods, practically no reduction of the maximum restoring force or inelastic increase of the displacement occur, so that we obtain $q = \mu_\Delta = 1$.

Generally, the behaviour factor q is a function of the construction type, the material used and also the target ductility class. Not only global ductility values μ_Δ defined by displacement rations at structure or substructure level but also the ductility values at section level are functions of q. The important task is to ensure that the implicitly or explicitly assumed ductility properties are actually present in the structure and that the cyclic plastic deformations to be expected for severe earthquakes cannot lead to collapse by causing dramatic degradation of structural strength or stiffness. A powerful tool that can be used in this context is the capacity design method originally developed by PAULAY and his co-workers [7.22]. As already mentioned, it ensures that plastic actions can only happen in special "critical" areas of the structure which are known beforehand and have been accordingly designed. Raising the target ductility of a structure, e.g. by choosing ductility class H or M instead of L, means lowering the design seismic loads for the price of having to observe more exacting dimensioning and detailing provisions.

In EC 8 the behaviour factor q is given by the expression:

$$q = q_0 \cdot k_D \cdot k_R \cdot k_W \geq 1{,}5 \qquad (7.8.4)$$

Here, q_0 is the basic value of q depending on the type of structure (its value lying between 2.0 for inverted pendulum systems and 5.0 for frame structures), k_D is a function of the target ductility class (with values between 0.5 for class L and 1.0 for class H), k_R takes the structural regularity in elevation into account (1.0 for regular, 0.8 for irregular systems) and finally k_W is a function of the failure mode of the pertinent lateral carrying system (1.0 for frames, $\frac{1}{2.5 - 0.5 \cdot \alpha_0} \leq 1$ for shear walls, with α_0 equal to the prevailing aspect ratio $\frac{H_W}{\ell_W}$ of wall height to wall length). We can see that q varies between 1.5 and 5.0, implying a reduction of the elastic spectrum with factors between 0.67 and 0.20. The design spectrum itself consists of four branches which are defined as follows:

$$S_d(T) = \alpha S \left[1 + \frac{T}{T_B}\left(\frac{\beta_o}{q} - 1\right)\right] \qquad \text{for } 0 \leq T \leq T_B \qquad (7.8.5)$$

$$S_d(T) = \alpha S \frac{\beta_o}{q} \qquad \text{for } T_B \leq T \leq T_C \qquad (7.8.6)$$

$$S_d(T) = \alpha S \frac{\beta_o}{q}\left[\frac{T_C}{T}\right]^{k_{d1}} \geq 0.20\alpha \quad \text{for } T_C \leq T \leq T_D \qquad (7.8.7)$$

$$S_d(T) = \alpha S \frac{\beta_o}{q}\left[\frac{T_C}{T_D}\right]^{k_{d1}}\left[\frac{T_D}{T}\right]^{k_{d2}} \geq 0.20\alpha \quad \text{for } T_D < T \qquad (7.8.8)$$

The following notation has been used:

$S_d(T)$ Ordinate of the design spectrum in units of g,

α Ratio between the design value of the ground acceleration a_g and the acceleration of gravity g, $\alpha = a_g / g$

k_{d1}, k_{d2} Exponents influencing the form of the design spectrum for natural periods greater than T_C or T_D .

Values of the parameters S, β_0, T_B, T_C and T_D as well as k_{d1} and k_{d2} are given in the following table:

Subsoil class	S	β_0	k_{d1}	k_{d2}	T_B in s	T_C in s	T_D in s
A	1.0	2.5	2/3	5/3	0.10	0.40	3.0
B	1.0	2.5	2/3	5/3	0.15	0.60	3.0
C	0.9	2.5	2/3	5/3	0.20	0.80	3.0

Table 7.8-1: Parameters of the design spectrum

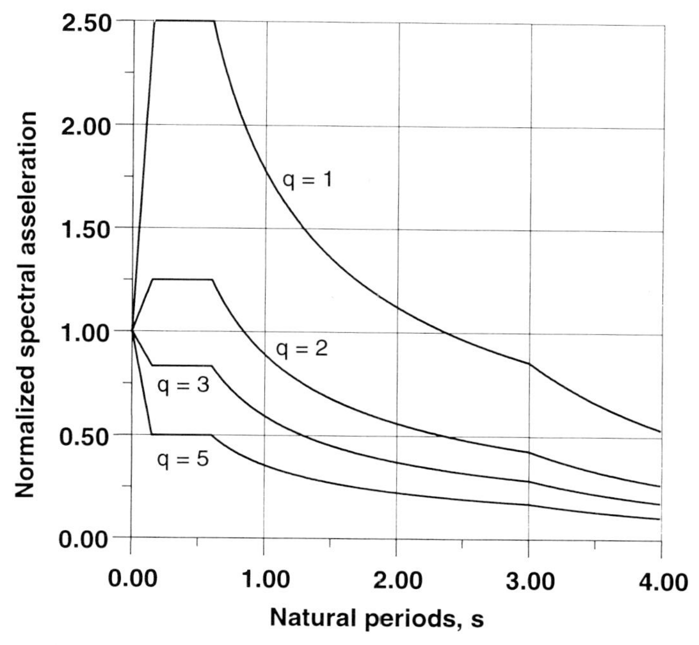

Fig. 7.8-3: Normalised design spectra according to EC 8 for different values of q

Fig. 7.8-3 shows as an example various design spectra for subsoil class B for different values of q. The ordinates are the values S_d (T) of the spectral acceleration in g units, divided by the product αS. These design spectra now represent non-linear spectra; they have, however, been derived from the corresponding elastic spectra simply by reducing their ordinates proportionally to the target ductility. In fact, the shape of true inelastic spectra, derived by subjecting non-linear SDOF systems to accelerograms as support motions, is generally different from the shape of the corresponding elastic spectra.

The relationship between the overall system ductility and the necessary local ductility supply strongly depends on the plastic mechanism inherently present in the structure. For safety reasons, high local accumulations of ductility demands should be avoided and the necessary plastic action distributed in several critical regions of the structure. The capacity design approach employed in order to reach this aim is not discussed here in detail because the pertinent rules are quite extensive and of course differ for different types of structures; for more details, the reader is referred to PAULAY et al. [7.22] and BACHMANN [7.23]. In the case of multi-story reinforced concrete frames, the contribution of many stories to the overall plastic action is considered favourable. This can be achieved by a "strong column-weak girder" design philosophy, which ensures that columns remain elastic (with the possible exception of the sections adjacent to the foundation) while plastic hinges form at the end sections of the girders. An example for the design of reinforced concrete shear walls is given in section 7.8.2. For other structural types the reader is referred to Part 1-3 of EC 8.

EC 8 distinguishes between regular and non-regular structures; this difference is crucial for the choice of the mathematical model of the structure, the algorithms to be employed and also the behaviour factor q. Generally, both 2D and 3D structural models may be employed. It is also possible to consider several modal contributions or just a single mode shape (normally the one corresponding to the lowest natural frequency).

The distinction between regular and non-regular structures is carried out separately for the plan and the elevation of a structure according to various regularity criteria. For regularity in plan it is important that the structure have approximately symmetrical stiffness and mass properties about two orthogonal axes, no re-entrant corners exceeding 25% of the plan dimensions, and sufficiently high in-plane stiffness of the floors. For regularity in elevation, the lateral load-carrying system should be continuous from the top of the building to the foundation and both mass and stiffness should not exhibit large variations between stories. The influence of structural regularity on the model to be employed is described in the following table:

Regularity conditions met		Type of structural model	Method of analysis
in plan	in elevation		
yes	yes	2D	First mode only
yes	no	2D	Multimodal
no	yes	3D	Multimodal
no	no	3D	Multimodal

Table 7.8-2: Influence of structural regularity on structural models and methods of analysis

For ordinary buildings, the regularity criteria both in plan and elevation are normally fulfilled. In this case, it is permitted to analyse the 3D system by means of two plane models while taking into account the actual or accidental eccentricities by means of approximate relationships, the contribution of the torsional response being relatively small. Being allowed to consider only the first mode shape further facilitates the analysis, leading in essence to a simple equivalent static force approach.

If the regularity criteria in elevation are not met, it is still possible to consider two independent plane models in the analysis, but more than one modal contribution must now be taken into account, leading to a multimodal response spectrum analysis of plane systems. If, additionally, the plan regularity criteria are not satisfied, it is not possible to consider independent plane systems in the analysis; instead, a full 3D analytical model must be employed.

The procedure of the simplified equivalent static force approach is summarised next. At the beginning, the total base shear force F_b must be determined for each principal direction; it is given as

$$F_b = S_d(T_1) \cdot W \qquad (7.8.9)$$

with

$S_d(T_1)$	design spectrum ordinate for the natural period T_1,
T_1	fundamental natural period of the building in the direction considered,
W	weight of the building.

The natural periods of vibration can be determined by means of finite element structural models or, approximately, through simplified expressions such as Eq. (7.8.14). The masses contributing to the building inertia forces are given by the following combination:

$$\sum G_{kj} + \sum \psi_{Ei} \cdot Q_{ki} \qquad (7.8.10)$$

Here, G_{kj} and Q_{ki} are the characteristic values of dead loads (action j) and live loads (action i), while ψ_{Ei} as combination coefficient quantifies the probability that only part of the live loads is actually present in the entire structure in the event of an earthquake. It is expressed as

$$\psi_{Ei} = \varphi \cdot \psi_2 \qquad (7.8.11)$$

with ψ_2 (according to Part 1 of EC 1) varying between 0.2 and 0.6 and φ (according to Table 3.2 of EC 8, Part 1-2) varying between 0.5 and 1.0.

The computed base shears are then distributed along the height of the building. If the displacements of the single stories in the fundamental vibration mode are known, this distribution is carried out according to the expression

$$F_i = F_b \frac{s_i \cdot W_i}{\sum_{j=1}^{n} s_j \cdot W_i} \qquad (7.8.12)$$

with the following notation:

F_i	horizontal force acting at story i,
F_b	base shear,
s_i, s_j	displacement ordinates of story masses m_i, m_j in the fundamental mode,
W_i, W_j	weights of the story masses m_i, m_j.

If the fundamental mode shape is not known, a triangular distribution of the seismic forces can be assumed. In this case, the story heights above ground z_i, z_j may be substituted for the displacements s_i, s_j in Eq. (7.8.12), leading to

$$F_i = F_b \frac{W_i z_i}{\sum_{j=1}^{n} W_j z_j} \qquad (7.8.13)$$

An approximate expression for the fundamental vibration period of buildings given in EC 8 is:

$$T_1 = C_t \cdot H^{0.75} \qquad (7.8.14)$$

H is the height of the building in m ($H < 80$m) and the coefficient C_t is set equal to 0.085 for moment-resisting spatial steel frames, 0.075 for moment-resisting spatial reinforced concrete frames and 0.050 for other systems. For buildings with reinforced concrete or masonry walls as load-carrying systems, C_t can also be computed according to the following expression given in Annex C2 of Part 1-2:

$$C_t = \frac{0.075}{\sqrt{A_c}} = \frac{0.075}{\sqrt{\sum \left[A_i \cdot \left(0.2 + \frac{\ell_{wi}}{H} \right)^2 \right]}}; \quad \frac{\ell_{wi}}{H} < 0.9 \qquad (7.8.15)$$

Here, A_c is the combined effective section of the ground floor walls, A_i the effective shear section of the wall i (both values in m^2) and ℓ_{wi} is the length of the ith ground floor wall parallel to the direction considered. Alternatively, the fundamental period T_1 may also be computed as

$$T_1 = 2 \cdot \sqrt{d} \qquad (7.8.16)$$

with d in m being the horizontal displacement at roof level due to the building's weight acting horizontally (Annex C3 in Part 1-2).

Two methods are given in EC 8 for approximately considering torsional effects in the context of the simplified equivalent static force approach. The first method multiplies the internal forces computed for each plane model with a factor δ given as

$$\delta = 1 + 0.60 \cdot \frac{x}{L_e}$$

In this expression, x is the distance between the structural component considered and the mass centre of the building, measured perpendicularly to the direction of the excitation, while L_e is the distance between the two outermost lateral load-carrying walls or frames, also measured perpendicularly to the direction of the excitation. This method is only applicable for buildings which satisfy both the plan and the elevation regularity conditions.

The second approximate method, described in Annex A of Part 1-2, is applicable for buildings which do not satisfy the regularity conditions in plan. First, additional accidental eccentricities $e_1 = 0.05 \cdot L_j$ are assumed, with L_j as building dimension perpendicular to the direction of the seismic excitation. Separate analyses are carried out on plane models for the two principal directions, in which it is further assumed that the horizontal story force F_i of the ith story is displaced from its location through the mass centre by an additional length e_2 (see Fig. 7.8-6). This additional eccentricity e_2 is the smaller of the two values given below:

$$e_2 = 0.1 \cdot (L + B) \cdot \sqrt{\frac{10 \cdot e_0}{L}} \leq 0.1 \cdot (L + B) \qquad (7.8.17)$$

$$e_2 = \frac{1}{2 \cdot e_0} \cdot \left[\ell_s^2 - e_0^2 - r^2 + \sqrt{\left(\ell_s^2 - e_0^2 - r^2\right)^2 + 4 \cdot e_0^2 \cdot r^2} \right] \qquad (7.8.18)$$

In this expression, we have

$$\ell_s^2 = \frac{L^2 + B^2}{12} \qquad (7.8.19)$$

with L and B as building dimensions perpendicular to and in direction of the seismic action, e_0 as distance between the mass and stiffness centres and r^2 as ratio of the torsional to the translational stiffness of the story. For more details, see the example in section 7.8.2. Since both horizontal seismic components must be considered, this leads to four load cases, with the shear forces in x and y direction being displaced both to the left and to the right of the mass centre in order to determine the most unfavourable combination for each single lateral load-carrying element.

Structures designed according to EC 8 must satisfy both the ultimate limit state (ULS) and the serviceability limit state (SLS) provisions. For the ultimate limit state, safety verifications include resistance and ductility conditions, global equilibrium and also the adequate behaviour of foundations and joints. The capacity design method is employed for ensuring the availability of adequate ductility; nevertheless, for important buildings (e.g. hospitals or communication centres), the continuous function of which in the case of emergencies is of prime importance, additional non-linear time history analyses of their actual behaviour under strong multi-component site-dependent seismic excitations is recommended.

For the serviceability limit state, verifications consider maximum permissible deformations, of which the inter-story drift is the most important. This takes the possible damage of (more or less brittle) partition walls and other non-structural elements into account For example, for buildings with brittle non-structural elements, the condition to be met is

$$\frac{d_s}{\nu} \leq 0.004\,h$$

with d_s as relative displacement of adjacent floors, h as the story height and ν as a factor (ranging between 2.0 and 2.5) considering the lower return period of the seismic event associated with the serviceability limit state.

A succinct overview of the conditions to be met according to EC 8 follows:

a) Ultimate limit state (ULS):
- Sufficient strength: $E_d \leq R_d$
- Sufficient ductility: Capacity design
- Global equilibrium (Overturning, sliding)
- Foundations and joints

b) Serviceability limit state (SLS):
- Limiting story drifts.

The overall design aim is to ensure that adequate ductility reserves are present in the structure, so that the plastic action expected to occur during strong earthquakes will not lead to critical deterioration of the structure's strength or stiffness.

According to Part 1-1 of EC 8 the following load case combination must be considered for seismic actions, with the summation sign denoting the combined influence of all pertinent actions:

$$\sum G_{kj} + \gamma_1 A_{ed} + \sum \psi_{2i} Q_{kj} \qquad (7.8.20)$$

Importance category	Buildings	Importance factor γ_1
I	Buildings, the integrity of which during earthquakes is of vital importance (e.g. hospitals)	1.4
II	Buildings, the resistance of which is important because of the consequences of a collapse (e.g. schools)	1.2
III	Buildings of intermediate size, intended for normal use (e.g. office buildings)	1.0
IV	Buildings of minor importance for public safety (e.g. agricultural buildings)	0.8

Table 7.8-3: Importance factors γ_1 according to EC 8

The single terms are defined as follows:

G_{kj}	Characteristic values of the permanent actions.
A_{ed}	Design seismic action.
γ_I	Importance factor according to Section 3.7, Part 1-2 of EC 8 (see Table 7.8-3).
Q_{ki}	Characteristic values of variable actions.
ψ_{2i}	Combination coefficient according to section 3.6, Part 1-2 of EC 8.

If prestressed structural elements are also present, this combination must additionally contain the prestressing action. More details are shown in the context of the example presented in the following section.

7.8.2 Example

7.8.2.1 Description of the building

The building considered as an example for designing according to EC 8 is a nine-story office building with two additional basement levels. It is a reinforced concrete structure, the lateral load-carrying system of which consists entirely of shear walls. The floors are reinforced concrete slabs cast in situ and the columns between floors carry only axial loads, not being part of moment-resisting frames. The reinforced concrete shear walls are symmetric in plan and connected to each other through the rigid floor diaphragms. The floor slab thickness is 26 cm, with the exception of the roof slab with a thickness of 24 cm. Some further assumptions follow:

- Design value for the ground acceleration: $a_g = 2.50 \frac{m}{s^2}$
- Importance category III
- Ductility class M
- Subsoil category A.

A relatively high ground acceleration value has been chosen in order to better exhibit the non-linear effects to be expected. Generally, the choice of a higher ductility class does not offer any advantage for low ground acceleration levels, since the resulting relatively low horizontal forces can usually be readily accomodated by the structure without causing marked plastic actions to happen.

7.8.2.2 Materials

The materials to be used in earthquake resistant construction must be chosen carefully, because of the implicit assumption that plastic actions are expected for strong seismic events. The magnitude of probable plastic actions increases for higher ductility classes (M or H), and, at the same time, the required quality characteristics of the materials become more important. As far as concrete is concerned, it should belong at least to class C20/25; higher strength concrete as given in EC 2 may also be used [7.24], [7.25].

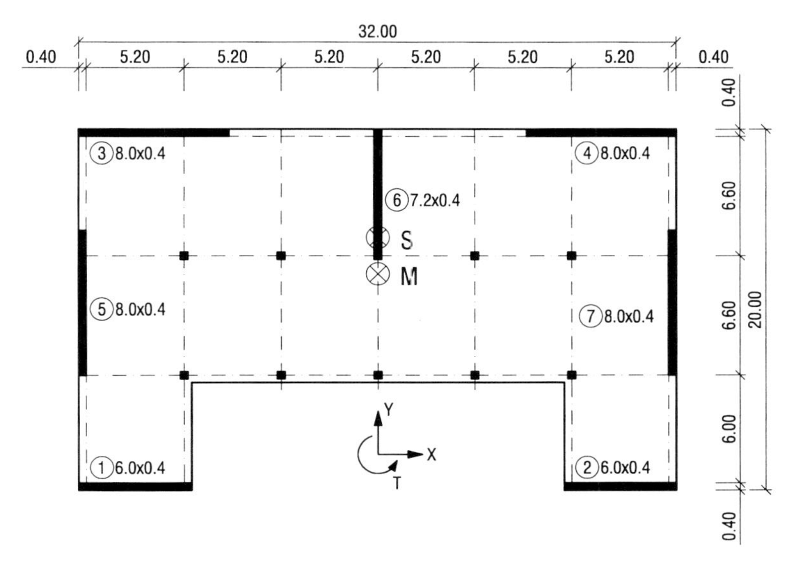

Fig. 7.8-4: Building plan

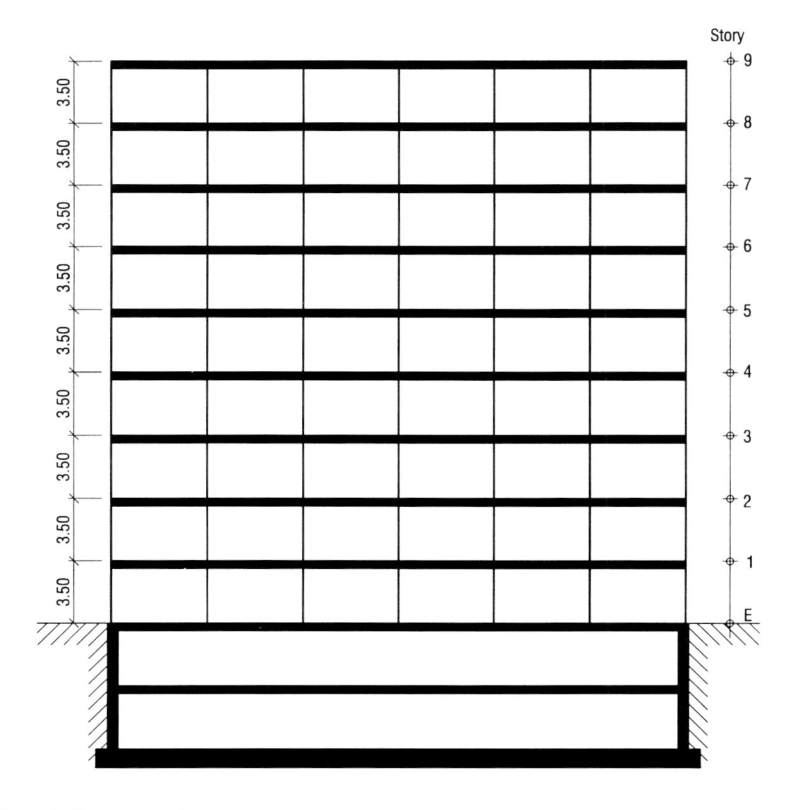

Fig. 7.8-5: Building elevation

Not all reinforcing steel qualities given in EC 2 are equally suitable for our purpose. High-ductility steel may be used for ductility class L structures; for ductility classes M and H the reinforcing steel must comply with the additional requirements given in Table 2.1, Part 1-3 of EC 8. These additional requirements refer to the ultimate yield strain $\varepsilon_{su,k}$ (which should be at least equal to 6% for ductility class M and 9% for ductility class H), the ratio f_t/f_y of the tensile to the yield strength (it should lie between 1.15 and 1.35 for ductility class M and 1.20 - 1.35 for ductility class H) and the ratio $f_{y,act}/f_{y,nom}$ of actual and nominal tensile yield strength (which should be less than or equal to 1.25 for ductility class M and 1.20 for ductility class H). These rules ensure adequate local rotation capacities and dimensions of plastic hinges.

Partial safety factors of $\gamma_c = 1.50$ for concrete and $\gamma_s = 1.15$ for steel were adopted according to EC 2, Table 2.3. Concrete class C30/37 was chosen, and the reinforcing steel was assumed to satisfy the requirements for ductility class M. The concrete's Young's modulus E_{cm} is equal to 32,000 MN/m², its Poisson's ratio ν equal to 0.20.

7.8.2.3 Structural model and equivalent seismic forces

A suitable analysis model for the building must be chosen with the help of Table 2.1, Part 1-2 of EC 8. The criteria for regularity in plan are not satisfied, the structure complies, however, with the criteria of Annex A, Part 1-2, regarding regularity in elevation. It is thus permitted to use the equivalent static force procedure for plane models in two mutually orthogonal planes. The following assumptions are also made:

- All floors are assumed to be rigid in their own plane and to offer no resistance to bending moments the vectors of which lie in their plane.
- The bending stiffness of shear walls about their weak axis is neglected.
- Uncracked concrete sections are used for the computation of bending stiffness values.
- The columns do not contribute to the horizontal stiffness, being considered hinged at both ends.
- No foundation flexibility is considered.

The assumption of uncracked concrete sections may lead to results lying on the unsafe side. In the present case the fundamental natural frequencies for both plane models happen to lie in the range of decreasing spectral ordinates (for increasing natural periods), so that the assumption of cracked sections (implying longer periods) leads to lower seismic forces. BACHMANN and PENELIS/KAPPOS recommend the assumption of an effective bending stiffness of 60% of the value calculated for the uncracked section [7.23], [7.19].

Eq. (7.8.10) must be applied in order to compute the story masses contributing to the inertia forces. The coefficient ψ_{Ei} is given by

$$\psi_{Ei} = \varphi \cdot \psi_2$$

as discussed in section 7.8.1. The combination coefficient ψ_2 from EC 1 is equal to 0.3 for all floors, φ is set equal to 1 for the top floor and 0.5 for all other floors according to Table 3.2 in Part 1-2 of EC 8. Table 7.8-4 summarises the results for permanent and live loads and presents the story masses thus obtained. The differences in the permanent loads are due to different equivalent distributed loads representing vertical structural components.

Story	Permanent load [kN/m²]	Live load [kN/m²]	Total mass [t]
9	10.26	2.00	575.78
8	12.15	3.25	695.89
7	12.17	3.25	671.14
6	12.17	3.25	671.14
5	12.17	3.25	671.14
4	12.17	3.25	671.14
3	12.17	3.25	671.14
2	12.17	3.25	671.14
1	14.80	3.25	810.28
Σ			6108.79

Table 7.8-4: Story masses

The behaviour factor q is next calculated from the expression

$$q = q_0 \cdot k_D \cdot k_R \cdot k_W \geq 1.50 \tag{7. 8. 21}$$

with the following terms:

q_0 Basic value of q as function of the lateral load-carrying system (here $q_0 = 4.0$ for uncoupled shear walls),

k_D Factor considering the ductility class (here 0.75 for ductility class M),

k_R Factor considering the structural regularity in elevation (here 1.0, since the corresponding criteria are satisfied),

k_W Factor considering the expected failure mode (bending or shear), given by $k_W = \dfrac{1}{2.50 - 0.5 \cdot \alpha_0}$ with $\alpha_0 = \dfrac{\sum H_{wi}}{\sum \ell_{wi}}$

Here the sum of wall heights is equal to $\sum H_{wi} = 7 \cdot 3.50 \cdot 9 = 220.50\text{m}$

The sum of wall lengths is equal to

$\sum \ell_{wi} = 2 \cdot (6.0 + 8.0 + 8.0) + 7.20 = 51.20\text{ m}$

$$k_W = \frac{1}{2.50 - 0.5 \cdot 4.31} = 2.90 \geq 1.00 \Rightarrow k_W = 1.00$$

The behaviour factor is thus equal to:

$$q = 4.00 \cdot 0.75 \cdot 1.00 \cdot 1.00 = 3.00 \tag{7. 8. 22}$$

Table 7.8-5 summarises all input data necessary for evaluating the spectral ordinates.

The computed natural periods lie in the range $T_C \leq T_{x/y} \leq T_D$ of the EC 8 design spectrum. The spectral ordinate is given by the expression:

$$S_d(T) = \alpha \cdot S \cdot \frac{\beta_0}{q} \cdot \left[\frac{T_C}{T}\right]^{K_{d1}} \text{ resp. } S_d(T) \geq 0.20 \cdot \alpha \tag{7. 8. 23}$$

Subsoil class A	
$a_g = 2.50 \frac{m}{s^2}$	
q	3.00
α	0.25
S	1.00
β_0	2.50
k_{d1}	2/3
k_{d2}	5/3
T_B	0.10 s
T_C	0.40 s
T_D	3.00 s
$T_{1,x}$	0.627 s
$T_{1,y}$	0.616 s

Table 7.8-5: Input data for the spectral analysis

This leads to

$$S_{d,x} = 0.1546$$

and

$$S_{d,y} = 0.1562$$

Table 7.8-5 contains values for the fundamental periods $T_{1,x}$ and $T_{1,y}$ that have been computed by means of a finite element model (see section 8.2). The equivalent static loads are computed as products of the spectral ordinates and the weight of the building. The latter is equal to $W = 6108.7 \cdot 9{,}81 = 59927$ kN (see Table 7.8-4). The equivalent static loads are thus equal to:

$$F_{b,x} = 9264.8 \text{ kN}$$

and

$$F_{b,y} = 9360.6 \text{ kN}$$

The distribution of the equivalent static loads along the height of the building in both principal planes is carried out by using the mode shapes computed by the aforesaid finite element model. For comparison purposes, the equivalent static loads are also evaluated for a triangular distribution law after Eq. (7.8.13). Results are summarised in Table 7.8-6.

The consideration of the torsional effects is carried out according to Annex A in Part 1-2 of EC 8. A separate analysis in two mutually orthogonal planes is admissible here. An additional accidental eccentricity must be considered and the overall eccentricity consists of three terms as given below:

$$e_{max} = e_0 + e_1 + e_2 \quad (7.8.24)$$

$$e_{min} = e_0 - e_1 \quad (7.8.25)$$

Fig. 7.7-3 showed the approach adopted in DIN 4149 which is quite similar to the one of EC 8, Part 1-2. The notation used is slightly different, though, and Fig. 7.8-6 depicts the eccentricities as employed in EC 8.

Story	Mass [t]	Weight [kN]	h_i [m]	$F_{i,x}$ [kN]	$F_{i,x}^{Triangle}$ [kN]	$F_{i,y}$ [kN]	$F_{i,y}^{Triangle}$ [kN]
9	575.78	5648.4	31.5	1986.8	2151.2	1997.2	2173.4
8	695.89	6826.7	28.0	2042.1	1583.8	2056.9	1600.2
7	671.14	6583.9	24.5	1624.5	1358.8	1639.6	1400.2
6	671.14	6583.9	21.0	1286.7	1187.9	1301.7	1200.1
5	671.14	6583.9	17.5	964.4	989.9	978.3	1000.1
4	671.14	6583.9	14.0	668.7	791.9	680.2	800.1
3	671.14	6583.9	10.5	410.9	593.9	419.4	600.1
2	671.14	6583.9	7.0	204.6	410.6	209.6	414.8
1	810.28	7948.8	3.5	76.2	169.9	77.9	171.6
			Σ	**9264.8**	**9264.8**	**9360.6**	**9360.6**

Table 7.8-6: Vertical distribution of the equivalent static forces

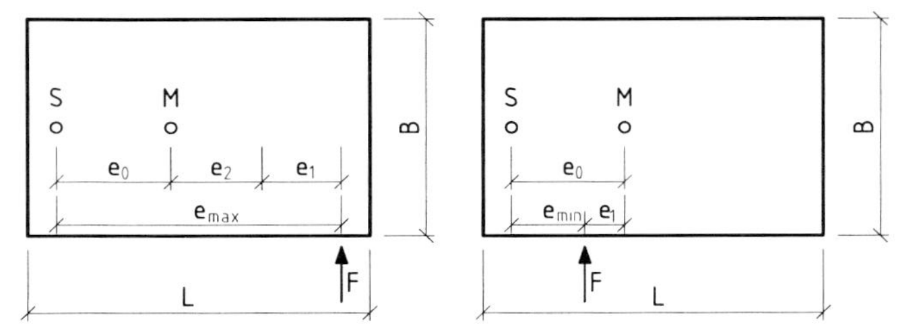

Fig. 7.8-6: Eccentricities according to EC 8

The three eccentricity terms are defined as follows:

Actual eccentricity e_0:
This is the distance between the mass and the stiffness centre of the floor; it should be as small as possible.

Accidental eccentricity e_1:
This is due to inaccuracies when considering the location of the permanent loads or the distribution of the live loads.

Additional eccentricity e_2 according to Annex A, EC 8, Part 1-2:
This additional term takes the contribution of torsional mode components into account, which cannot be directly considered in a plane model. It depends primarily on the ratio of the torsional to the translational rigidity of the system. In the present case, the corresponding expressions of Annex A yielded the result that an additional eccentricity must be considered in both principal planes. For an overview of the different concepts for approximately considering torsional actions the reader is referred to [7.26].

Starting with the eccentricities in y direction, we obtain the following results:

$$e_{0,y} = 8.29 - 6.01 = 2.28\ \text{m}$$
$$e_{1,y} = 0.05 \cdot L_i = 0.05 \cdot 20.0 = 1.00\ \text{m}$$

The eccentricity e_2 can be calculated by means of the Eqs. (7.8.17) to (7.8.19) which are repeated here for convenience:

$$e_2 = \min \begin{bmatrix} 0.10 \cdot (L+B) \cdot \sqrt{\dfrac{10 \cdot e_0}{L}} \leq 0.10 \cdot (L+B) \\ \dfrac{1}{2 \cdot e_0}\left(\ell_s^{\,2} - e_0^{\,2} - r^2 + \sqrt{\left(\ell_s^{\,2} + e_0^{\,2} - r^2\right)^2 + 4 \cdot e_0^{\,2} \cdot r^2}\right) \end{bmatrix} \tag{7. 8. 26}$$

with

$$\ell_s^{\,2} = \frac{L^2 + B^2}{12} \tag{7. 8. 27}$$

As already mentioned, r^2 expresses the ratio of the system's torsional stiffness to its translational stiffness. For buildings with shear walls as lateral load-carrying systems, it can be evaluated by means of Eq. (7.8.28) [7.9]. In this expression, I_{ei} stands for the equivalent moment of inertia of wall i about its strong axis and r_i is the distance of the wall to the stiffness centre. $I_{ei,x}$ and $I_{ei,y}$ are moments of inertia of the wall i about the x and y axes through the wall's centre of gravity. For more details the reader is referred to [7.27], where expressions for other building types may also be found.

$$\begin{aligned} r^2 &= \frac{\sum I_{ei} \cdot r_i^{\,2}}{\sum I_{ei,x}} \quad (\text{x direction}) \\ r^2 &= \frac{\sum I_{ei} \cdot r_i^{\,2}}{\sum I_{ei,y}} \quad (\text{y direction}) \end{aligned} \tag{7. 8. 28}$$

By means of these expressions we obtain

$$e_2 = \min \begin{bmatrix} 5.20\ \text{m} \\ 1.846\ \text{m} \end{bmatrix} \tag{7. 8. 29}$$

This finally leads to maximum eccentricities of

$$e_{y,max} = 2.28 + 1.00 + 1.846 = 5.126\ \text{m}$$
$$e_{y,min} = 2.28 - 1.00 = 1.280\ \text{m}$$

The evaluation of the eccentricities in the x direction is much simpler because no actual eccentricity e_0 is present. Therefore, e_2 is zero and we obtain:

$$e_{0,x} = 0.00\ \text{m}$$
$$e_{1,x} = 0.05 \cdot L_i = 0.05 \cdot 32.00 = 1.60\ \text{m}$$
$$e_{2,x} = 0.00\ \text{m}$$

The extreme values are then given by

$$e_{x,max} = e_1 = 1.60\ \text{m}$$
$$e_{x,min} = -e_1 = -1.60\ \text{m}$$

7.8.2.4 Design forces and verifications

In the last section the equivalent seismic forces were determined, together with their distribution along the height of the building, and also the torsional eccentricities. The next task is to evaluate the forces acting on each shear wall. Fig. 7.8-7 shows the four cases for eccentrically applying the story shear forces. As we shall see, in the course of combining the single seismic actions, eight different cases must actually be considered, because factors of both 1.0 and 0.3 must be associated with the equivalent static loads (see Eqs. (7.8.37) and (7.8.38) later in this section).

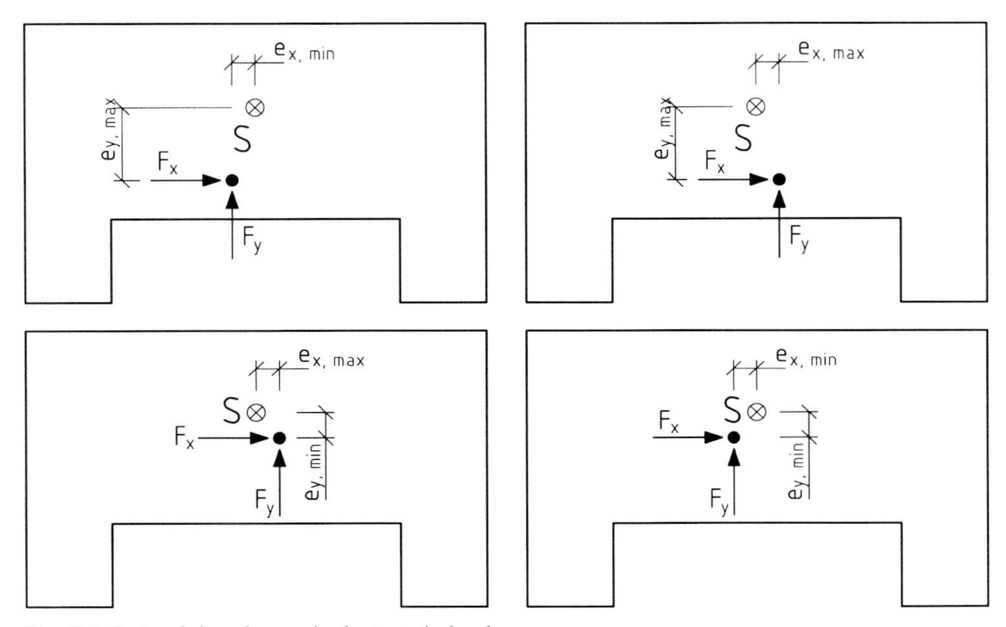

Fig. 7.8-7 Applying the equivalent static loads

The computation of the story forces acting on each single wall can be carried out by hand or by computer, using the programs already introduced. Both approaches will be demonstrated here and the results for a single shear wall (No. 7 in Fig. 7.8-4) presented in detail.

In the manual approach, the distribution of the story shears is carried out through calculating distribution coefficients, to be evaluated separately for shear forces and torsional moments. The pertinent expressions can be found in [7.23]. For the shear forces, the distribution coefficients are given in the form

$$V_{ix,q} = \frac{I_{iy}}{\sum I_{iy}} \tag{7.8.30}$$

Analogously, distribution coefficients for torsional moments are given as

$$V_{ix,t} = \frac{I_{yi} \cdot \overline{y}_i}{\sum \left(I_{iy} \cdot \overline{y}_i^{\,2} + I_{ix} \cdot \overline{x}_i^{\,2} \right)} \tag{7.8.31}$$

$$V_{iy,t} = \frac{I_{xi} \cdot \overline{x}_i}{\sum\left(I_{iy} \cdot \overline{y}_i^{\,2} + I_{ix} \cdot \overline{x}_i^{\,2}\right)} \tag{7. 8. 32}$$

In the last two equations, $\overline{x}_i$ and $\overline{y}_i$ are the co-ordinates of the shear centre of wall i relative to the stiffness centre of the floor. The calculations must be carried out separately for the x and y directions. As an example, we consider wall no. 7 in Fig. 7.8-4. For excitation in direction x, we obtain the expressions:

$$F_{7,min} = F_{ix} \cdot V_{ix,q} + e_{y,min} \cdot F_{ix} \cdot V_{ix,t} \tag{7. 8. 33}$$

$$F_{7,max} = F_{ix} \cdot V_{ix,q} + e_{y,max} \cdot F_{ix} \cdot V_{ix,t} \, . \tag{7. 8. 34}$$

For the y direction, we obtain:

$$F_{7,min} = F_{iy} \cdot V_{iy,q} + e_{x,min} \cdot F_{iy} \cdot V_{iy,t} \tag{7. 8. 35}$$

$$F_{7,max} = F_{iy} \cdot V_{iy,q} + e_{x,max} \cdot F_{iy} \cdot V_{iy,t} \tag{7. 8. 36}$$

The following table summarises the computed story shear forces for wall no. 7 for all four load cases. The last column in this table presents the combination leading to the extreme story shears for this wall. These extreme values are attained for full seismic action in the y direction (column 5 with factor 1.0) and reduced seismic action in the x direction (column 3 with factor 0.3 according to Eq. (7.8.38) in this section). Here, indices x and y denote the direction of the seismic action, while indices min and max refer to the story shear force having its minimum or maximum eccentricity.

Story	$F_{7,x,min}$ [kN]	$F_{7,x,max}$ [kN]	$F_{7,y,min}$ [kN]	$F_{7,y,max}$ [kN]	$F_{7,extr}$ [kN]
9	55.26	221.30	662.52	801.40	867.79
8	56.80	227.47	682.34	825.37	893.61
7	45.19	180.95	543.90	657.91	712.20
6	35.79	143.32	431.82	522.33	565.33
5	26.82	107.42	324.52	392.54	424.77
4	18.60	74.48	225.64	272.94	295.28
3	11.43	45.77	139.13	168.29	182.02
2	5.69	22.79	69.52	84.10	90.94
1	2.12	8.49	25.84	31.25	33.80

Table 7.8-7: Story shear forces for wall no. 7, manual evaluation

In the most general case, all three translational seismic components must be considered, and the most unfavourable combinations determined. This is clearly a major undertaking, so simple combination rules are very important.

Regarding the vertical component, EC 8, Part 1-1, states that it can be modelled by the same spectrum used for the horizontal components with its ordinates reduced by factors between 0.5 and 0.7, depending on the natural periods (0.7 for $T < 0.15$ s, 0.5 for $T > 0.5$ s and linear interpolation in-between). According to Part 1-2 of EC 8, the effects of the vertical component may be ignored altogether if the structure does not contain any of the following:

- Horizontal (or almost horizontal) members with spans over 20 m
- Horizontal (or almost horizontal) cantilevers
- Horizontal (or almost horizontal) pre-stressed members
- Beams supporting columns.

In the example considered no such structural member is present, so that it suffices to consider only the horizontal seismic components. Their combined effect may be considered in the form

$$E_{Ed} = E_{edx} \text{ "+" } 0.30 \cdot E_{Edy} \tag{7. 8. 37}$$

$$E_{Ed} = E_{edy} \text{ "+" } 0.30 \cdot E_{Edx} \tag{7. 8. 38}$$

Relevant for design purposes is the maximum of these two combinations, with the + sign in the two preceding expressions meaning "to be combined with" rather than an algebraic addition. Alternatively, a standard SRSS combination of the modal internal forces can be carried out for both horizontal seismic components.

The evaluation of the design internal forces and the necessary reinforcement in the single walls is omitted here. The distribution of the story shear forces to the single shear walls can also be carried out with the help of the programs already discussed based on the pseudo-spatial model (see section 7.6). In this model, the only active kinematic degrees of freedom are the two mutually orthogonal horizontal displacements of each floor and its rotation about a vertical axis. The load combination leading to the most unfavourable story shears for wall no. 7 is the one shown at the right of the upper row in Fig. 7.8-7; the equivalent loads in the x direction are simultaneously considered with a factor of 0.3. The system displacements are evaluated with the program DISP3D and the story shear forces of each shear wall are then computed by the program FORWND. The results are compared in Table 7.8-8 with the corresponding values of the manual approach.

Story	$F_{7,extr}$ [kN] electronic solution	$F_{7,extr}$ [kN] solution by hand
9	867.31	867.79
8	893.80	893.61
7	712.97	712.20
6	565.47	565.33
5	424.25	424.77
4	295.93	295.28
3	181.84	182.02
2	91.06	90.94
1	33.84	33.80

Table 7.8-8: Story shear forces of wall no. 7, computed electronically and by hand

8 Examples and special topics

This section deals with some special topics and application examples from the field of seismic structural design and analysis. Section 8.1 centres on the non-linear seismic behaviour of reinforced concrete buildings, section 8.2 presents some results of a study of the seismic behaviour of the Cologne Cathedral and section 8.3 deals with the modelling of a complex refinery container.

8.1 Non-linear behaviour of seismically excited reinforced concrete buildings

In view of the high damage levels associated with a strong seismic event hitting densely populated urban centres, the further development of design rules towards a minimisation of damages is becoming increasingly important. Meanwhile, it is generally accepted that the aseismic design of new buildings and also the seismic strengthening and retrofitting of existing structures (which may have suffered some degree of damage from past earthquakes) should focus on critical deformation states and explicitly consider inelastic damage. The so-called "Performance Based Seismic Design" philosophy poses the central question "How can a desired functionality level of a structure be guaranteed for a certain seismic excitation level"? The design process must match the structural damage levels to the excitation intensities in order to avoid inadequate structural performance at relatively low seismic intensity levels. A simple relationship connecting structural damage states (defined by functional criteria) to excitation levels (defined by their occurrence probability) according to [8.1] is shown in the following table:

Structural damage state	*50-year-event occurrence probability*
Operational	about 50%
Occupancy	10-20%
Life Safety	5-10%
Near Collapse	< 5%

Fig. 8.1-1 graphically depicts this relationship between ground excitation levels and structural damage for different building classes according to their importance

The verification of the continuing functionality of the building is carried out by developing fragility curves of the building as a whole and/or of its sub-assemblages, structural fragility being defined as the probability of exceeding a certain structural response threshold as a function of the seismic intensity level. A quantitative safety estimation requires information on the structural damage corresponding to the "life safe" level, with a 50 year occurrence probability about 5%. The available safety of an existing building can then be estimated by means of a relationship

$$\text{available safety} = \frac{D_{ult}}{D_{target}} \tag{8. 1. 1}$$

Here, D_{ult} is the damage indicator for the "life safe" level and D_{target} the corresponding value for the target seismic excitation level. The structural damage information can be supplied only by non-linear tine history analyses of the building; the remainder of this section discusses methods for conducting such analyses for buildings with reinforced concrete shear walls as lateral load-carrying systems.

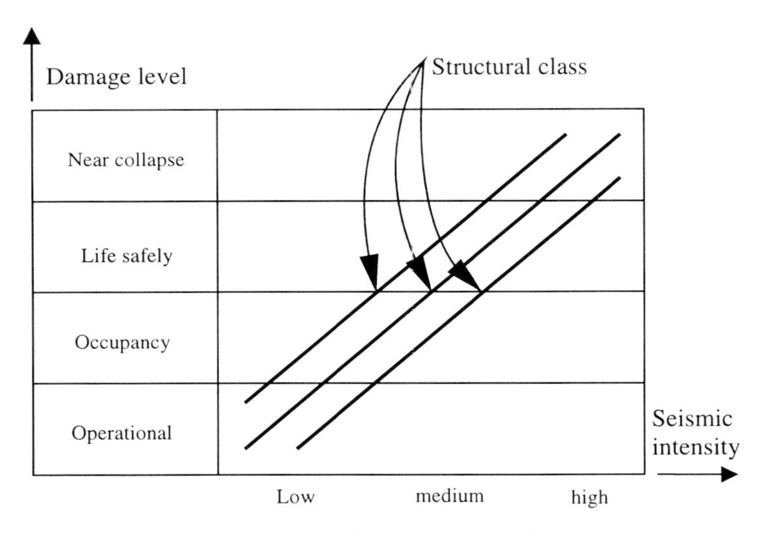

Fig. 8.1-1: Relationship between structural damage and seismic intensity

The modelling of reinforced concrete shear walls can be carried out by means of suitable finite element "micro-models" (using non-linear multi-layer reinforced concrete plate elements) [8.3] or by using half-empirical so-called „macro-models" [8.2]. The finite elements used in micro-models include sophisticated constitutive laws for the cyclic behaviour of the concrete, the reinforcing steel and the bond between the two components; they are much more accurate than macro-models but accordingly complex to apply. Macro-models are, on the other hand, much easier to use but must be calibrated carefully in order to yield dependable results. They are mostly descendants of the classical „Three-Vertical-Line" model introduced by KABEYASAWA et al. [8.4].

Classical damage indicators are the maximum and cumulative ductility values. They are shown in Figs. 7.8-2 and 7.8-3 as defined at section level with curvatures κ as variables. The maximum ductility μ_κ is defined as the ratio

$$\mu_\kappa = \frac{\kappa_m}{\kappa_y} \tag{8. 1. 2}$$

The cumulative ductility μ_{cu} is given by:

$$\mu_{cu} = \frac{\sum_i \kappa_{pl,i}}{\kappa_y} \tag{8. 1. 3}$$

Maximum ductility values are quite suitable for describing failure types induced by a single excessive loading peak, but they are unable to mirror the progressive failure type caused by gradual stiffness and strength deterioration after many loading cycles. The latter case requires damage indicators resembling the cumulative ductility μ_{cu} defined above. An inherent disadvantage of both μ_κ and μ_{cu} is the lack of normalising factors so that the range from undamaged structure to the state prior to collapse is projected in the 0 to 1 (or 0 to 100%) interval. Such a normalised damage indicator for reinforced concrete structures has been given by MEYER and GARSTKA and is termed the D_Q damage indicator ([8.5], [8,6]). It is based on dissipated energy and it is normalised by the maximum work E_u introduced in the structure under monotonic loading (Fig. 8.1-4). The following equations hold:

$$D_Q = D_Q^+ + D_Q^- - D_Q^+ \cdot D_Q^- \tag{8. 1. 4}$$

with

$$D_Q^+ = \frac{\sum_i E_{S,i}^+ + \sum_i E_i^+}{E_u^+ + \sum_i E_i^+}, \quad D_Q^- = \frac{\sum_i E_{S,i}^- + \sum_i E_i^-}{E_u^- + \sum_i E_i^-} \tag{8. 1. 5}$$

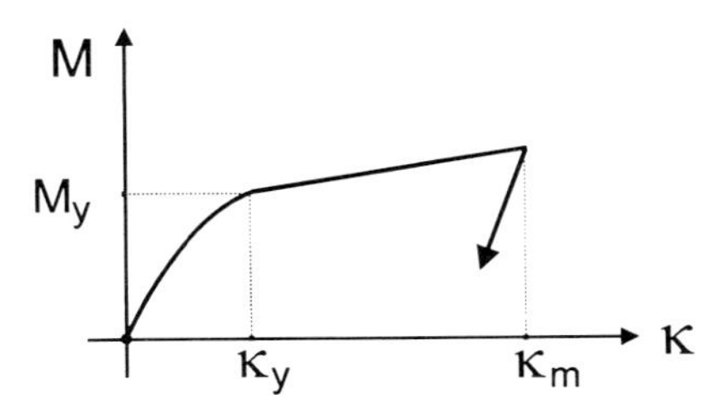

Fig. 8.1-2: Maximum curvature ductility at section level

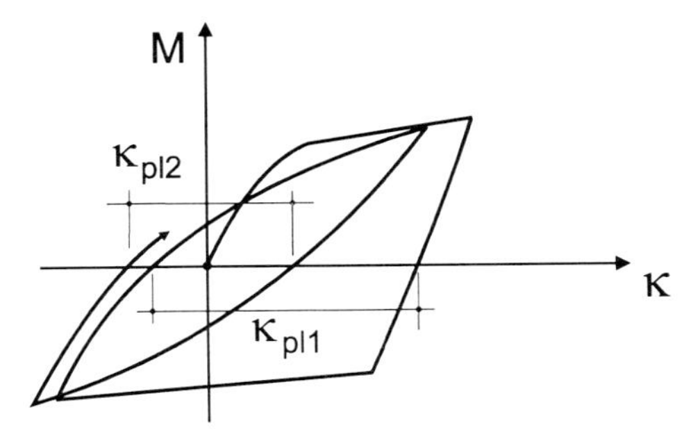

Fig. 8.1-3: Cumulative ductility at section level

The first term in Eq. (8.1-5) is valid for positive, the second for negative deformations. $E_{S,i}$ is the energy absorbed during the primary half-cycle, which is the loading curve leading to a new maximum deformation, E_i the energy absorbed in the following half-cycles (the maximum amplitudes of which do not exceed the maximum deformation reached by the primary half-cycle) and E_u^+, E_u^- are the normalizing factors for positive and negative deformations.

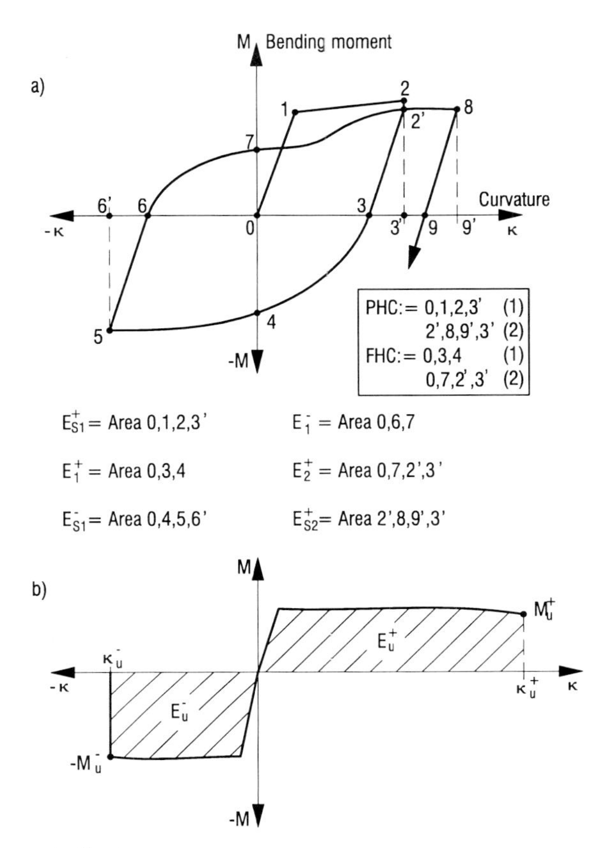

Fig. 8.1-4: D_Q damage indicator

A well-known damage indicator consisting of both a deformation and an energy term has been given by PARK and ANG [8.7]; for the section level it is given by:

$$D = \frac{\theta_{max}}{\theta_u} + \frac{\beta}{M_y \theta_u} \int dE \tag{8. 1. 6}$$

Here θ_u is the ultimate rotation of the section, M_y the section yield moment, θ_{max} the maximum rotation attained during the loading history and β is an empirical coefficient with values between 0.05 and 0.25. Damage levels of $D \approx 0 - 0{,}30$ are typical for no or slight damage, $D \approx 0{,}40 - 1{,}0$ means heavy damage up to collapse.

For high-rise buildings, damage indicators at the story level are most suitable. These can be maximum or cumulative ductilities based on deformation parameters or energy-based indicators. The independent monitoring of the story drifts and the story shears is an additional effective tool for checking the overall structural behaviour.

The non-linear behaviour of the building already introduced in section 7.8 will now be checked as an example. Fig. 8.1-5 shows the finite element idealisation employed.

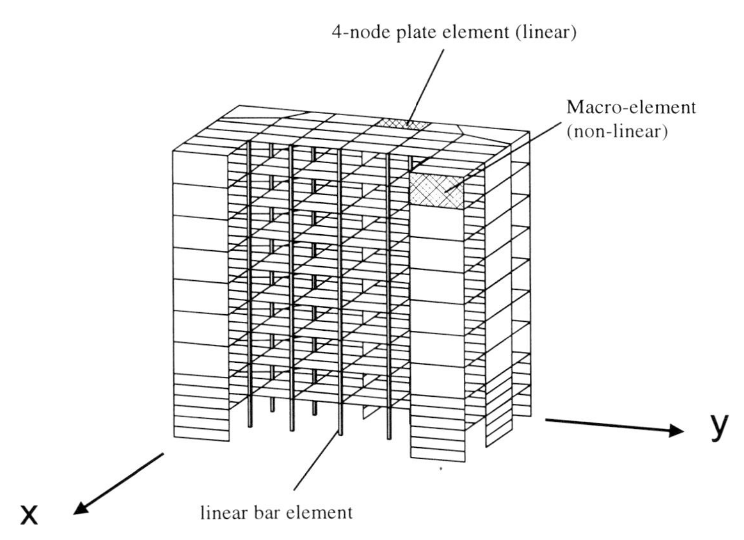

Fig. 8.1-5: Finite element model of the building

The building floors were discretised by standard rectangular plate elements, the columns between floors by bar elements and the shear walls by suitable macro elements. The latter were basically of the element type proposed by LINDE [8.8]. A rigid foundation was assumed throughout.

The macro-model utilised for discretising the shear walls consists of three vertical springs which account for the bending and axial deformation behaviour and of a horizontal spring to account for the shear deformation. The central vertical spring stays linear, but all other springs are considered to have bi- or trilinear hysteretic laws. Fig. 8.1-6 shows the macro-element and the hysteretic laws of its springs.

In [8.13] it is shown that this macro-element can reproduce the non-linear behaviour of seismically excited wall-stiffened buildings with sufficient accuracy as compared to the discretisation with non-linear plate elements. Added advantages of the macro-model are its simplicity and the ease of evaluating quantities such as ductilities and the section-level PARK/ANG damage indicator. The more accurate micro-model which uses non-linear layered plate elements is much more difficult to apply and requires large amounts of CPU time; its advantage lies in being able to describe damage right down to the material point level and thus explain the initiation of damage accumulation phenomena.

The input data needed for the calibration of macro-elements (spring stiffness values, limit spring deformations etc.) are evaluated prior to the actual non-linear time-history analysis by special computer programs based on layer models of the shear wall sections [8.13]. These programs evaluate also the parameters needed for the PARK/ANG damage indicator and the moment-axial force interaction diagrams for the wall section which are required later for the actual section design.

The damage indicators evaluated with the macro-model idealisation are story drifts, rotation and curvature ductilities and the PARK/ANG indicator given by Eq. (8.1.6) [8.9]. An important step is the evaluation of the ultimate curvatures of the wall sections; they are computed separately by means of layer models as mentioned above. The empirical factor β has been chosen to be 0.10; recent investigations [8.10] have shown that the significance of the energy term of the PARK/ANG indicator is not very high for well-designed structures. It must be remarked, however, that well-designed structures are mostly of recent origin and that this term may be quite important for in the context of investigating the seismic vulnerability of older buildings.

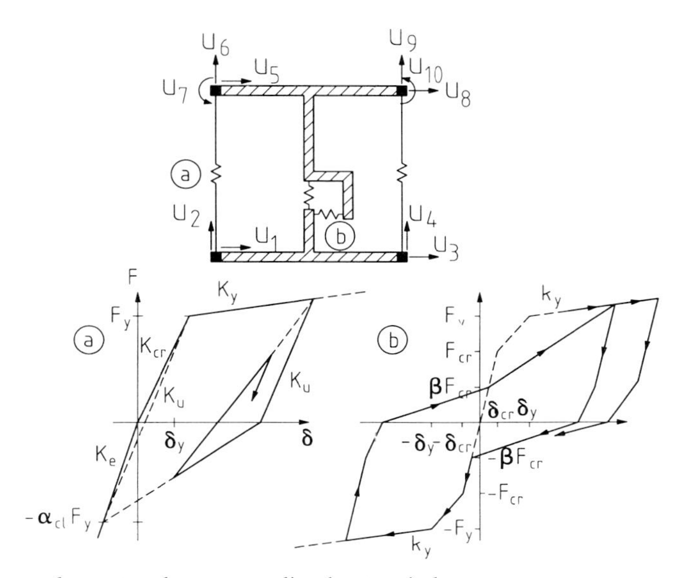

Fig. 8.1-6: Macro-element and corresponding hysteretic laws

The non-linear time-history investigations of the building were conducted for the measured ground motion records of the Kalamata earthquake of 1986 and the JMA record of the 1995 Kobe earthquake. Both horizontal excitation components were applied, with the N-S components acting along the Y axis and the E-W components along the X axis of the structure. Additionally, a synthetic target-compatible accelerogram was generated by the program SYNTH, matching the design spectrum of the EC 8. The synthetic accelerogram and the Kalamata records were scaled to a peak ground acceleration of 3.0 m/s^2 whereas the JMA record was used as recorded. Fig. 8.1-7 depicts a component of the JMA record together with a component of the Kalamata event scaled to 3.0 m/s^2 peak acceleration; its original PGA value was 2.35 m/s^2 .

The permanent loads were considered as static loads during the time-history analysis; as stipulated in the EC 8, no vertical ground motion component was taken into account.

In a first step, the results of a linear time-history analysis of the "exact" model (taking into account the true stiffness values of the floors) were compared to the results of a pseudo-spatial idealisation which requires only three active kinematic degrees of freedom per floor by assuming floors to be rigid in their own plane and disregarding vertical compatibility.

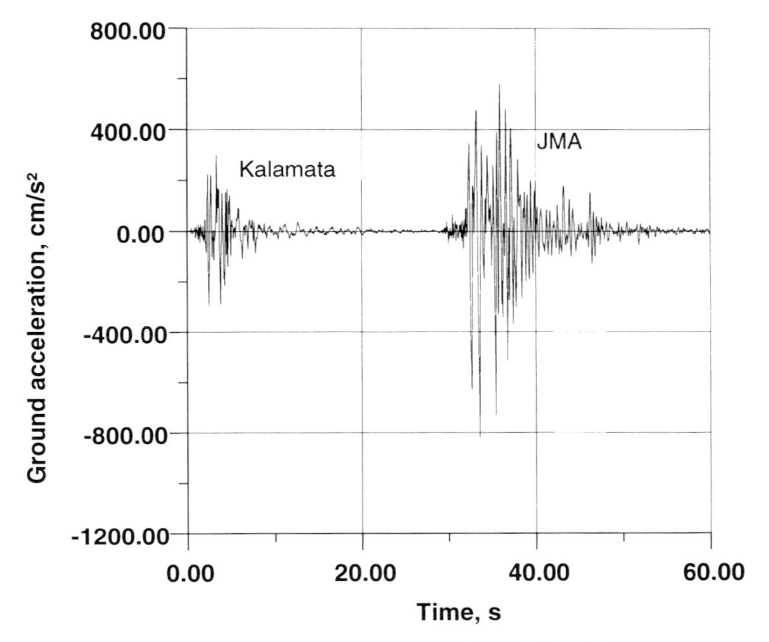

Fig. 8.1-7: Comparison of a JMA-Kobe component and a scaled Kalamata component

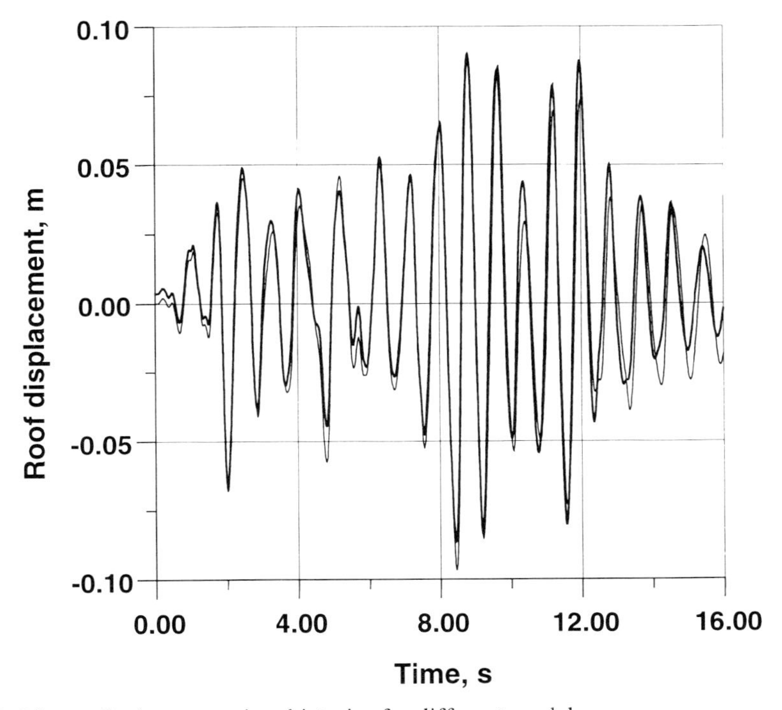

Fig. 8.1-8: Linear displacement time histories for different models

Fig. 8.1-8 shows the resulting roof displacement time histories in the Y direction for the synthetic accelerogram as excitation; the initial value present at t=0 for the exact model (bold curve) is due to the action of the permanent loads. The spatial finite element model can take into account the full uncracked stiffness values of the reinforced concrete sections or consider the reduced cracked section stiffness values. The following table compares the resulting values for the fundamental period of the structure:

Idealisation	T [s]
Pseudo-spatial	0.709
Spatial with full stiffness (uncracked)	0.640
Spatial with reduced stiffness (cracked)	0.840

The reduced bending stiffness was set to 60 % of its uncracked value. As can be seen from the fundamental mode shape shown in Fig. 8.1-9, torsional components are present due to the lack of symmetry. Figs. 8.1-10 and 8.1-11 depict two further mode shapes of the spatial idealisation (reduced stiffness), the natural periods of which are $T_2 = 0.83$ s and $T_3 = 0.61$ s.

Fig. 8.1-12 shows the time history of the bending moment at the encastré section of wall no. 7 due to the scaled Kalamata record, this being the most severely affected section of the building. The slight exceedance of the design bending moment of $1.0 \cdot 10^5$ kNm is as expected, and the damage values (D = 0.318 for the PARK/ANG indicator, $\mu_\kappa = 2.318$, $\mu_{cu} = 4.515$) that were actually attained pose no safety problems.

The non-linear time history analyses yielded for the single damage indicators the values summarised in the following table. The response spectra of both the scaled Kalamata records and the synthetic accelerogram are similar to the EC 8 design spectrum, while the JMA record is a much stronger excitation which can lead to structural failure. This is indeed supported by the results of the time history analyses, which yield damage values D > 1 for this seismic record.

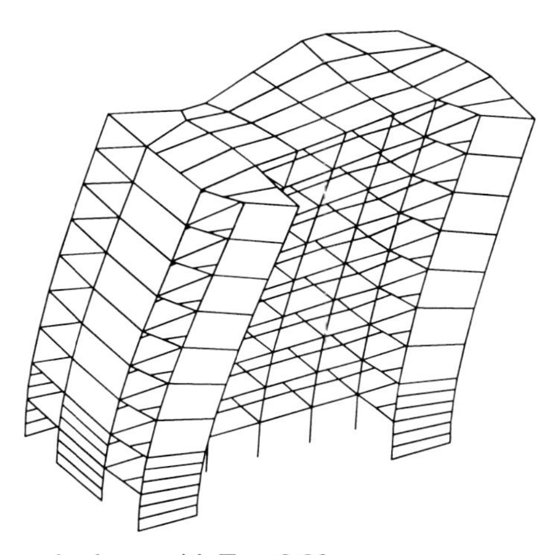

Fig. 8.1-9: Fundamental mode shape with $T_1 = 0.83$ s

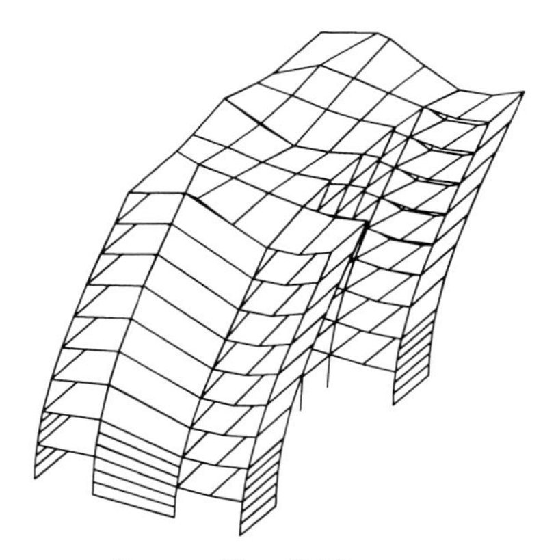

Fig. 8.1-10: Mode shape corresponding to $T_2 = 0.85$ s

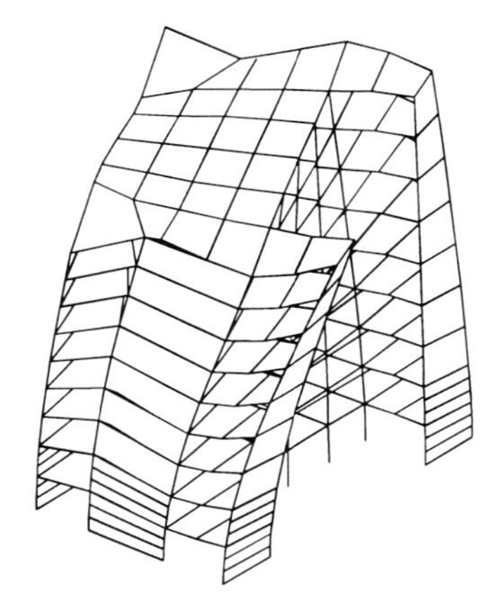

Fig. 8.1-11: Mode shape corresponding to $T_3 = 0.61$ s

Wall No.		1	2	3	4	5	6	7
Record 1	$\mu_{\theta,max}$	2.07	2.17	< 1.0	1.03	1.55	1.31	1.69
(synthetic)	D	0.183	0.190	0.088	0.092	0.238	0.178	0.269
Record 2	$\mu_{\theta,max}$	3.04	3.25	2.77	2.63	2.69	2.43	2.61
(Kalamata)	D	0.294	0.301	0.257	0.254	0.376	0.342	0.418
Record 2	$\mu_{\theta,max}$	3.04	3.25	2.77	2.63	2.69	2.66	3.10
(pre-damaged)	D	0.338	0.346	0.281	0.276	0.399	0.413	0.586
Record 2	$\mu_{\theta,max}$	3.23	3.34	2.59	2.61	2.36	2.47	2.55
(Wall 5 damage)	D	0.310	0.315	0.255	0.252	0.481	0.334	0.399
Record 3	$\mu_{\theta,max}$	7.01	7.46	7.29	7.06	6.57	6.27	8.25
(JMA Kobe)	D	0.849	0.869	0.786	0.788	1.081	1.161	1.802

The table contains results of two further investigations in rows 3 and 4. The first one (row 3) deals with subjecting the structure which has been already damaged by the Kalamata earthquake to the same earthquake once more. In the second investigation, the encastré section of wall no. 5 was assumed to be pre-damaged in such a way that its stiffness was reduced to 60% (with its strength being diminished accordingly). Such analyses are quite useful for appraising the impact of weak spots in the structure on the overall safety level.

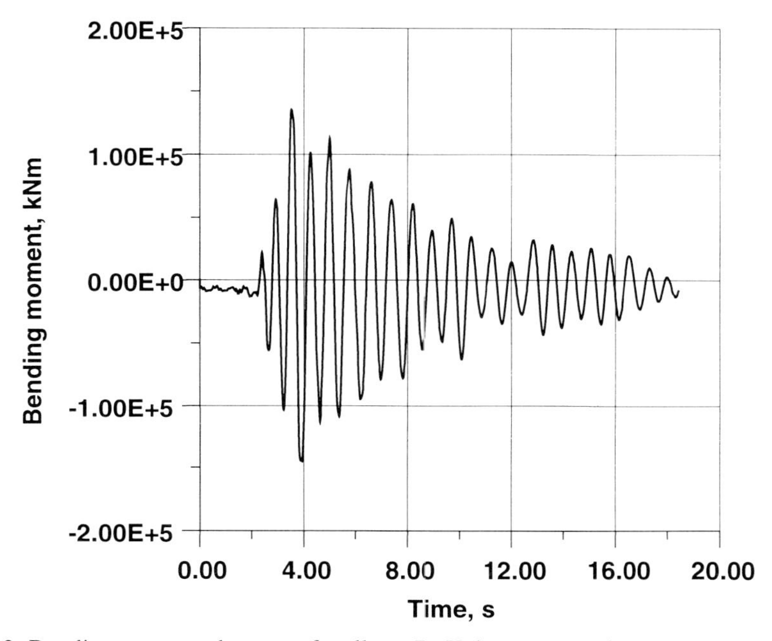

Fig. 8.1-12: Bending moment, bottom of wall no. 7, Kalamata record

Fig. 8.1-13 shows the story drifts in both the X and Y directions for the scaled Kalamata record. The figure presents the values both for the linear and the non-linear models and shows also the (non-reduced) limit value of 0.4% given in EC 8. The JMA-Kobe record induces extremely high drifts (Fig. 8.1-14) which are, in fact, tantamount to collapse.

Fig. 8.1-15 depicts the roof displacement time histories for the structure being subjected sequentially twice to the Kalamata earthquake. A slight damage accumulation occurs at the pre-damaged sections, but it can be seen that the structure as a whole, having been designed in accordance with the EC 8 code provisions, is remarkably resistant to progressive deterioration.

The analysis results show that earthquakes, the spectra of which more or less correspond to the design spectrum, can cause some inelastic action to happen in the structure but are not in a position to seriously impair the overall structural safety. In the case of the much stronger Kobe event, which would have caused structural collapse, a marked variation of the degree of damage throughout the structure was evident. In [8.11] a possible redistribution of the elastically evaluated equivalent static loads of the single shear walls in the range of up to 30% is mentioned. Non-linear time history analyses can be of help in determining a pattern of wall shear forces that leads to a better spatial distribution of plastic actions in the structure and thus to enhanced safety levels.

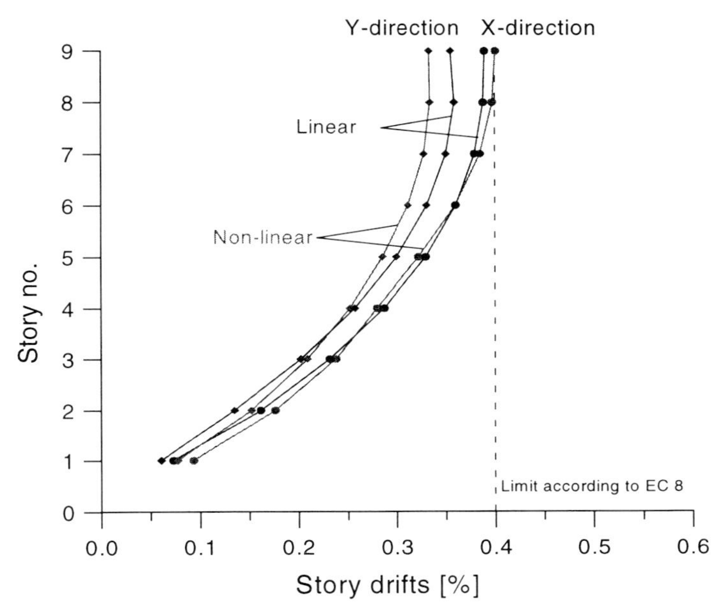

Fig. 8.1-13: Linear and non-linear story drifts for the Kalamata earthquake

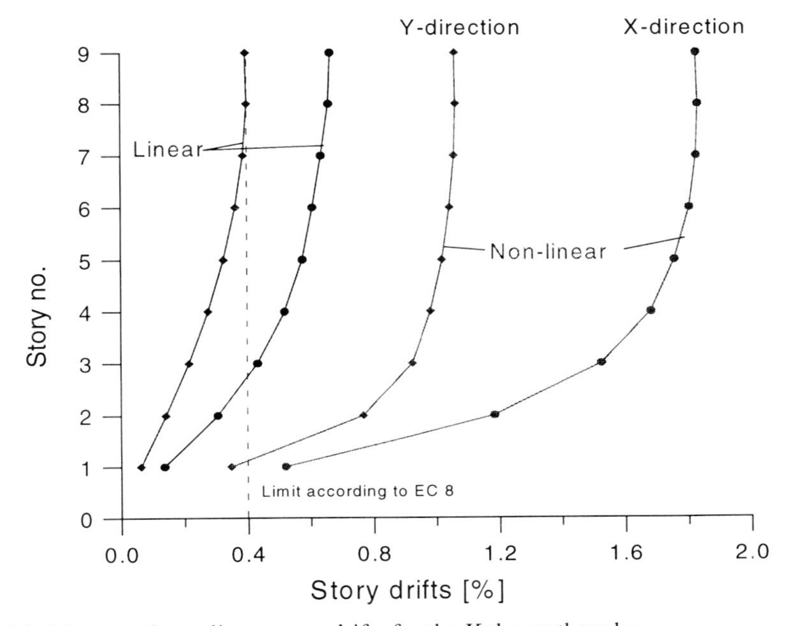

Fig. 8.1-14: Linear and non-linear story drifts for the Kobe earthquake

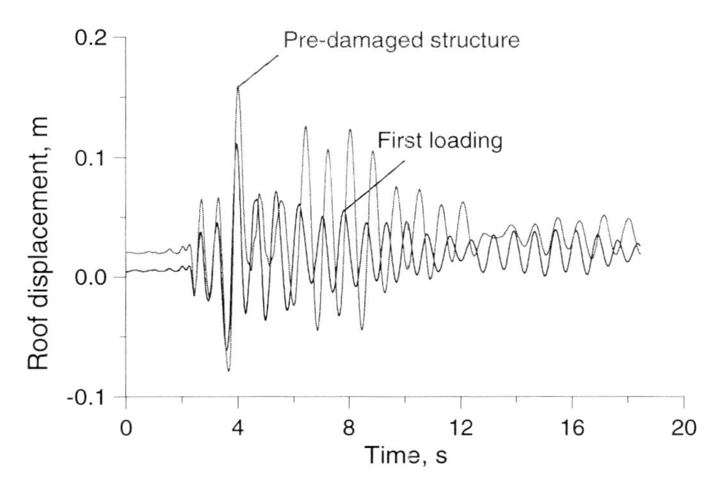

Fig. 8.1-15: Roof displacement time histories, Kalamata earthquake

Another point concerns the relationship between damage indicators at the section level (based on curvature ductility) and (partly or wholly) energy-based indicators of the PARK/ANG type. Figs. 8.1-16, 8.1-17 and 8.1-18 show some results for the encastré sections of the seven shear walls of this example structure. While the correlation between maximum curvature ductility and the PARK/ANG indicator shown in Fig. 8.1-16 is more or less satisfactory, this is certainly not the case for the relationship between the cumulative curvature ductility and the PARK/ANG indicator (Fig. 8.1-17). On the other hand, the cumulative ductility correlates quite satisfactorily with the energy term of the PARK/ANG indicators, as depicted in Fig. 8.1-18.

It must be stressed that these results cannot be extended to damage indicators at the substructure level; they simply point out that more than one damage indicator is needed in order to describe the complex damage evolution process.

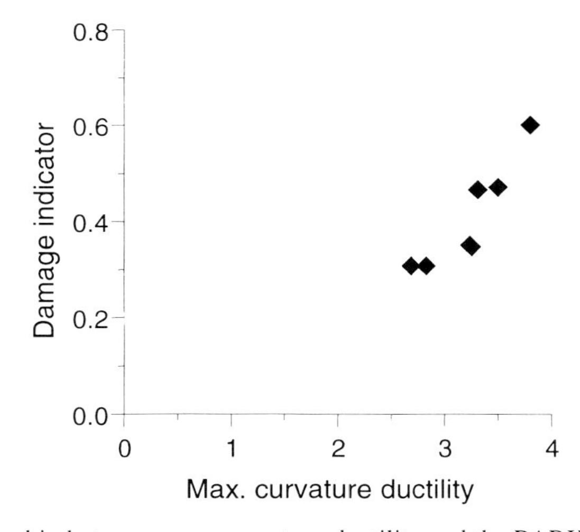

Fig. 8.1-16: Relationship between max. curvature ductility and the PARK/ANG indicator

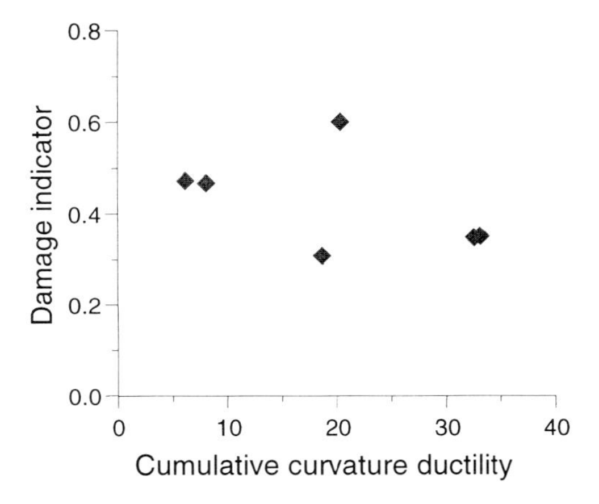

Fig. 8.1-17: Relationship between cumulative curvature ductility and the PARK/ANG indicator

To summarise the results of this section, it was shown that the example structure designed according to EC 8 has exhibited a non-linear behaviour completely compatible with expectations. Such non-linear time history analyses can serve for enhancing seismic safety by pointing the way to a more balanced distribution of plastic actions throughout the structure and discovering weak points, where damage accumulation is liable to take place. They can also be utilised in the context of seismic retrofitting measures for determining which strengthening interventions lead to the best results in terms of structural performance and safety enhancement.

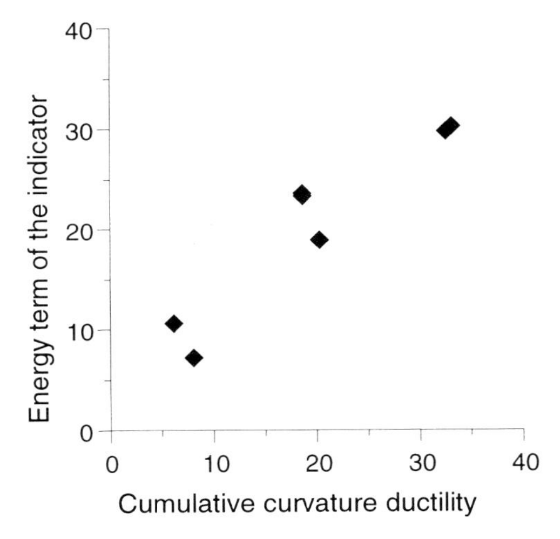

Fig. 8.1-18: Relationship between cumulative curvature ductility and the energy term of the PARK/ANG indicator

8.2 Seismic investigation of the towers of Cologne cathedral

On the occasion of the 750th anniversary (in 1998) of the laying of the foundation stone of the Cologne cathedral, the Chair of Structural Statics and Dynamics of the Aachen Technical University and the Bensberg Seismic Observatory of the Cologne University (Head: Dr. Klaus-Günter Hinzen) have jointly carried out an investigation in order to determine whether the 157.3 m high towers of the cathedral are able to withstand possible future earthquakes. To this effect, suitable accelerograms were produced as described in section 7.4 and used as input for time history analyses of a complex finite element model of the cathedral's two towers. The accuracy of the finite element model was checked against measured values of the natural frequencies of the towers which were determined as described in section 6.2. Here some further aspects of this investigation are presented, concerning the finite element model of the towers and the resulting system displacements.

Fig. 8.2-1: Finite element model of the Cologne cathedral towers

Standard plate/shell and three-dimensional beam/column elements were used for the discretisation of the structure, the former being used for modelling the foundation slabs and the arches connecting the columns. In all, 3955 degrees of freedom were introduced, with 48 plate/shell and 958 three-dimensional beam-column elements. Fig. 8.2-1 depicts the finite element model which was set up and analysed by the computer program FEMAS90. Two different types of rock material were employed in the towers of the Cologne cathedral, namely Oberkirchener sandstone weighing 2.2 kg/dm^3 with a YOUNG's modulus of

$1.9 \cdot 10^4\ MN/m^2$ and also Drachenfelstrachyt weighing 2.3 kg/dm^3 with a YOUNG's modulus of $4 \cdot 10^4\ MN/m^2$. A detailed modelling of the nave of the cathedral was not attempted in view of the many uncertainties concerning the mechanical properties of its components; in some discretisation variants, an approximate consideration of its influence was carried out by introducing spring supports restraining the motion of the towers in the EW direction. The corresponding results showed that this influence was not very pronounced because of the large stiffness of the lower parts of the towers. These springs were therefore omitted and only the two towers considered in the final analyses, an additional argument being the better transparency and reproducibility of the results.

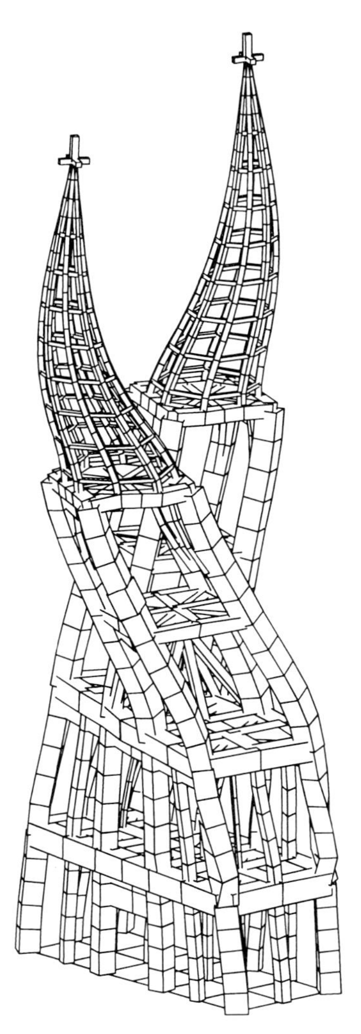

Fig. 8.2-2: Mode shape for T = 1.3 s

The section properties (areas and moments of inertia) of the beams were determined numerically by means of the computer program AREMOM (see section 2.4) because of their

complicated contours. The masses corresponding to the single degrees of freedom were then determined by hand; the total mass of a single tower amounts to approximately 23 500 t.
Solving the eigenproblem for the finite element model of the two towers shown in Fig. 8.2-1 yielded natural frequencies at 0.60, 0.74, 0.80 and 1.02 Hz, the corresponding periods being 1.67, 1.35, 1.25 and 0.98 s. The measurements conducted at a height of approx. 100 m in the southern tower yielded naturalfrequencies between 0.74 and 1.10 Hz (Fig. 6.2-3). This agrees quite well with the numerical results, because the mode shape corresponding to the natural frequency of 0.60 Hz mostly concerns the northern tower. Two calculated mode shapes are shown in Figs. 8.2-2 and 8.2-3.

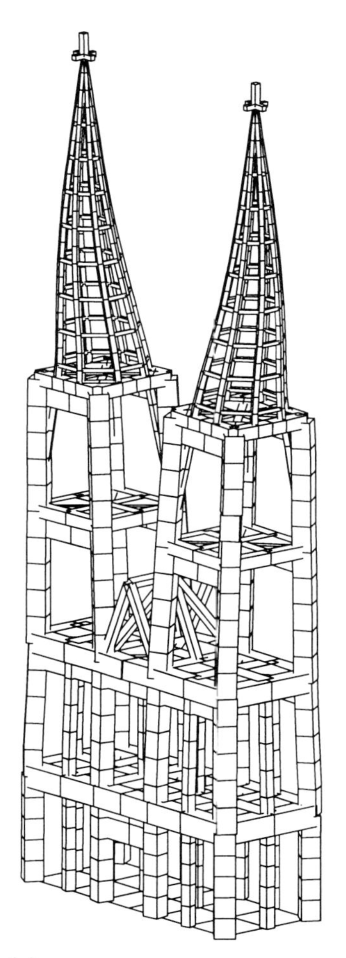

Fig. 8.2-3: Mode shape for T = 1.0 s

The time history analyses were conducted both for medium intensity seismic events such as the 1992 Roermond earthquake (adjusted to the cathedral site) and for the maximum credible accelerogram (TOL60) with an underlying moment magnitude of 6.0; the determination of the latter has been described in section 7.4 (Fig. 7.4-8). The following table summarises the maximum calculated horizontal displacements for the two towers at different heights:

Level	North tower		South tower	
	$max\ u_{N\text{-}S}$	$max\ u_{O\text{-}W}$	$max\ u_{N\text{-}S}$	$max\ u_{O\text{-}W}$
150 m	4.0 cm	7.0 cm	3.7 cm	4.0 cm
95 m	1.7 cm	5.8 cm	1.5 cm	2.5 cm

The results of this analysis show that the two towers are in a position to withstand the maximum credible seismic event without suffering significant damage. The maximum horizontal displacement at the 150 m level of the north tower is only 0.07 m or 0.047 % drift. These displacements are about three times as large as the values calculated for the 1992 Roermond earthquake (the epicentre of which lay 70 km away) adjusted to the Cathedral site; it is known that only very slight damage occurred at that time. A simple explanation of the high seismic resistance of the towers lies in the position of the fundamental natural period in respect to the peaks of the spectra shown in Fig. 8.2-4. It is obvious that the acceleration spectra of the TOL60 and the Roermond records (the latter adjusted to the cathedral site), obtain their maximum values for periods about 0.2 s, a safe distance away from the structure's lowest natural periods.

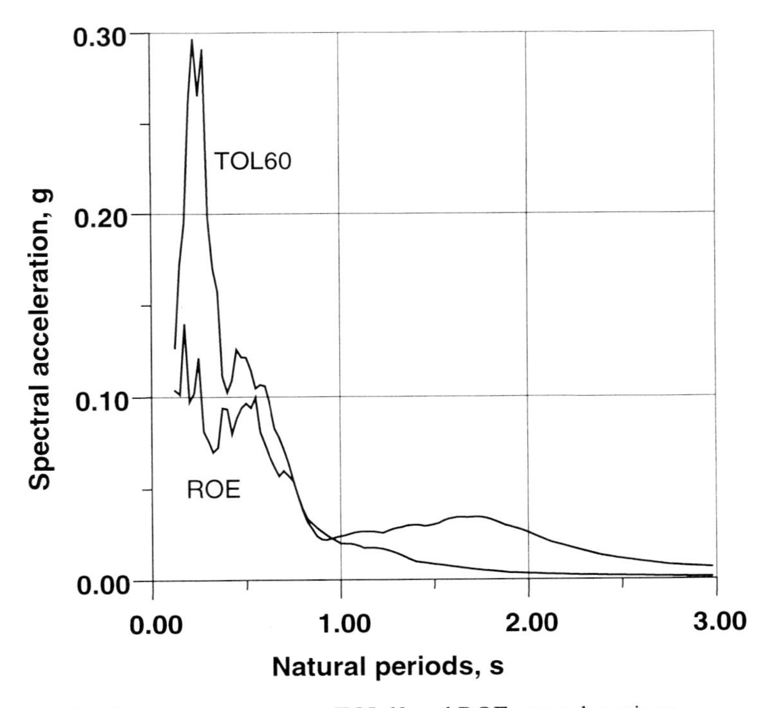

Fig. 8.2-4: Acceleration response spectra, TOL60 and ROE ground motions

8.3 Natural frequencies and mode shapes of a refinery vessel

In this section, the natural frequencies and mode shapes of a refinery vessel are computed both by means of a detailed spatial finite element model and by means of simple stick-model idealisations, and the results compared. This work was conducted as part of a diploma thesis [8.12].

The refinery vessel in question is an approximately 30 m high steel cylinder bolted to a reinforced concrete platform with a height of approximately 10 m. The cylindrical vessel itself contains only very little crude oil, the mass of which may be neglected in the computation. This also holds for the various pipes connecting the subassemblage to other components.

The finite element analysis was carried out by the computer program FEMAS90 on a HP9000-J210 workstation. The structure was discretised using a rather fine mesh of plate/shell elements as well as 3D beams elements for the platform supports. A total of 29346 degrees of freedom were introduced, which required about 20 MB core storage for the system stiffness matrix. The finite element mesh used is shown in Fig. 8.3-1; consistent mass matrices were used for the description of the system's mass properties.

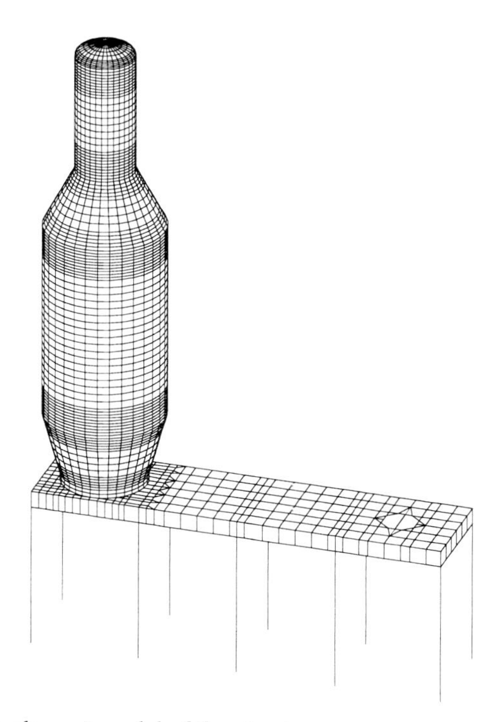

Fig. 8.3-1: Spatial finite element model of the structure

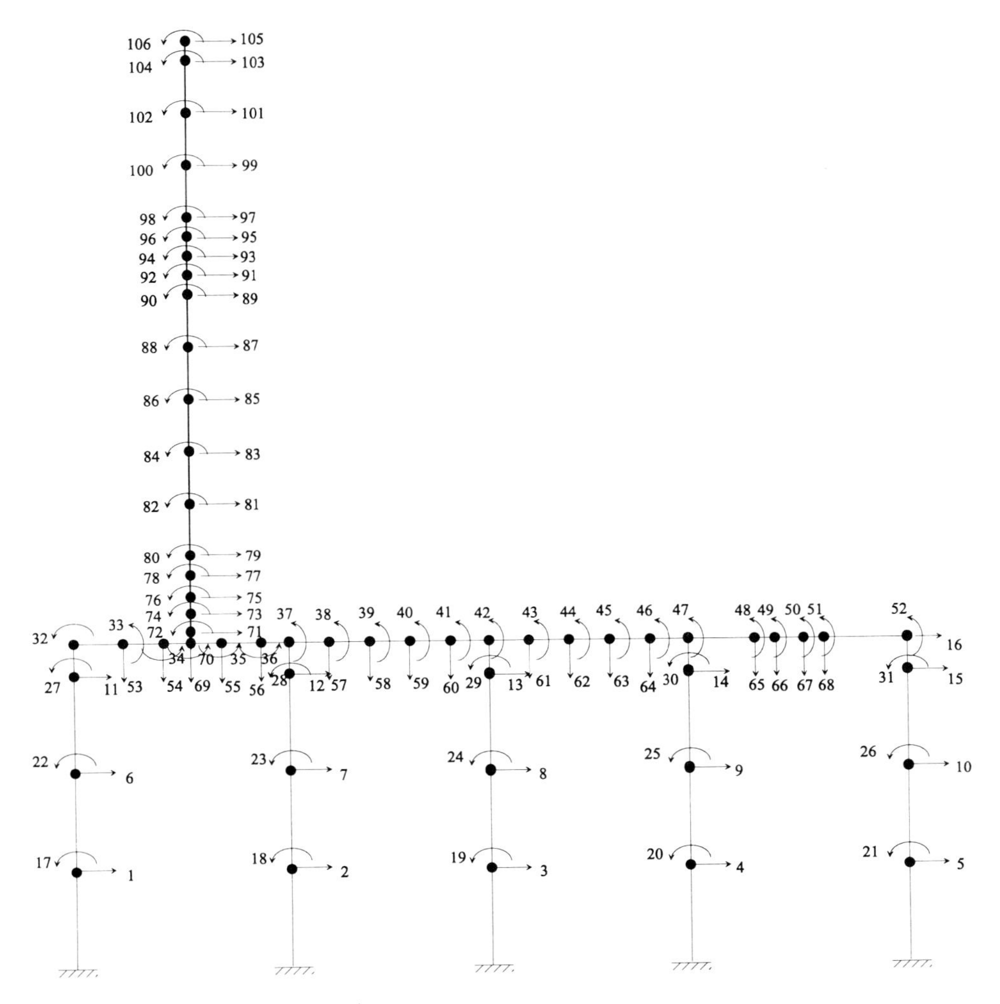

Fig. 8.3-2: Plane structural model

The simplified plane model of the structure consists of massless beam elements with constant stiffness properties along their length and discrete point masses. In order to keep the overall number of degrees of freedom small, inextensional members are modelled by introducing the same degree of freedom at both element end sections (see for example degree-of-freedom no. 16 for the horizontal displacement of the table or degree-of-freedom no. 69 for the vertical displacement of the tower as shown in Fig. 8.3-2). The plane model of Fig. 8.3-2 consists of 59 beam elements with a total number of 106 active kinematic degrees of freedom, these being also the master degrees of freedom. It is of course much easier to handle than the detailed spatial finite element model; on the other hand, the effort needed for computing the stiffness values of the single beam elements should not be underestimated. The mass matrix of the simplified plane model system is a diagonal matrix which contains the masses corresponding to these 106 degrees of freedom.

The calculation of the natural frequencies and mode shapes of the simplified model is carried out by the JACOBI computer program. The fundamental mode shape is dominated by a translation of the table with the vessel simply "riding along", while the second mode shape corresponds to the standard form of a cantilever beam. These important mode shapes can also be identified in the results of the detailed spatial finite element model; they are shown in Fig. 8.3-3.

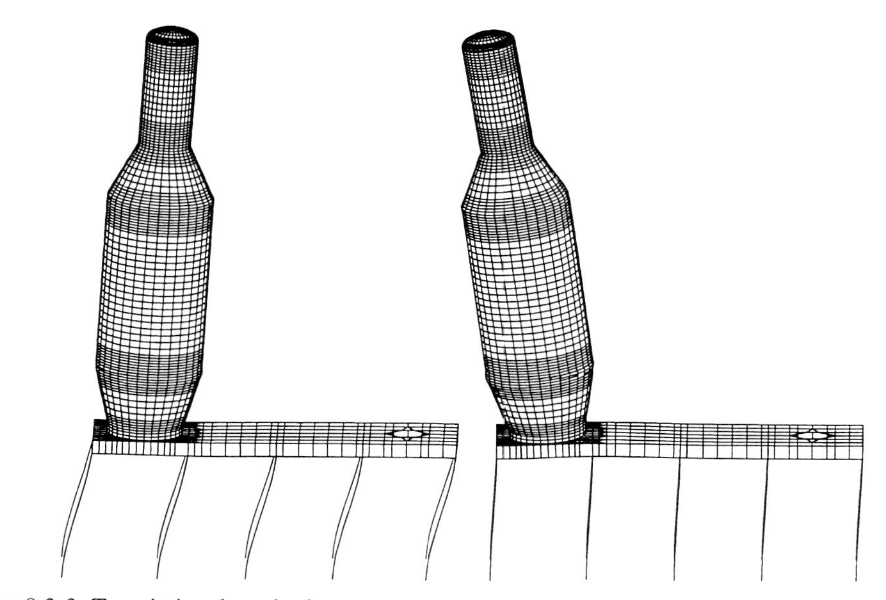

Fig. 8.3-3: Translational mode shapes

A comparison of the natural frequencies calculated with the simplified plane model and the spatial finite element model is given in the following table for these two translational modes. The degree of accuracy of the simple model is quite satisfactory.

Mode No.	**Natural frequency [Hz] spatial model**	**Natural frequency [Hz] plane model**
1	2.46	2.11
2	4.78	4.05

One reason for the differences is that with a small number of plane beam elements with constant stiffness it is not possible to model the stiffness properties of a complex structure with numerous changes in geometry with great accuracy. Furthermore, the beam theory yields only an approximation for the load-carrying behaviour of a cylindrical shell.

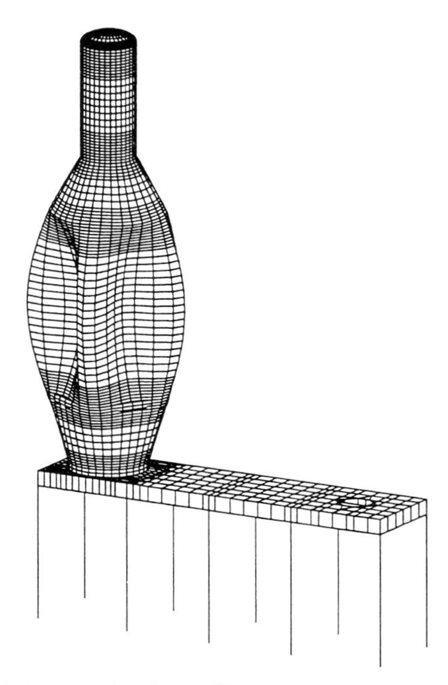

Fig. 8.3-4: Mode shape of the vessel at f = 11 Hz

Of course, the detailed spatial finite element model is also capable of yielding mode shapes which correspond to buckling-type deformations of the cylindrical vessel. These mode shapes clearly cannot be identified by means of the plane stick model; however, their natural frequencies are quite high so that their importance for seismic analyses is not very pronounced. Fig 8.3-4 shows the first of these mode shapes; the natural frequency of the mode shape depicted is approximately 11 Hz.

9 Fluid containers subject to seismic loading

Structures which are in contact with fluids exhibit a markedly different behaviour under seismic loads from other structures. The seismically induced motion of the structure causes dynamic pressure in the fluid, which in turn acts upon the structure and modifies its response. We are dealing therefore with a complex „fluid-structure interaction problem“, which arises not only for fluid containers but also for large water reservoirs and their containing structures. The underlying mathematical theory is quite involved, and since no closed-form solutions are generally available, numerical methods are usually employed, which make use of a series of simplifying assumptions in order to reduced the overall effort for determining design forces and deformations with engineering accuracy. This section presents some useful formulas for standard fluid containers.

9.1 General

During the horizontal seismic excitation of a tank containing fluid, the inertia of the liquid exerts a horizontal force (or pressure per unit area) on the tank walls, the so-called impulsive pressure p_i. Additionally, the prior level water surface is disturbed, leading to "sloshing", which causes an additional "convective" pressure p_s on the tank walls, p_s being independent of p_i. Fig. 9.1-1 depicts qualitatively the hydrodynamic pressure distributions acting on the tank wall, with the impulsive pressure due to the inertia forces being further divided into two components. These are the pressure p_{i1} acting on a rigid tank wall and the additional pressure p_{i2} due to the flexibility of the tank wall. Clearly, the latter part diminishes in proportion to the stiffness of the tank walls.

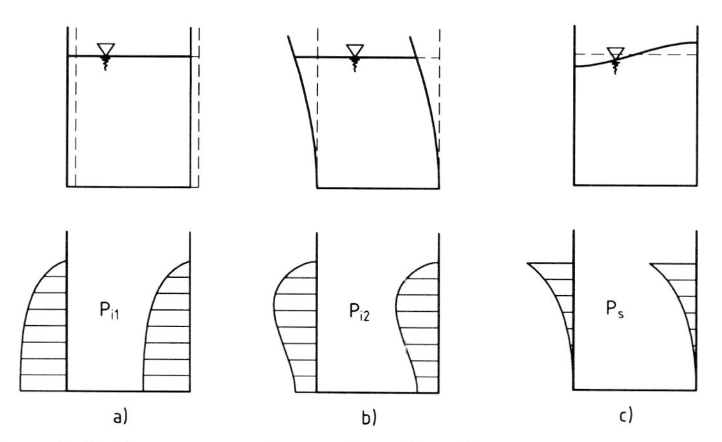

Fig. 9.1-1: Dynamic fluid pressures acting on the tank wall

In the past, most seismically excited liquid storage tanks were analysed by means of the well-known simplified HOUSNER approach [9.1], which considers only the hydrodynamic pressure on a rigid wall and neglects p_{i2} as shown in Fig. 9.1-1b. In this case, the impulsive pressure is independent of the deformation of the tank wall and the corresponding force is simply proportional to the ground acceleration $\ddot{u}_B$. Since some tanks designed according to this concept suffered serious damage under seismic loading, it was determined that the actual hydrodynamic pressure was larger than the one computed according to HOUSNER without considering the tank wall flexibility. Nevertheless, the classical HOUSNER approach is still a part of some codes of practice [9.2], which consider the tank wall flexibility by means of additional corrective factors.

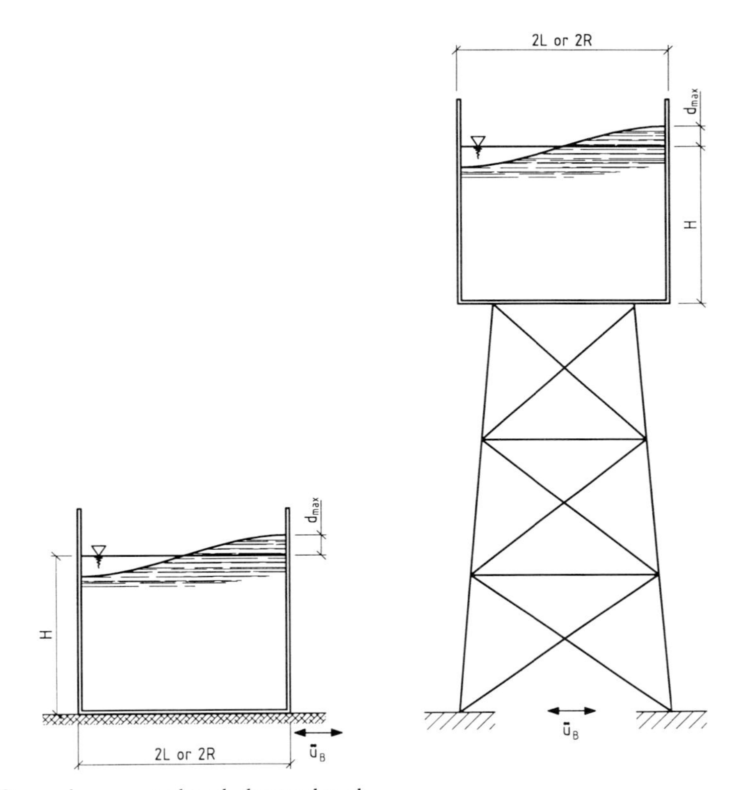

Fig. 9.2-1: Ground-supported and elevated tank

The most important methods for considering the influence of the wall flexibility are the finite element method and the Assumed Mode method. In the latter, deformation modes corresponding to the degrees of freedom of the tank are assumed and the mass of the liquid is taken into account as an added mass [9.3, 9.4, 9.5]. While the simplest model idealises the tank as a cantilever beam, finite element models are usually employed if a greater degree of accuracy is desired. Finite elements can also be used for modelling the liquid inside the tank [9.6][9.7]. This approach is not pursued further in this context, apart from some remarks in section 9.3.

9.2 HOUSNER's approximate method

9.2.1 Assumptions

The main assumptions inherent in the approximate method of HOUSNER are the following:

- Tanks with constant rectangular or circular sections and flat bottoms.
- Purely horizontal seismic excitation.
- Rigid tank walls.

Fig. 9.2-1 depicts two such tanks subject to a horizontal ground acceleration $\ddot{u}_B$.

Both a tank supported on the ground and an elevated tank are shown. In both cases we are interested in the maximum vertical wave displacement d_{max} and the resulting seismic forces.

As already mentioned, the horizontal ground acceleration induces both an impulsive force due to the inertia of the liquid that opposes the movement of the tank wall and a convective force due to sloshing. The liquid mass is accordingly split into two parts, the impulsive mass m_0, which is considered firmly attached to the tank wall, and the convective mass m_1, which is connected to the tank wall by means of longitudinal springs and can thus move relative to the wall (Fig. 9.2-2).

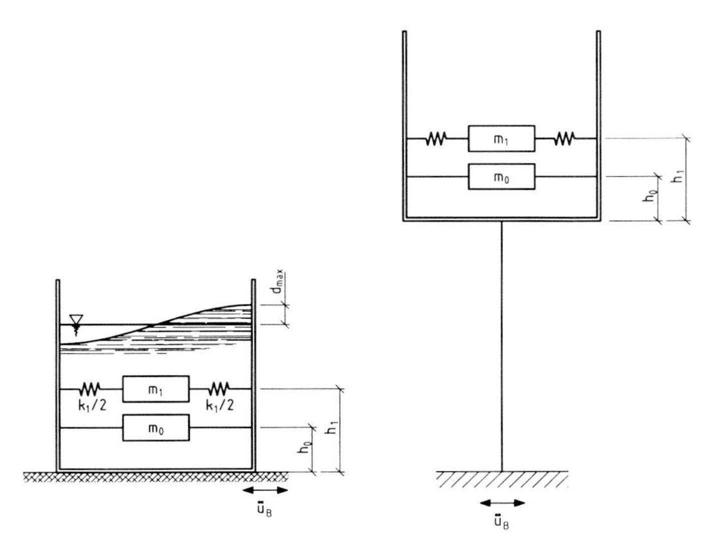

Fig. 9.2-2: Dividing the liquid mass in m_0 and m_1

The masses m_0 and m_1 are acted upon by the accelerations $\ddot{u}_0$ and $\ddot{u}_1$ and yield equivalent seismic forces P_0 , P_1 . Multiplied by h_0 and h_1, these forces cause moments M_0 and M_1 on the tank bottom, as seen in Fig. 9.2-3. The masses m_0 and m_1 must be determined in such a way that the resulting stresses are similar to the true ones caused by the actual liquid in the seismically excited container.

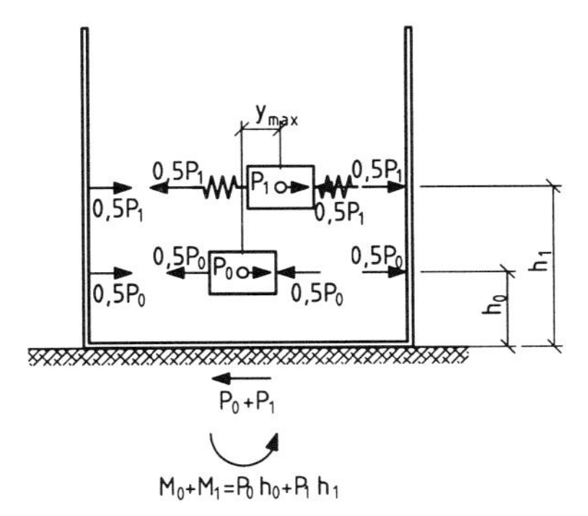

Fig. 9.2-3: Equilibrium equations for a ground-supported tank

In the case of a ground-supported tank with rigid walls, both the tank and the ground experience the same absolute displacement and acceleration, so that the maximum impulsive equivalent force P_0 of the mass m_0 is proportional to the peak acceleration of the tank.

The convective mass m_1 can move relatively to the tank wall, but since the latter is rigid, the maximum horizontal displacement y_{max} of the mass relatively to the wall is also proportional to the maximum convective equivalent force P_1 and the vertical wave height d_{max}. This implies that ground-supported tanks may be considered as single-degree-of-freedom systems.

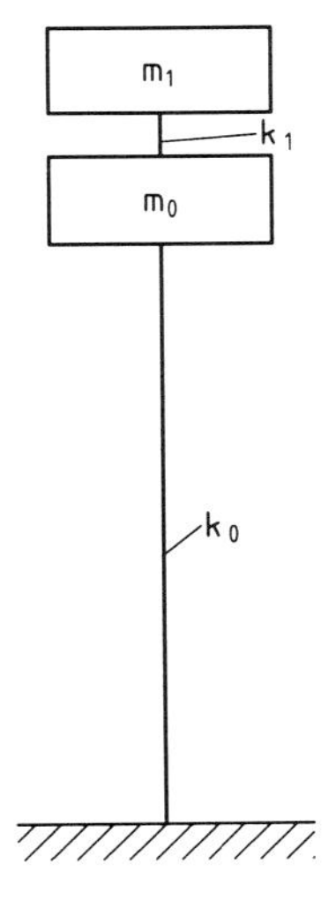

Fig. 9.2-4: Model for an elevated tank

In the case of elevated tanks it is not permitted to apply the ground acceleration directly to the impulsive mass m_0 of the liquid tank content. The flexibility of the supporting structure must be considered, which leads to a two-degree-of-freedom system as a suitable model (Fig. 9.2-4).

In the following sections some examples are given for the computation of liquid storage tanks using the formulas based on the HOUSNER approach as given in the Codes [9.8] and [9.9].

The site seismicity has been characterised by the acceleration spectrum given by HOUSNER in [9.8] and shown in Fig. 9.2-5. It is normalized to a unit ground acceleration, so that its ordinates must be multiplied by the corresponding site acceleration. As an example, the U.S. Coast and Geodetic Survey Seismic Probability Map of the United States [9.10] supplies a value of $\ddot{u}_B = 0{,}33 \cdot g$ for California.

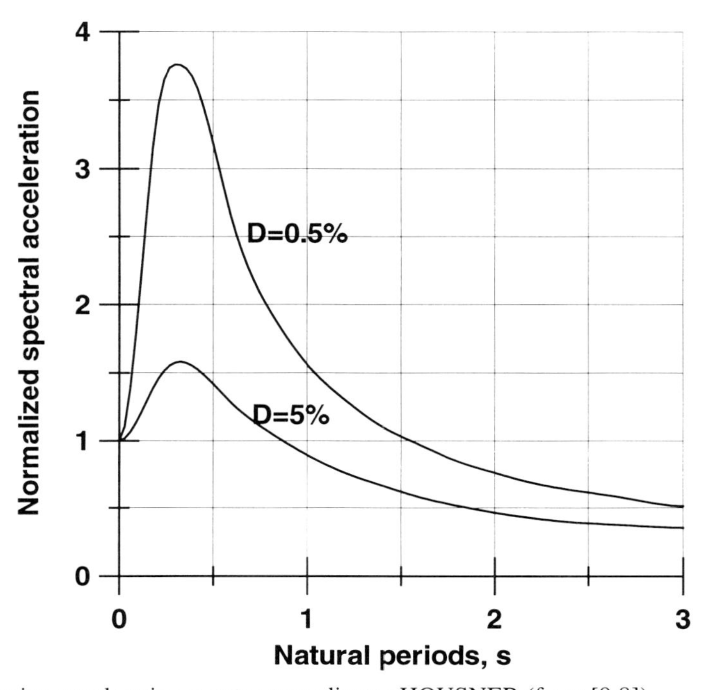

Fig. 9.2-5: Design acceleration spectra according to HOUSNER (from [9.8])

The following assumptions and remarks must be observed:

- The equivalent forces P_0 and P_1 with the corresponding heights h_0 and h_1 furnish the bending moments for the tank wall and the overturning moment for the tank bottom. For each one of the two heights h_0 and h_1 we must consider two values (h_0^g, h_0^k and h_1^g, h_1^k), where the values with the upper index g already contain the influence of the bottom pressure, in contrast to the ones with the upper index k. The smaller values (with the upper index k) serve for computing the bending moments directly above the tank bottom, whereas the values with the upper index g are used for computing the overturning moment (for which consideration of the bottom pressure is essential).
- The mass of the tank, including the mass of the support structure in the case of elevated tanks, is added to the impulsive mass m_0. The distance h_0 must be computed while considering the different positions of the gravity centres for the mass of the liquid and the mass of the structure.
- In the process of evaluating the convective and the impulsive masses, one must distinguish between slender tanks (with large height-to-diameter ratios) and squat tanks (see section 9.2.2).

9.2.2 Formulas for ground-supported tanks

A ground-supported rigid tank experiences the same absolute displacements as the ground. The sum of the tank mass m_T [t] and the impulsive mass m_0 [t] of the liquid together with the peak ground acceleration $\ddot{u}_B\,[m/s^2]$ furnish a force $P_0 = \ddot{u}_B \cdot (m_0 + m_T)$ acting on the tank wall. The evaluation of the impulsive mass m_0 depends on the tank geometry, since for a slender tank the percentage of the liquid contributing to the impulsive mass is larger than for a squat tank, the latter assumed to be the case if H/L is less than 1.5. Only the upper part of the liquid with a height of $\overline{H} = 1.5 \cdot R$ or $1.5 \cdot L$ is divided into an inpulsive and a convective part; the remaining liquid volume with a height of $H - \overline{H}$ is simply added to the impulsive mass.

The convective liquid mass m_1 is connected to the tank wall by means of a spring with a spring constant k_1. The force P_1 acting on the tank wall is determined from the maximum horizontal displacement y_{max} [m] (Fig. 9.2-3). The differences between a squat and a slender tank are further clarified in Fig. 9.2-6.

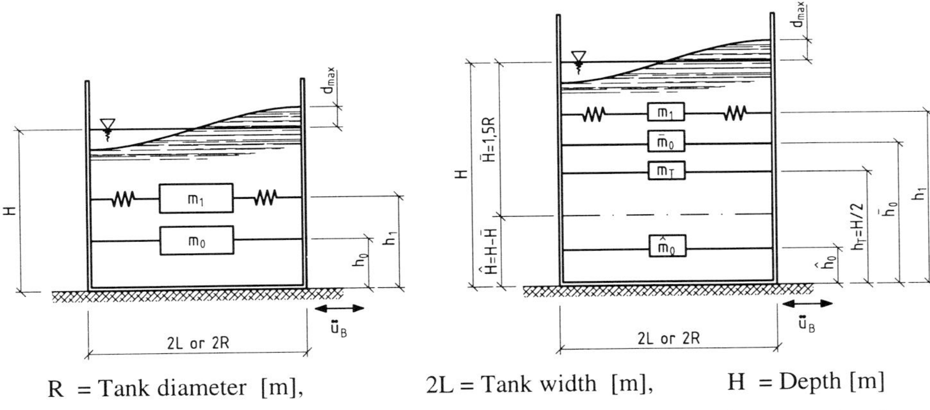

R = Tank diameter [m], 2L = Tank width [m], H = Depth [m]

Fig. 9.2-6: Impulsive and convective masses for squat and slender ground-supported tanks

The general approach is as follows:

- Determine the horizontal peak ground acceleration $\ddot{u}_B$.
- Evaluate the impulsive values m_0, h_0 and P_0 as well as the bending moment above the tank bottom and the overturning moment.
- Compute the fundamental natural frequency ω and corresponding period T of the liquid.
- Obtain the spectral acceleration ordinate S_a as a function of T and an assumed damping ratio for the fluid (about 0.5%).
- Compute the horizontal displacement y_{max} and the angle θ_h for the oscillation of the water surface.
- Compute the convective values m_1, h_1 and P_1 as well as the bending moment above the tank bottom and the overturning moment.
- Add the impulsive and convective contributions together.

9.2.2.1 Squat ground-supported tanks

The table below summarises the necessary formulas for the analysis of squat ground-supported cylindrical or rectangular tanks.

	Cylindrical tank	Rectangular tank
Liquid mass $\mathbf{m_w}$ [t]	$m_w = \rho \cdot H \cdot R^2 \pi$	$m_w = \rho \cdot H \cdot 2L \cdot B$ (B = other tank dimension)
impulsive liquid mass $\mathbf{m_0}$ [t]	$m_0 = m_w \cdot \left[\frac{\tanh(\sqrt{3} \cdot \frac{R}{H})}{\sqrt{3} \cdot \frac{R}{H}}\right]$	$m_0 = m_w \cdot \left[\frac{\tanh(\sqrt{3} \cdot \frac{L}{H})}{\sqrt{3} \cdot \frac{L}{H}}\right]$
impulsive level $\mathbf{h_0}$ [m] without bottom pressure	$h_0^k = \frac{3}{8} \cdot H$	$h_0^k = \frac{3}{8} \cdot H$
impulsive level $\mathbf{h_0}$ [m] with bottom pressure	$h_0^g = \frac{H}{8} \cdot \left[\frac{4}{\left[\frac{\tanh(\sqrt{3} \cdot \frac{R}{H})}{\sqrt{3} \cdot \frac{R}{H}}\right]} - 1\right]$	$h_0^g = \frac{H}{8} \cdot \left[\frac{4}{\left[\frac{\tanh(\sqrt{3} \cdot \frac{L}{H})}{\sqrt{3} \cdot \frac{L}{H}}\right]} - 1\right]$
impulsive equivalent force $\mathbf{P_0}$ [kN]	$P_0 = \ddot{u}_0 \cdot (m_0 + m_T)$	$P_0 = \ddot{u}_0 \cdot (m_0 + m_T)$
convective liquid mass $\mathbf{m_1}$ [t]	$m_1 = m_w \cdot 0.18 \cdot \frac{R}{H} \tanh(1.84 \cdot \frac{H}{R})$	$m_1 = m_w \cdot 0.527 \cdot \frac{L}{H} \tanh(1.58 \cdot \frac{H}{L})$
convective level $\mathbf{h_1}$[m] without bottom pressure	$h_1^k = H \cdot \left[1 - \frac{\cosh(1.84\frac{H}{R}) - 1}{1.84\frac{H}{R} \cdot \sinh(1.84\frac{H}{R})}\right]$	$h_1^k = H \cdot \left[1 - \frac{\cosh(1.58\frac{H}{L}) - 1}{1.58\frac{H}{L} \cdot \sinh(1.58\frac{H}{L})}\right]$
convective level $\mathbf{h_1}$[m] with bottom pressure	$h_1^g = H \cdot \left[1 - \frac{\cosh(1.84\frac{H}{R}) - 2.01}{1.84\frac{H}{R} \cdot \sinh(1.84\frac{H}{R})}\right]$	$h_1^g = H \cdot \left[1 - \frac{\cosh(1.58\frac{H}{L}) - 2}{1.58\frac{H}{L} \cdot \sinh(1.58\frac{H}{L})}\right]$
natural circular frequency ω [rad/s]	$\omega^2 = \frac{1.84 \cdot g}{R} \cdot \tanh(1.84 \cdot \frac{H}{R})$	$\omega^2 = \frac{1.58 \cdot g}{L} \cdot \tanh(1.58 \cdot \frac{H}{L})$
max. horizontal diaplacement $\mathbf{y_{max}}$ [m]	$y_{max} = \frac{S_v}{\omega}$	$y_{max} = \frac{S_v}{\omega}$
Angle θ_h [rad]	$\theta_h = 1.534 \cdot \frac{y_{max}}{R} \cdot \tanh(1.84 \cdot \frac{H}{R})$	$\theta_h = 1.58 \cdot \frac{y_{max}}{L} \cdot \tanh(1.58 \cdot \frac{H}{L})$
convective equivalent force $\mathbf{P_1}$ [kN]	$P_1 = 1.2 \cdot m_1 \cdot g \cdot \theta_h \cdot \sin(\omega \cdot t)$	$P_1 = m_1 \cdot g \cdot \theta_h \cdot \sin(\omega \cdot t)$
max. vertical displacement $\mathbf{d_{max}}$ [m]	$d_{max} = \frac{0.408 \cdot R \cdot \coth(1.84 \cdot \frac{H}{R})}{\frac{g}{\omega^2 \cdot \theta_h \cdot R} - 1}$	$d_{max} = \frac{0.527 \cdot L \cdot \coth(1.58 \cdot \frac{H}{L})}{\frac{g}{\omega^2 \cdot \theta_h \cdot L} - 1}$

The following points must be considered:

- The sum of the impulsive and convective masses $m_0 + m_1$ is not equal to the total liquid mass m_w.
- It can be generally assumed that the centre of gravity of the tank mass m_T lies at the same level as the centre of gravity of the impulsive mass m_0, the level h_0 being thus the same in both cases.
- The spectral ordinates for displacement d, velocity v and acceleration a are interrelated according to $S_v = S_d \cdot \omega = \frac{S_a}{\omega}$.

Example

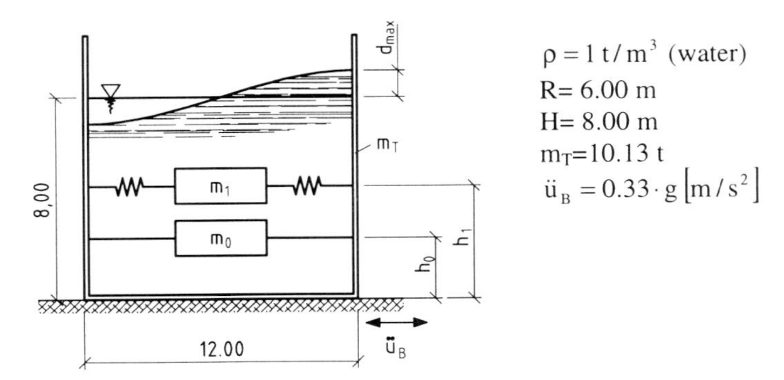

Fig. 9.2-7: Example structure (squat ground-supported tank)

a) Basic quantities

$$\frac{H}{R} = 1.33 < 1.5 \Rightarrow \text{squat tank}$$

Water mass:

$$m_w = \rho \cdot V = 1 \cdot \pi \cdot 6.00^2 \cdot 8.00 = 904.78\,t$$

Auxiliary terms:

$$\left[\frac{\tanh(\sqrt{3} \cdot \frac{R}{H})}{\sqrt{3} \cdot \frac{R}{H}}\right] = \frac{\tanh(\sqrt{3} \cdot 0.75)}{\sqrt{3} \cdot 0.75} = 0.663$$

$$1.84 \cdot \frac{H}{R} = 2.453$$

b) Impulsive part

Masses:

$$m_0 = m_w \cdot \left[\frac{\tanh(\sqrt{3} \cdot \frac{R}{H})}{\sqrt{3} \cdot \frac{R}{H}}\right] = 904.78 \cdot 0.663 = 599.87\,t$$

$$m_{0+T} = 599.87 + 10.13 = 610\,t$$

Levels:

$$h_0^k = \frac{3}{8} \cdot H \overset{!}{=} h_{0+T}^k = 3.00\,m$$

$$h_0^g = \frac{H}{8} \cdot \left[\frac{4}{\left[\frac{\tanh(\sqrt{3} \cdot \frac{R}{H})}{\sqrt{3} \cdot \frac{R}{H}} \right]} - 1 \right] = \frac{8.00}{8} \cdot \left[\frac{4}{0.663} - 1 \right] = 5.033\,m$$

Equivalent forces with $\ddot{u}_B = \ddot{u}_0$:

$$P_0 = \ddot{u}_0 \cdot m_0 = 0.33 \cdot g \cdot 610 = 1974.75\,kN$$

Moments:

$$M_0^k = 1974.75 \cdot 3.00 = 5924.25\ kNm$$

$$M_0^g = 1974.75 \cdot 5.033 = 9938.92\ kNm$$

c) Convective part

Masses:

$$m_1 = m_w \cdot 0.318 \cdot \frac{R}{H} \cdot \tanh(1.84 \cdot \tfrac{H}{R}) = 904.78 \cdot 0.318 \cdot \frac{6.00}{8.00} \cdot \tanh(2.453)$$
$$= 212.62\ t$$

Levels:

$$h_1^k = H \cdot \left[1 - \frac{\cosh(1.84 \cdot \frac{H}{R}) - 1}{1.84 \cdot \frac{H}{R} \cdot \sinh(1.84 \cdot \frac{H}{R})} \right] = 8.00 \cdot \left[1 - \frac{\cosh(2.453) - 1}{2.453 \cdot \sinh(2.453)} \right] = 5.255\,m$$

$$h_1^g = H \cdot \left[1 - \frac{\cosh(1.84 \cdot \frac{H}{R}) - 2.01}{1.84 \cdot \frac{H}{R} \cdot \sinh(1.84 \cdot \frac{H}{R})} \right] = 8.00 \cdot \left[1 - \frac{\cosh(2.453) - 2.01}{2.453 \cdot \sinh(2.453)} \right] = 5.826\ m$$

Natural circular frequency:

$$\omega^2 = \frac{1.84 \cdot g}{R} \cdot \tanh(1.84 \cdot \tfrac{H}{R}) = \frac{1.84 \cdot 9.81}{6.00} \cdot \tanh(2.453) = 2.964\ (rad/s)^2 \Rightarrow \omega = 1.722\ \ rad/s$$

$$T = \frac{2 \cdot \pi}{\omega} = 3.649\ s$$

The damping ratio for water is assumed equal to 0.5%. From the spectrum of Fig. 9.2-5 with D = 0.5% and $\ddot{u}_B = 0.33 \cdot g$ we obtain

$$S_a = f(T) \approx 0.45 \cdot 0.33 \cdot g = 1.457\ m/s^2$$

$$S_v = \frac{S_a}{\omega} = \frac{1.457}{1.722} = 0.846\ m/s$$

Oscillation of the water surface:

$$y_{max} = \frac{S_v}{\omega} = \frac{0.846}{1.722} = 0.492\,m$$

$$\theta_h = 1.534 \cdot \frac{y_{max}}{R} \cdot \tanh(1.84 \cdot \tfrac{H}{R}) = 1.534 \cdot \frac{0.492}{6.00} \cdot \tanh(2.453) = 0.124\ rad$$

Vertical displacement:

$$d_{max} = \frac{0.408 \cdot R \cdot \coth(1.84 \cdot \frac{H}{R})}{\frac{g}{\omega^2 \cdot \theta_h \cdot R} - 1} = \frac{0.408 \cdot 6.00 \cdot \coth(2.453)}{4.449 - 1} = 0.72 \text{ m}$$

Equivalent forces:

$$P_1 = 1.20 \cdot m_1 \cdot g \cdot \theta_h \cdot \sin(\omega \cdot t) = 1.20 \cdot 212.62 \cdot 9.81 \cdot 0.124 \cdot \sin(1.722 \cdot t) =$$
$$= 310.37 \text{ kN} \cdot \sin(1.722 \cdot t)$$

$$\max P_1 = 310.37 \text{ kN}$$

Moments:

$$M_1^k = 310.37 \cdot 5.255 = 1630.99 \text{ kNm}$$
$$M_1^g = 310.37 \cdot 5.826 = 1808.22 \text{ kNm}$$

d) Adding the impulsive and the convective contributions

$$P = P_0 + P_1 = 1974.75 + 310.37 = 2286 \text{ kN}$$
$$M^k = M_0^k + M_1^k = 5924.25 + 1630.9 = 7556 \text{ kNm}$$
$$M^g = M_0^g + M_1^g = 9938.92 + 1808.22 = 11748 \text{ kNm}$$

The convective force causes an oscillation of the water surface with $\omega = 1.722$ rad/s, leading to a sloshing height of 0.72 m relative to the undisturbed surface. The design values are given by:

$P =$	2286 kN	max. shear force at bottom level,
$M^k =$	7556 kNm	Bending moment in the tank wall at bottom level,
$M^g =$	11748 kNm	Overturning moment.

9.2.2.2 Slender ground-supported tanks

As can be seen in Fig. 9.2-6, the overall depth of the liquid H is divided into two parts in the case of a slender tank. The first one is the "fixed" level $\hat{H}$ while the second is the "free" depth $\overline{H}$. The following formulas hold for slender, ground-supported tanks.

	Cylindrical tank	Rectangular tank
Liquid mass $\mathbf{m_w}$ [t]	$m_w = \rho \cdot H \cdot R^2 \pi$	$m_w = \rho \cdot H \cdot 2L \cdot B$ (B = other tank dimension)
Impulsive part		
"Free" depth $\overline{H}$ [m]	$\overline{H} = 1.5 \cdot R$	$\overline{H} = 1.5 \cdot L$
"Fixed" level $\hat{H}$ [m]	$\hat{H} = H - \overline{H}$	$\hat{H} = H - \overline{H}$
„fixed" liquid mass $\hat{\mathbf{m}}_0$ [t]	$\hat{m}_0 = \rho \cdot \hat{H} \cdot R^2 \pi$	$\hat{m}_0 = \rho \cdot \hat{H} \cdot 2L \cdot B$
„fixed" level $\hat{\mathbf{h}}$ [m]	$\hat{h}_0 = \frac{\hat{H}}{2}$	$\hat{h}_0 = \frac{\hat{H}}{2}$
„free" liquid mass $\overline{\mathbf{m}}$ [t]	$\overline{m} = \rho \cdot \overline{H} \cdot R^2 \pi$	$\overline{m} = \rho \cdot \overline{H} \cdot 2L \cdot B$
corresponding impulsive liquid mass $\overline{\mathbf{m}}_0$ [t]	$\overline{m}_0 = \overline{m} \cdot \left[\frac{\tanh(\sqrt{3} \cdot 0.667)}{\sqrt{3} \cdot 0.667}\right] = \overline{m} \cdot 0.7095$	$\overline{m}_0 = \overline{m} \cdot \left[\frac{\tanh(\sqrt{3} \cdot 0.667)}{\sqrt{3} \cdot 0.667}\right] = \overline{m} \cdot 0.7095$
impulsive level $\overline{\mathbf{h}}_0$ [m] without bottom pressure	$\overline{h}_0^k = \frac{3}{8} \cdot 1.5 \cdot R + \hat{H}$	$\overline{h}_0^k = \frac{3}{8} \cdot 1.5 \cdot L + \hat{H}$
impulsive level $\overline{\mathbf{h}}_0$ [m] with bottom pressure	$\overline{h}_0^g = \frac{1.5 \cdot R}{8} \cdot \left[\frac{4}{0.7095} - 1\right] + \hat{H}$	$\overline{h}_0^g = \frac{1.5 \cdot L}{8} \cdot \left[\frac{4}{0.7095} - 1\right] + \hat{H}$
impulsive liquid mass $\mathbf{m_0}$ [t]	$m_0 = \hat{m}_0 + \overline{m}_0$	$m_0 = \hat{m}_0 + \overline{m}_0$
impulsive level $\mathbf{h_0}$ [m] without bottom pressure	$h_0^k = \frac{\hat{m}_0 \cdot \hat{h}_0 + \overline{m}_0 \cdot \overline{h}_0^k}{m_0}$	$h_0^k = \frac{\hat{m}_0 \cdot \hat{h}_0 + \overline{m}_0 \cdot \overline{h}_0^k}{m_0}$
impulsive level $\mathbf{h_0}$ [m] with bottom pressure	$h_0^g = \frac{\hat{m}_0 \cdot \hat{h}_0 + \overline{m}_0 \cdot \overline{h}_0^g}{m_0}$	$h_0^g = \frac{\hat{m}_0 \cdot \hat{h}_0 + \overline{m}_0 \cdot \overline{h}_0^g}{m_0}$
impulsive equivalent force $\mathbf{P_0}$ [kN]	$P_0 = \ddot{u}_0 \cdot m_0$	$P_0 = \ddot{u}_0 \cdot m_0$

Convective part		
convective liquid mass $\mathbf{m_1}$ [t]	$m_1 = m_w \cdot 0.318 \cdot \frac{R}{H} \cdot \tanh(1.84 \cdot \frac{H}{R})$	$m_1 = m_w \cdot 0.527 \cdot \frac{L}{H} \cdot \tanh(1.58 \cdot \frac{H}{L})$
convective level $\mathbf{h_0}$ [m] without bottom pressure	$h_1^k = H\left[1 - \frac{\cosh(1.84 \cdot \frac{H}{R}) - 1}{1.84 \cdot \frac{H}{R} \cdot \sinh(1.84 \cdot \frac{H}{R})}\right]$	$h_1^k = H\left[1 - \frac{\cosh(1.58 \cdot \frac{H}{L}) - 1}{1.58 \cdot \frac{H}{L} \cdot \sinh(1.58 \cdot \frac{H}{L})}\right]$
convective level $\mathbf{h_0}$ [m] with bottom pressure	$h_1^g = H \cdot \left[1 - \frac{\cosh(1.84 \cdot \frac{H}{R}) - 2.01}{1.84 \cdot \frac{H}{R} \cdot \sinh(1.84 \cdot \frac{H}{R})}\right]$	$h_1^g = H \cdot \left[1 - \frac{\cosh(1.58 \cdot \frac{H}{L}) - 2}{1.58 \cdot \frac{H}{L} \cdot \sinh(1.58 \cdot \frac{H}{L})}\right]$
Natural circular frequency ω [rad/s]	$\omega^2 = \frac{1.84 \cdot g}{R} \cdot \tanh(1.84 \cdot \frac{H}{R})$	$\omega^2 = \frac{1.58 \cdot g}{L} \cdot \tanh(1.58 \cdot \frac{H}{L})$
max. horizontal displacement $\mathbf{y_{max}}$ [m]	$y_{max} = \frac{S_v}{\omega}$	$y_{max} = \frac{S_v}{\omega}$
Angle θ_h [rad]	$\theta_h = 1.534 \cdot \frac{y_{max}}{R} \cdot \tanh(1.84 \cdot \frac{H}{R})$	$\theta_h = 1.58 \cdot \frac{y_{max}}{L} \cdot \tanh(1.58 \cdot \frac{H}{L})$
convective equivalent force $\mathbf{P_1}$ [kN]	$P_1 = 1.2 \cdot m_1 \cdot g \cdot \theta_h \cdot \sin(\omega \cdot t)$	$P_1 = m_1 \cdot g \cdot \theta_h \cdot \sin(\omega \cdot t)$
max. vertical displacement $\mathbf{d_{max}}$ [m]	$d_{max} = \frac{0.408 \cdot R \cdot \coth(1.84 \cdot \frac{H}{R})}{\frac{g}{\omega^2 \cdot \theta_h \cdot R} - 1}$	$d_{max} = \frac{0.527 \cdot L \cdot \coth(1.58 \cdot \frac{H}{L})}{\frac{g}{\omega^2 \cdot \theta_h \cdot L} - 1}$

Example

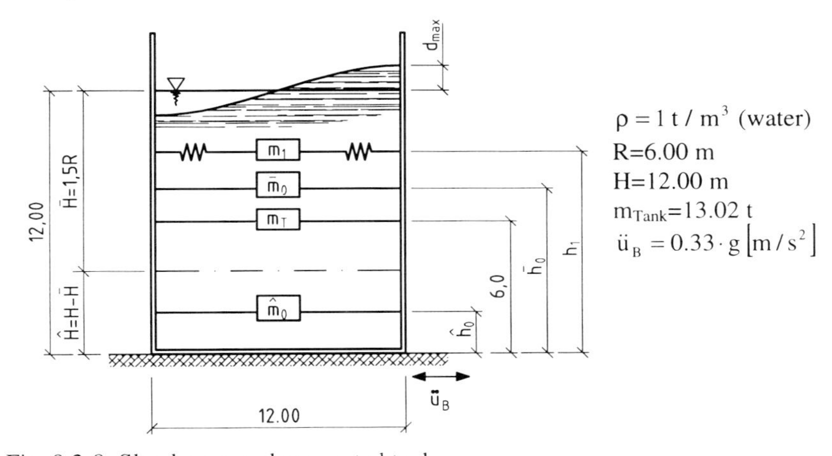

$\rho = 1\ t/m^3$ (water)
R=6.00 m
H=12.00 m
m_{Tank}=13.02 t
$\ddot{u}_B = 0.33 \cdot g\ [m/s^2]$

Fig. 9.2-8: Slender ground-supported tank

a) Basic quantities

$$\frac{H}{R} = 2 > 1.5 \Rightarrow \text{slender tank}$$

Water mass:

$$m_w = \rho \cdot V = 1 \cdot \pi \cdot 6.00^2 \cdot 12.00 = 1357.17 \text{ t}$$

Auxiliary term:

$$1.84 \cdot H / R = 3.68$$

"Fixed" and "free" depths:

$$\overline{H} = 1.5 \cdot R = 9.00 \text{ m}$$

$$\hat{H} = 12.00 - 9.00 = 3.00 \text{ m}$$

b) Impulsive part

„fixed" water mass:

$$\hat{m}_0 = \rho \cdot V = 1 \cdot \pi \cdot 6.00^2 \cdot 3.00 = 339.29 \text{ t}$$

$$\hat{h}_0 = \frac{3.00}{2} = 1.50 \text{ m}$$

„free" water mass:

$$\overline{m} = 1 \cdot \pi \cdot 6.00^2 \cdot 9.00 = 1017.88 \text{ t}$$

$$\overline{m}_0 = \overline{m} \cdot 0.7095 = 1017.88 \cdot 0.7095 = 722.69 \text{ t}$$

$$\overline{h}_0^k = \frac{3}{8} \cdot 1.5 \cdot R + \hat{H} = 3.375 + 3.00 = 6.375 \text{ m}$$

$$\overline{h}_0^g = \frac{1.5 \cdot R}{8} \cdot \left[\frac{4}{0.7095} - 1 \right] + \hat{H} = 0.1875 \cdot 6.00 \cdot 4.634 + 3.00 = 8.213 \text{ m}$$

$$m_0 = \hat{m}_0 + \overline{m}_0 + m_{Tank} = 339.29 + 722.69 + 13.02 = 1075 \text{ t}$$

$$h_0^k = \frac{\hat{m}_0 \cdot \hat{h}_0 + \overline{m}_0 \cdot \overline{h}_0^k + m_{Tank} \cdot \frac{H}{2}}{m_0}$$

$$= \frac{339.29 \cdot 1.50 + 722.69 \cdot 6.375 + 13.02 \cdot 6.00}{1075} = 4.83 \text{ m}$$

$$h_0^g = \frac{\hat{m}_0 \cdot \hat{h}_0 + \overline{m}_0 \cdot \overline{h}_0^g + m_{Tank} \cdot \frac{H}{2}}{m_0}$$

$$= \frac{339.29 \cdot 1.50 + 722.69 \cdot 8.213 + 13.02 \cdot 6{,}00}{1075} = 6.07 \text{ m}$$

Equivalent force with $\ddot{u}_B = \ddot{u}_0$:

$$P_0 = \ddot{u}_0 \cdot m_0 = 0.33 \cdot g \cdot 1075 = 3480.10 \text{ kN}$$

Moments:

$$M_0^k = 3480.10 \cdot 4.83 = 16808.88 \text{ kNm}$$

$$M_0^g = 3480.10 \cdot 6.07 = 21124.21 \text{ kNm}$$

c) Convective part

$$m_1 = m_w \cdot 0.318 \cdot \frac{R}{H} \cdot \tanh(1.84 \cdot \tfrac{H}{R}) = 1357.17 \cdot 0.318 \cdot 0.50 \cdot \tanh(3.68) = 215.52 \text{ t}$$

$$h_1^k = H \cdot \left[1 - \frac{\cosh(1.84 \cdot \frac{H}{R}) - 1}{1.84 \cdot \frac{H}{R} \cdot \sinh(1.84 \cdot \frac{H}{R})}\right] = 12.00 \cdot \left[1 - \frac{\cosh(3.68) - 1}{3.68 \cdot \sinh(3.68)}\right] = 8.90 \text{ m}$$

$$h_1^g = H \cdot \left[1 - \frac{\cosh(1.84 \cdot \frac{H}{R}) - 2.01}{1.84 \cdot \frac{H}{R} \cdot \sinh(1.84 \cdot \frac{H}{R})}\right] = 12.00 \cdot \left[1 - \frac{\cosh(3.68) - 2.01}{3.68 \cdot \sinh(3.68)}\right] = 9.07 \text{ m}$$

Natural circular frequency:

$$\omega^2 = \frac{1.84 \cdot g}{R} \cdot \tanh(1.84 \cdot \tfrac{H}{R}) = \frac{1.84 \cdot 9.81}{6.00} \cdot \tanh(3.68) = 3.01\,(\text{rad/s})^2 \Rightarrow \omega = 1.735 \text{ rad/s}$$

$$T = \frac{2 \cdot \pi}{\omega} = 3.62 \text{ s}$$

The damping ratio for water is assumed equal to 0.5%. From the spectrum of Fig. 9.2-5 with D = 0.5% and $\ddot{u}_B = 0.33 \cdot g$ we obtain

$$S_a = f(T) \approx 0.45 \cdot 0{,}33 \cdot g = 1.457 \text{ m/s}^2$$

$$S_v = \frac{Sa}{\omega} = \frac{1.457}{1.735} = 0.840 \text{ m/s}$$

Motion of the water surface:

$$y_{max} = \frac{S_v}{\omega} = \frac{0.840}{1.735} = 0.484 \text{ m}$$

$$\theta_h = 1.534 \cdot \frac{y_{max}}{R} \cdot \tanh(1.84 \cdot \tfrac{H}{R}) = 1.534 \cdot \frac{0.484}{6.00} \cdot \tanh(3.68) = 0.124 \text{ rad}$$

Vertical motion:

$$d_{max} = \frac{0.408 \cdot R \cdot \coth(1.84 \cdot \frac{H}{R})}{\frac{g}{\omega^2 \cdot \theta_h \cdot R} - 1} = \frac{0.408 \cdot 6.00 \cdot \coth(3.68)}{4.381 - 1} = 0.725 \text{ m}$$

Equivalent force:

$$P_1 = 1.2 \cdot m_1 \cdot g \cdot \theta_h \cdot \sin(\omega \cdot t) = 1.2 \cdot 215.52 \cdot 9.81 \cdot 0.124 \cdot \sin(1.735 \cdot t) =$$
$$= 314.60 \text{ kN} \cdot \sin(1.735 \cdot t)$$
$$\max P_1 = 314.60 \text{ kN}$$

Moments:

$$M_1^k = 314.60 \cdot 8.90 = 2799.94 \text{ kNm}$$
$$M_1^g = 314.60 \cdot 9.07 = 2853.42 \text{ kNm}$$

d) Adding the impulsive and the convective contributions

$$M^k = M_0^k + M_1^k = 16808.88 + 2799.94 = 19609 \text{ kNm}$$

$$M^g = M_0^g + M_1^g = 21124.21 + 2853.42 = 23978 \text{ kNm}$$

$$P = P_0 + P_1 = 3480.10 + 314.60 = 3795 \text{ kN}$$

The convective force causes an oscillation of the water surface with $\omega = 1.735$ rad/s, leading to a sloshing height of 0.725 m relative to the undisturbed surface. The design values are given by:

$P =$	3795 kN	max. shear force at bottom level
$M^k =$	19609 kNm	Bending moment in the tank wall at bottom level
$M^g =$	23978 kNm	Overturning moment.

9.2.3 Formulas for elevated tanks

In the case of elevated tanks, the stiffness properties of the support structure must be considered, since the tanks are able to move relatively to the ground and thus experience a different absolute acceleration than $\ddot{u}_B$. The simplest possible model is the two-degree-of-freedom system shown in Fig. 9.2-9.

The impulsive and the convective liquid mass are determined as functions of the slenderness ratio of the tank. The convective mass m_1 is connected to the walls by means of longitudinal springs; the impulsive mass m_0 contains in addition to the liquid mass m_0^* shown in Fig. 9.2-9 the mass of the tank and its support construction. The analysis supplies the maximum horizontal shear forces and bending moments as well as the sloshing amplitude. Since the convective properties of the liquid are not influenced by the support structure, the relevant expressions stay the same.

	Cylindrical tank	Rectangular tank
Liquid mass $\mathbf{m_w}$ [t]	$m_w = \rho \cdot H \cdot R^2 \cdot \pi$	$m_w = \rho \cdot H \cdot 2L \cdot B$ (B= other tank dimension)
convective mass $\mathbf{m_1}$ [t]	$m_1 = m_w \cdot 0.318 \cdot \frac{R}{H} \cdot \tanh(1.84 \cdot \frac{H}{R})$	$m_1 = m_w \cdot 0.527 \cdot \frac{L}{H} \cdot \tanh(1.58 \cdot \frac{H}{L})$
Natural circular frequency ω [rad/s]	$\omega^2 = \frac{1.84 \cdot g}{R} \cdot \tanh(1.84 \cdot \frac{H}{R})$	$\omega^2 = \frac{1.58 \cdot g}{L} \cdot \tanh(1.58 \cdot \frac{H}{L})$

The natural circular frequency ω is directly connected to the spring stiffness k_1:

$$\omega^2 = \frac{k}{m} \Rightarrow \omega^2 = \frac{k_1}{m_1}\left[\frac{kN/m}{t}\right] \Rightarrow k_1 = \omega^2 \cdot m_1$$

The spring stiffness of the support structure can be determined by applying a horizontal unit load to the centre of gravity of the tank and computing the resulting horizontal displacement f_0. The spring stiffness k_0 is equal to the reciprocal, $k_C = 1/f_0$.

Fig. 9.2-10 shows the two-degree-of-freedom shear beam model with the discrete masses m_0, m_1 and the springs with stiffness values k_0 and k_1. For this shear beam idealisation it is assumed that the masses can only move horizontally, without additional rotation.

The two-degree-of-freedom shear beam model has two natural frequencies and mode shapes. Usually, the fundamental mode is associated with sloshing, while the second mode is a mode of the supporting structure. The sloshing mode can be assumed to have a damping of D=0.5% for water tanks, while the damping of the second mode depends on the type of the support construction.

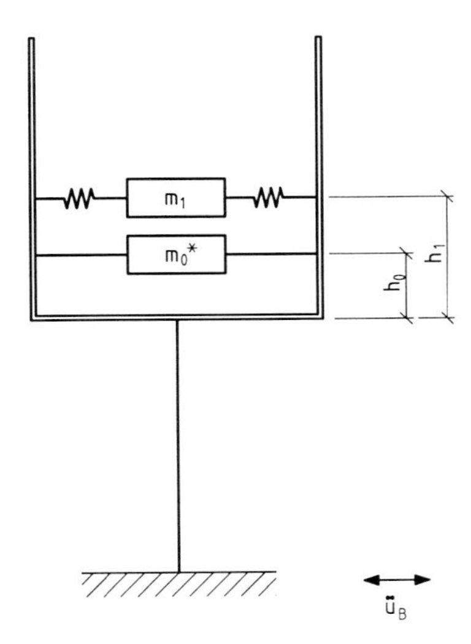

Fig. 9.2-9: Elevated tank

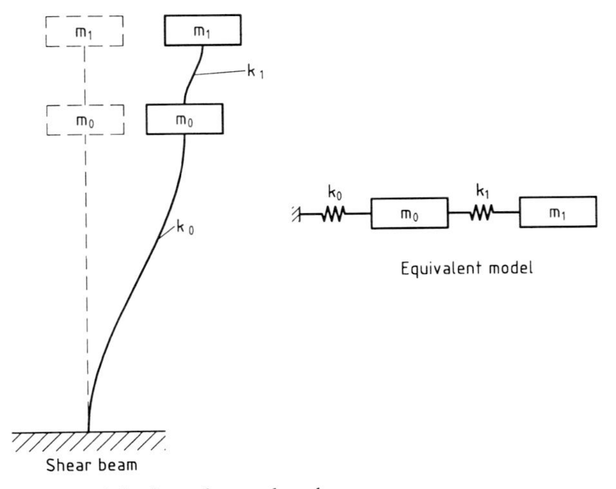

Fig. 9.2-10: Shear beam model of an elevated tank

The stiffness and mass matrices for the system depicted in Fig. 9.2-10 are given by:

$$\underline{K} = \begin{bmatrix} k_0 + k_1 & -k_1 \\ -k_1 & k_1 \end{bmatrix} \qquad \underline{M} = \begin{bmatrix} m_0 & 0 \\ 0 & m_1 \end{bmatrix}$$

In this case, the eigenproblem can be solved analytically in order to obtain closed-form expressions for the natural circular frequencies and the mode shapes:

a) Natural circular frequencies:

$$\det \left| \underline{K} - \omega^2 \cdot \underline{M} \right| = 0$$

$$\left((k_0 + k_1) - \omega^2 \cdot m_0\right) \cdot \left(k_1 - \omega^2 \cdot m_1\right) - \left(k_1\right)^2 = 0$$

$$\omega^4 - \omega^2 \cdot \left(\frac{k_0 + k_1}{m_0} + \frac{k_1}{m_1} \right) + \frac{(k_0 + k_1) \cdot k_1 - k_1^2}{m_1 \cdot m_0} = 0$$

As a result, we obtain the natural circular frequencies from the expression:

$$\omega_{1,2}^2 = \frac{1}{2} \cdot \left(\frac{(k_0 + k_1)}{m_0} + \frac{k_1}{m_1} \right) \pm \sqrt{\frac{1}{4} \left(\frac{(k_0 + k_1)}{m_0} - \frac{k_1}{m_1} \right)^2 + \frac{k_1^2}{m_1 \cdot m_0}}$$

The algebraically smaller natural frequency corresponds to the sloshing mode, the larger one to the support vibration mode.

b) Mode shapes $\underline{u}_i$:

$$\left\{ \begin{bmatrix} k_0 + k_1 & -k_1 \\ -k_1 & k_1 \end{bmatrix} - \omega^2 \cdot \begin{bmatrix} m_0 & 0 \\ 0 & m_1 \end{bmatrix} \right\} \cdot \begin{bmatrix} u_0 \\ u_1 \end{bmatrix} = \begin{bmatrix} 0 \\ 0 \end{bmatrix}$$

We arbitrarily set the modal component u_1 equal to unity for both mode shapes. The u_0 components are given by

$$u_0 = 1 - \frac{\omega_{Slosh}^2 \cdot m_1}{k_1} \quad \text{for the sloshing mode}$$

$$u_0 = 1 - \frac{\omega_{vibr}^2 \cdot m_1}{k_1} \quad \text{for the support vibration mode}$$

The following steps c) through f) must be carried out separately for both modal contributions. As the result, we obtain the equivalent forces for both modes (sloshing and support vibration):

$$\underline{P}_{slosh} = \begin{bmatrix} P_{0slosh} \\ P_{1slosh} \end{bmatrix}$$

$$\underline{P}_{vibr} = \begin{bmatrix} P_{0\,vibr} \\ P_{1\,vibr} \end{bmatrix}$$

c) Normalizing the mode shapes $\underline{u}$:

Normalizing factor $a = u_0^2 \cdot m_0 + u_1^2 \cdot m_1$

normalized mode shape: $\underline{\Phi} = \frac{1}{\sqrt{a}} \cdot \underline{u}$

d) Maximum displacements:

The participation factors β_i are determined first:

$\beta = \underline{\Phi}^T \cdot \underline{M} \cdot \underline{r}$
where $\underline{r}$ is the vector containing the displacements in the degrees of freedom for a unit support displacement, $\underline{r}^T = [1,1]$,

$$\beta = \underline{\Phi}^T \cdot \begin{bmatrix} m_0 & 0 \\ 0 & m_1 \end{bmatrix} \cdot \begin{bmatrix} 1 \\ 1 \end{bmatrix} = \frac{1}{\sqrt{a}} \cdot \begin{bmatrix} u_0 & 1 \end{bmatrix} \cdot \begin{bmatrix} m_0 & 0 \\ 0 & m_1 \end{bmatrix} \cdot \begin{bmatrix} 1 \\ 1 \end{bmatrix} = \frac{u_0 \cdot m_0 + m_1}{\sqrt{u_0^2 \cdot m_0 + m_1}}$$

The maximum displacements are then given by:

$$\max \underline{y} = \beta \cdot S_d \cdot \underline{\Phi}$$

$$\max \underline{y} = \begin{bmatrix} y_0 \\ y_1 \end{bmatrix} = \beta \cdot S_d \cdot \underline{\Phi} = \beta \cdot S_d \cdot \frac{1}{\sqrt{a}} \cdot \begin{bmatrix} u_0 \\ 1 \end{bmatrix}$$

with

$$S_d = S_a \cdot \frac{1}{\omega^2} \text{ and } S_d = S_v \cdot \frac{1}{\omega}$$

The spectral ordinates are functions of T and thus different for the two modes.

e) Equivalent static forces:

These are equal to the masses multiplied by the corresponding absolute accelerations:

$$\max \underline{\ddot{u}} = \beta \cdot S_a \cdot \underline{\Phi}$$

$$\max \underline{\ddot{u}} = \begin{bmatrix} \ddot{u}_0 \\ \ddot{u}_1 \end{bmatrix} = \beta \cdot S_a \cdot \underline{\Phi} = \beta \cdot S_a \cdot \frac{1}{\sqrt{a}} \cdot \begin{bmatrix} u_0 \\ 1 \end{bmatrix}$$

$$P_0 = \ddot{u}_0 \cdot m_0$$

$$P_1 = \ddot{u}_1 \cdot m_1$$

Alternatively, they can be computed as

$$P_0 = (k_0 + k_1) \cdot y_0 - k_1 \cdot y_1$$

$$P_1 = -k_1 \cdot y_0 + k_1 \cdot y_1$$

f) Internal forces:

They are computed from the equivalent static loads P_0 and P_1 separately for the two modal contributions and the single components are combined by means of the SRSS rule.

g) Maximum sloshing height d_{max}:

The oscillation angle of the fluid surface is first computed for each mode:

$$\theta_h = 1.534 \cdot \frac{y_1 - y_0}{R} \cdot \tanh\left(1.84 \cdot \frac{H}{R}\right)$$

The maximum sloshing height can then be determined from the expression

$$d_{max,slosh} = \frac{0.408 \cdot R \cdot \coth(1.84 \cdot \frac{H}{R})}{\dfrac{g}{\omega_{slosh}^2 \cdot \theta_{h,slosh} \cdot R} - 1} \quad \text{and} \quad d_{max,vibr} = \theta_{h,vibr} \cdot R$$

$$d_{max} = \sum d_{max,i}$$

The single calculation steps are summarised in the following table.

Liquid mass $\mathbf{m_w}$ [t]	
convective liquid mass $\mathbf{m_1}$ [t]	
Natural circular frequency ω [rad/s]	
Spring stiffness $\mathbf{k_1}$ [kN/m]	
impulsive liquid mass $\mathbf{m_0}$ [t]	
Structure stiffness $\mathbf{k_0}$ [kN/m]	
$\omega_{1,2}^2 = \frac{1}{2} \cdot \left(\frac{(k_0 + k_1)}{m_0} + \frac{k_1}{m_1} \right) \pm \sqrt{\frac{1}{4} \left(\frac{(k_0 + k_1)}{m_0} - \frac{k_1}{m_1} \right)^2 + \frac{k_1^2}{m_1 \cdot m_0}}$ with $\omega_{slosh} = \omega_{min}$ and $\omega_{vibr} = \omega_{max}$	
Support vibration mode	Sloshing mode
$u_0 = 1 - \frac{\omega_{vibr}^2 \cdot m_1}{k_1}$	$u_0 = 1 - \frac{\omega_{slosh}^2 \cdot m_1}{k_1}$
$\beta = \frac{u_0 \cdot m_0 + m_1}{\sqrt{u_0^2 \cdot m_0 + m_1}}$	$\beta = \frac{u_1 \cdot m_0 + m_1}{\sqrt{u_0^2 \cdot m_0 + m_1}}$
$a = u_0^2 \cdot m_0 + m_1$	$a = u_0^2 \cdot m_0 + m_1$
$\max \underline{y} = \beta \cdot S_d \cdot \frac{1}{\sqrt{a}} \cdot \begin{bmatrix} u_0 \\ 1 \end{bmatrix}$	$\max \underline{y} = \beta \cdot S_d \cdot \frac{1}{\sqrt{a}} \cdot \begin{bmatrix} u_0 \\ 1 \end{bmatrix}$
$\max \underline{\ddot{u}} = \beta \cdot S_a \cdot \frac{1}{\sqrt{a}} \cdot \begin{bmatrix} u_0 \\ 1 \end{bmatrix}$	$\max \underline{\ddot{u}} = \beta \cdot S_a \cdot \frac{1}{\sqrt{a}} \cdot \begin{bmatrix} u_0 \\ 1 \end{bmatrix}$
$P_0 = \ddot{u}_0 \cdot m_0$ $P_1 = \ddot{u}_1 \cdot m_1$	$P_0 = \ddot{u}_0 \cdot m_0$ $P_1 = \ddot{u}_1 \cdot m_1$
Computation of the internal forces due to the equivalent static forces and combination by means of the SRSS rule	
$\theta_{h,slosh} = 1.534 \cdot \frac{y_1 - y_0}{R} \cdot \tanh\left(1.84 \cdot \frac{H}{R}\right)$	$\theta_{h,vibr} = 1.534 \cdot \frac{y_1 - y_0}{R} \cdot \tanh\left(1.84 \cdot \frac{H}{R}\right)$
Cylindrical tank: $d_{max,slosh} = \frac{0.408 \cdot R \cdot \coth(1.84 \cdot \frac{H}{R})}{\frac{g}{\omega^2_{slosh} \cdot \theta_{h,slosh} \cdot R} - 1}$ Rectangular tank: $d_{max,slosh} = \frac{0.527 \cdot L \cdot \coth(1.58 \cdot \frac{H}{L})}{\frac{g}{\omega^2_{slosh} \cdot \theta_{h,slosh} \cdot L} - 1}$	Cylindrical tank: $d_{max,vibr} = \theta_{h,vibr} \cdot R$ Rectangular tank: $d_{max,vibr} = \theta_{h,vibr} \cdot L$
Total $d_{max} = d_{max,vibr} + d_{max,slosh}$	

Example

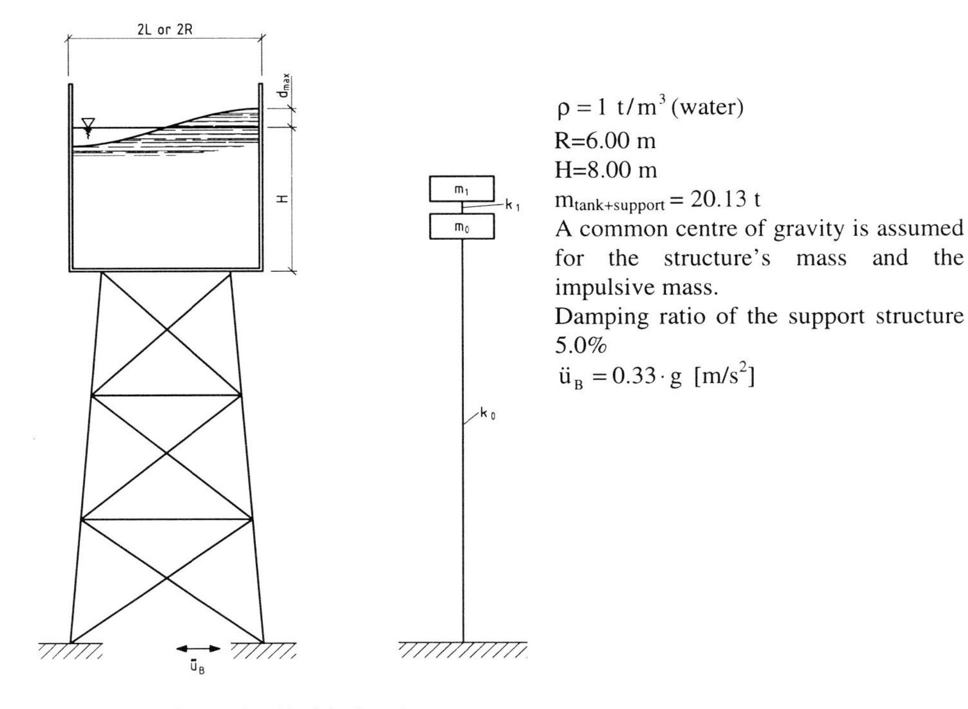

$\rho = 1\ t/m^3$ (water)
R=6.00 m
H=8.00 m
$m_{tank+support} = 20.13$ t
A common centre of gravity is assumed for the structure's mass and the impulsive mass.
Damping ratio of the support structure 5.0%
$\ddot{u}_B = 0.33 \cdot g\ [m/s^2]$

Fig. 9.2-11: Elevated cylindrical tank

a) Basic quantities

Checking whether the tank is slender or squat:

$$\frac{H}{R} = 1.33 < 1.5 \Rightarrow \text{squat tank}$$

Water mass:

$$m_w = \rho \cdot V = 904.78\ t$$

Auxiliary terms:

$$\left[\frac{\tanh(\sqrt{3} \cdot \frac{R}{H})}{\sqrt{3} \cdot \frac{R}{H}}\right] = 0.663$$

$$1.84 \cdot \frac{H}{R} = 2.453$$

b) Convective part

Mass:

$$m_1 = m_w \cdot 0.318 \cdot \frac{R}{H} \cdot \tanh(1.84 \cdot \tfrac{H}{R}) = 212.62 \ t$$

Natural circular frequency:

$$\omega^2 = \frac{1.84 \cdot g}{R} \cdot \tanh(1.84 \cdot \tfrac{H}{R}) = 2.964 \, (rad/s)^2 \Rightarrow \omega = 1.722 \ rad/s$$

Spring stiffness k_1 (sloshing):

$$k_1 = m_1 \cdot \omega^2 = 212.62 \cdot 2.964 = 630.21 \text{ kN/m}$$

c) Impulsive part

Masses:

$$m_0 = m_w \cdot 0.663 = 599.87 \text{ t}$$
$$m_{0+T} = 599.87 + 20.13 = 620 \text{ t}$$

The spring constant k_0 of the support structure is assumed to be known and to be equal to $k_0 = 14369.79$ kN/m

Stiffness and mass matrix:

$$\underline{K} = \begin{bmatrix} 14369.79 + 630.21 & -630.21 \\ -630.21 & 630.21 \end{bmatrix} \qquad \underline{M} = \begin{bmatrix} 620 & 0 \\ 0 & 212.62 \end{bmatrix}$$

Natural circular frequencies:

$$\omega^2 = 0.5 \cdot \left(\frac{15000}{620} + \frac{630.21}{212.62} \right) \pm \sqrt{0.25 \cdot \left(\frac{15000}{620} - \frac{630.21}{212.62} \right)^2 + \frac{630.21 \cdot 630.21}{620 \cdot 212.62}}$$

For the support vibration mode:

$$\omega_0 = \sqrt{13.58 + \sqrt{115.69}} = 4.93 \, rad/s \Rightarrow T_0 = \frac{2\pi}{\omega_0} = 1.27 \, s$$

For the sloshing mode:

$$\omega_1 = \sqrt{13.58 - \sqrt{115.69}} = 1.68 \ rad/s \Rightarrow T_1 = \frac{2\pi}{\omega_1} = 3.74 \ s$$

	Sloshing mode	Support vibration mode
Natural circular frequencies	$\omega_1 = 1.68 \frac{\text{rad}}{\text{s}}$	$\omega_0 = 4.93 \frac{\text{rad}}{\text{s}}$
Mode shapes $\begin{bmatrix} u_0 \\ u_1 \end{bmatrix} = \begin{bmatrix} u_0 \\ 1 \end{bmatrix}$	$u_0 = 1 - \frac{1{,}68^2 \cdot 212.62}{630.21} = 0.048$	$u_0 = 1 - \frac{4.93^2 \cdot 212.62}{630.21} = -7.20$
Participation factors	$\beta = \frac{0.048 \cdot 620 + 212.62}{\sqrt{0.048^2 \cdot 620 + 212.62}}$ $= 16.57$	$\beta = \frac{-7.2 \cdot 620 + 212.62}{\sqrt{7.2^2 \cdot 620 + 212.62}}$ $= -23.64$
Normalizing factors	$a = 0.048^2 \cdot 620 + 212.62$ $= 214.05$	$a = 7.2^2 \cdot 620 + 212.62$ $= 32353.42$
Spectral ordinates for 5.0% and 0.5% damping	D=0.5% $S_a = f(T = 3.74) \approx 0.48 \cdot 0.33 \cdot g$ $= 1.55\,\text{m/s}^2$	D=5.0% $S_a = f(T = 1.27) \approx 0.7 \cdot 0.33 \cdot g$ $= 2.266\,\text{m/s}^2$
Spectral displacements	$S_d = \frac{S_a}{\omega^2} = \frac{1.55}{1.68^2} = 0.549$	$S_d = \frac{S_a}{\omega^2} = \frac{2.266}{4.93^2} = 0.0932$
max. displacements $\max \underline{y} = \begin{bmatrix} y_0 \\ y_1 \end{bmatrix}$ [m]	$\begin{bmatrix} y_0 \\ y_1 \end{bmatrix} = 16.57 \cdot \frac{0.549}{\sqrt{214.05}} \cdot \begin{bmatrix} 0.048 \\ 1 \end{bmatrix}$ $= \begin{bmatrix} 0.0298 \\ 0.6218 \end{bmatrix}$	$\begin{bmatrix} y_0 \\ y_1 \end{bmatrix} = -23.64 \cdot \frac{0.0932}{\sqrt{32353.42}} \cdot \begin{bmatrix} -7.2 \\ 1 \end{bmatrix}$ $= \begin{bmatrix} 0.0882 \\ -0.0122 \end{bmatrix}$
max. accelerations $\max \underline{\ddot{u}} = \begin{bmatrix} \ddot{u}_0 \\ \ddot{u}_1 \end{bmatrix}$	$\begin{bmatrix} \ddot{u}_0 \\ \ddot{u}_1 \end{bmatrix} = 16.57 \cdot \frac{1.55}{\sqrt{214.05}} \cdot \begin{bmatrix} 0.048 \\ 1 \end{bmatrix}$ $= \begin{bmatrix} 0.0843 \\ 1.7555 \end{bmatrix}$	$\begin{bmatrix} \ddot{u}_0 \\ \ddot{u}_1 \end{bmatrix} = -23.64 \cdot \frac{2.266}{\sqrt{32353.42}} \cdot \begin{bmatrix} -7.2 \\ 1 \end{bmatrix}$ $= \begin{bmatrix} 2.1443 \\ -0.2978 \end{bmatrix}$
Equivalent forces [kN]	$P_0 = 0.0843 \cdot 620 = 52.27$ $P_1 = 1.7555 \cdot 212.62 = 373.25$	$P_0 = 2.1443 \cdot 620 = 1329.47$ $P_1 = -0.2979 \cdot 212.62 = -63.32$

The internal forces are then computed for each mode from the equivalent static forces and the results combined by the SRSS rule. The free oscillation of the water surface is computed as follows:

Horizontal displacement of the water surface	$y_{max} = y_1 - y_0$ $= 0.6218 - 0.0298$ $= 0.592\ m$	$y_{max} = y_1 - y_0$ $= -0.0122 - 0.0882$ $= -0.1004\ m$
Oscillation angle of the surface	$\theta_h = 1.534 \cdot \frac{y_{max}}{6.00} \cdot \tanh(2.543)$ $= 0.1495$	$\theta_h = 1.534 \cdot \frac{y_{max}}{6.00} \cdot \tanh(2.543)$ $- 0{,}0254$
Vertical displacement of the surface	$d_{max} = \dfrac{0.408 \cdot R \cdot \coth\left(1.84 \frac{H}{R}\right)}{\frac{g}{\omega^2 \cdot \theta_h \cdot R} - 1}$ $= \frac{2.47844}{3.87488 - 1}$ $= 0.862\ m$	$d_{max} = \theta_h \cdot R$ $= -0.0254 \cdot 6.00$ $= -0.152\ m$

The max. total vertical displacement of the water is equal to:

$$d_{max} = 0.862 - 0.152 = 0.71\ m$$

The idealised system model has masses of 620 t for the sum of impulsive mass and the mass of the structure and 212 t for the convective water mass. The spring constants are equal to 14370 kN/m and 630 kN/m, and the circular natural frequencies are 1.68 rad/s for the sloshing mode and 4.93 rad/s for the structural vibration mode. The maximum wave height is 0.71 m relative to the surface, with the fundamental mode being predominant.

9.3 Numerical analysis of the structure-fluid interaction problem

The consideration of the flexibility of the tank's walls leads to more complex models than the approximate HOUSNER model of section 9.2. According to FISCHER [9.3], [9.11], there is no marked interaction between sloshing and structural vibration, since the sloshing natural frequencies usually lie well below the natural frequencies of the liquid-filled tank. This suggests that computing the influence of the convective mass under the assumption of rigid walls is a valid approximation. On the other hand, computing the influence of the impulsive mass under such a rigid-wall-assumption might lead to serious discrepancies on the unsafe side, because the wall displacements are in reality not identical to the ground displacement. In order to remedy this, the computed impulsive liquid mass is divided into two parts, one of which, $m_{0\ rigid}$, corresponds to the rigid-wall assumption and is much larger than the second part, $m_{0\ rel}$, which corresponds to the relative displacement of the wall and the tank bottom:

$$m_0 = m_{0rigid} + m_{0\ rel} \qquad (9.\ 3.\ 1)$$

Even though $m_{0\ rel}$ is not large, it has a marked influence because the corresponding acceleration $\ddot{u}_{rel}$ can become much larger than the ground acceleration $\ddot{u}_B$.

The HOUSNER formula

$$P_0 = \ddot{u}_B \cdot m_0 \tag{9. 3. 2}$$

may now be substituted by the expression according to FISCHER:

$$P_0 = \ddot{u}_B \cdot m_{0\,rigid} + \ddot{u}_{rel} \cdot m_{0\,rel} \tag{9. 3. 3}$$

FISCHER [9.3],[9.11] presents a method for taking the hydrodynamic effect of the liquid into account by adding suitable masses to the empty tank. According to [9.6], [9.12], this method leads to conservative results.

A more detailed numerical analysis of liquid-structure interaction effects can be carried out by using finite elements for the liquid mass, which can be derived in an Eulerian or Lagrangian formulation. In the former, a velocity potential function is assumed and the behaviour of the liquid described through pressure or velocity variables at the element nodes. On the other hand, since the configuration of the structure is described by displacement variables, two separate problems must be solved in order to consider the interaction effects [9.15]. This complication is circumvented by applying Lagrangian elements, which can use displacements as fluid element variables. Usual idealisations are plane 4-node or spatial 8-node elements [9.6, 9.12 through 9.14]. Both the Langrangian and the Eulerian approach lead to essentially the same results as long as the displacements of the liquid stay small, which is normally the case for liquid-filled tanks.

Two mode shapes for a rectangular water tank are shown in the following for illustration purposes. The underlying finite elements were derived according to the Lagrangian formulation, and the results were verified by the approximate method of HOUSNER. The geometrical data of the example were a water depth H equal to 1.905 m and a tank width of 2L = 5.08 m.

The analytic solution yields the fundamental natural circular frequency according to

$$\omega^2 = \frac{1.58 \cdot g}{L} \cdot \tanh\left(1.58 \cdot \frac{H}{L}\right) = \frac{1.58 \cdot g}{2.54} \cdot \tanh\left(1.58 \cdot \frac{1.905}{2.54}\right) = 5.059\ (\mathrm{rad/s})^2$$

$$\Rightarrow \omega = 2.249\ \mathrm{rad/s}$$

A very similar value was also determined in the finite element investigation, which furnished the results shown in the following figures 9.3-1 and 9.3-2 for the first two mode shapes.

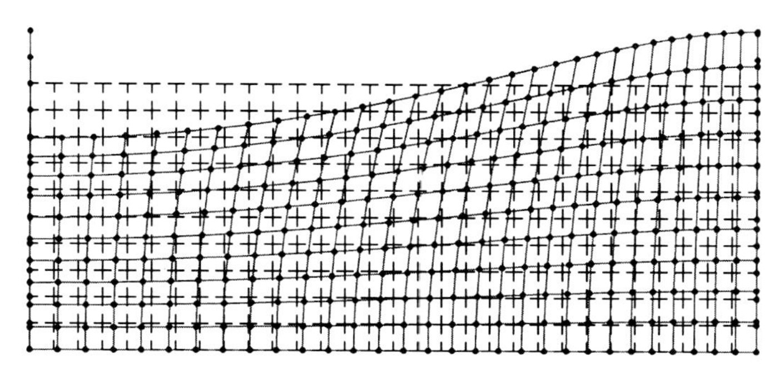

Fig. 9.3-1: Fundamental mode shape for $\omega = 2.24$ rads/s

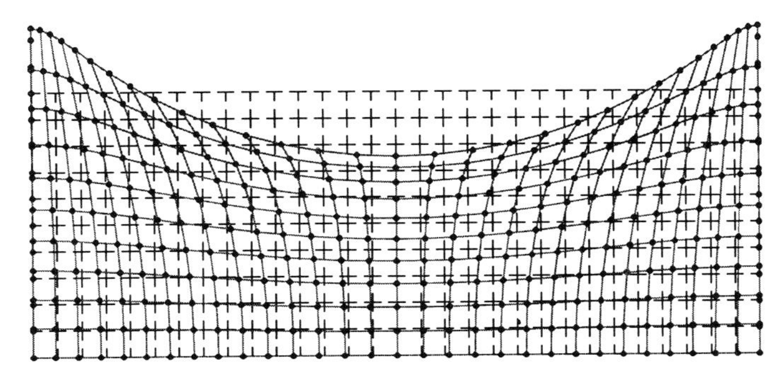

Fig. 9.3-2: Mode shape for $\omega = 3.49$ rad/s

References

General References

Bachmann, H. et al.: Vibration problems in structures-Practical Guidelines. Birkhäuser, 1995.

Beards, C.F.: Structural Vibration-Analysis and Damping. Edward Arnold, 1996.

Buchholdt, H.: Structural dynamics for engineers. Thomas Telford, 1997.

Chopra, A.K.: Dynamics of Structures: Theory and Applications to Earthquake Engineering. Prentice Hall, 1995.

Clough, R.W., Penzien, J.: Dynamics of Structures. 2nd ed., McGraw-Hill,1993.

Dowding, C.H.: Construction Vibration. Prentice Hall 1996

Fertis, D.G.: Mechanical and Structural Vibration. J. Wiley & Sons, 1995.

Géradin, M., Rixen, D.: Mechanical vibrations. J. Wiley & Sons, 1994.

Ghosh, S.K., Domel Jr., A.W., Fanella, D.A.: Design of concrete buildings for earthquake and wind forces. 2nd ed., Portland Cement Association, 1995.

Paz, M.: Structural Dynamics: Theory and Computation. Chapman & Hall, 4th ed., 1997.

References Chapter 2

[2.1] Fleßner, H.: Ein Beitrag zur Ermittlung von Querschnittswerten mit Hilfe elektronischer Rechenanlagen. Der Bauingenieur 37 (1962), No. 4, p. 146.

[2.2] H. Prediger.: Zur Berechnung von Massenträgheitsmomenten durch ein Computer-Programm. Der Stahlbau 50 (1981), p. 21-24.

[2.3] Petersen, Chr.: Dynamik der Baukonstruktionen. Braunschweig / Wiesbaden: F. Vieweg & Sohn, 1996.

[2.4] Cooley, J.W., Tukey, J.W.: An Algorithm for the Machine Calculation of Complex Fourier Series. Mathematics of Computation 19 (1965), S. 297-301.

References Chapter 3

[3.1] Newmark, N. M.: A Method of Computation for Structural Dynamics. ASCE Journal of the Engineering Mechanics Division 85 (1959), p. 67-94.

References Chapter 4

[4.1] Krätzig, W.B., Meskouris, K., Link, M.: Chapter „Baudynamik und Systemidentifikation“ in „Der Ingenieurbau - Grundwissen in 9 Bänden“, volume „Baustatik/ Baudynamik“, ed. G. Mehlhorn, W. Ernst & Sohn (Wiley) 1996, p. 365-518.

[4.2] Guyan, R.J.: Reduction of Stiffness and Mass Matrices. AIAA Journal 3 (1965), p. 380.

[4.3] Irons, B.M.: Structural Eigenvalue Problems: Elimination of Unwanted Variables. AIAA Journal 3 (1965), p. 961-962.

[4.4] Caughey, T.K.; O'Kelly, M.E.J.: Classical Normal Modes in Damped Linear Dynamic Systems. Transactions of the ASME, Journal of Applied Mechanics, 12 (1965), p. 583-588.

[4.5] Hurty, W.C., Rubinstein, M.F.: Dynamics of Structures. Englewood Cliffs: Prentice-Hall, 1964

[4.6] Itoh, T.: Damped Vibration Mode Superposition Method for Dynamic Response Analysis. Earthquake Engineering and Structural Dynamics 2 (1973), p. 47-57.

[4.7] Roesset, J.M. et al.: Modal Analysis for Structures with Foundation Interaction. ASCE, Journal of the Structural Division 99 (1973), p. 399-416.

[4.8] Mojtahedi, S., Clough, R.W.: Earthquake Response Analysis Considering Non-proportional Damping. Earthquake Engineering and Structural Dynamics 4 (1976), p. 489-496.

[4.9] Newmark, N. M.: A Method of Computation for Structural Dynamics. ASCE Journal of the Engineering Mechanics Division 85 (1959), p. 67-94.

References Chapter 5

[5.1] Paz, M.: Structural Dynamics. New York: Van Nostrand Reinhold Company, 1980.

[5.2] Kolousek, V.: Dynamics in Engineering Structures. London: Butterworth 1973

[5.3] Akesson, B.A.: Byggnadsdynamik. In: BYGG Handbok för hus-, väg- och vattenbyggnad 1B (Allmänna grunder). Stockholm: AB Byggmästarens förlag 1972.

References Chapter 6

[6.1] Petersen, Chr.: Dynamik der Baukonstruktionen. Braunschweig/Wiesbaden: F. Vieweg & Sohn, 1996.

[6.2] Uhrig, R.: Kinetik der Tragwerke - Baudynamik. Mannhein: Bibliographisches Institut & F.A. Brockhaus AG, 1992.

References Chapter 7

[7.1] Lay, T., Wallace, T. G.: Modern Global Seismology. San Diego: Academic Press 1995.

[7.2] Gubbins, D.: Seismology and plate tectonics. Cambridge: Cambridge University Press 1990.

[7.3] Bolt, B.A.: Earthquakes and Geological Discovery. Scientific American Library, New York/Oxford: Freeman & Co., 1993

[7.4] Tselentis, Akis: Synchroni Seismologia (Modern Seismology, in Greek). Athens: Papasotiriou 1997

[7.5] Kanamori H.: The energy release in great earthquakes. Journal of geophysical research 82 (1977), p. 2981-2987

[7.6] Arias, A.: A Measure of Earthquake Intensity. Seismic Design for Nuclear Power Plants, ed. R. Hansen, MIT Press, Cambridge Mass. 1970.

[7.7] Bolt, B.A.: Duration of Strong Ground Motion. Proceedings 5th WCEE, Vol. 1, Rome 1973, p. 1304-1313.

[7.8] Trifunac, M.D., Brady, A.G.: A Study on the Duration of Strong Earthquake Ground Motion. Bulletin of the Seismological Society of America 65 (1975), p. 581-626.

[7.9] Novikova, E.I., Trifunac, M.D.: Duration of Strong Ground Motion in Terms of Earthquake Magnitude, Epicentral Distance, Site Conditions and Site Geometry. Earthquake Engineering and Structural Dynamics 23 (1994), p. 1023-1043.

[7.9] Sucukoglu, H., Nurtug, A.: Earthquake Ground Motion Characteristics and Seimic Energy Dissipation. Earthquake Engineering and Structural Dynamics 24 (1995), p. 1195-1213.

[7.10] Uang, C., Bertero, V.V.: Evaluation of Seismic Energy in Structures. Earthquake Engineering and Structural Dynamics 19 (1990), p. 77-90.

[7.11] Newmark, N.M., Hall, W.J.: Procedures and Criteria for Earthquake Resistant Design. Building Science Series 46, Building Practices For Disaster Mitigation, National Bureau of Standards, Feb. 1973, p. 209-236.

[7.12] User's Manual for SHAKE91 (Original version by Schnabel, Lysmer & Seed, modified by I.M. Idriss and J. I. Sun). Center for Geotechnical Modelling, Department of Civil & Environmental Engineering, University of California, Davis, California, November 1992.

[7.13] Nakamura, Y.: A method for dynamic characteristics estimation of subsurface using microtremor on the ground surface. QR Railway Tech. Res. Inst. 30 (1989), p. 25-33.

[7.14] Coutel, F., Mora, P.: Simulation-Based Comparison of Four Site-response Estimation techniques. Bulletin of the Seismological Society of America 88 (1998), p. 30-42.

[7.15] Wilson, E.L., Der Kiureghian, A., Bayo, E.P.: A replacement for the SRSS method in seismic analysis. Earthquake Engineering and Structural Dynamics 9 (1981), p. 187-192

[7.16] Gupta, A.K., Singh, M.P.: Design of column sections subjected to three components of earthquakes. Nuclear Engineering and Design 41 (1977), p. 129-133.

[7.17] Anastassiadis, K.: Antiseismikes Kataskeves (Seismic Resistant Structures, in Greek), Thessaloniki: CT Computer Technics 1989.

[7.18] Rosenblueth, E., Contreras, H.: Approximate Design for multicomponent earthquakes. ASCE, Journal of the Engineering Mechanics Division 103 (1977).

[7.19] Penelis, G.G., Kappos, A.J.: Earthquake-Resistant Concrete Structures. London: E & FN Spon, 1997.

[7.20] Cook, R.D.: Concepts and Applications of Finite Element Analysis. London: John Wiley 1974.

[7.21] European Committee for Standardization (CEN). Eurocode 8 - Design Provisions for Earthquake Resistance of Structures. European prenorm, Oct. 1994.

[7.22] Paulay, Th., Bachmann, H. und Moser, K.: Erdbebensicherung von Stahlbeton-hochbauten. Basel: Birkhäuser Verlag 1990.

[7.23] Bachmann, H.: Erdbebensicherung von Bauwerken. Basel: Birkhäuser Verlag, 1995.

[7.24] DIN V ENV 1991-1-1: Eurocode 2 - Planung von Stahlbeton- und Spannbeton-tragwerken; Teil 1-1: Grundlagen und Anwendungsregeln für den Hochbau, 1992 edition.

[7.25] Deutscher Auschuß für Stahlbeton. Richtlinie zur Anwendung von Eurocode 2 - Planung von Stahlbeton- und Spannbetontragwerken; Part 1: Grundlagen und Anwendungsregeln für den Hochbau, April 1993.

[7.26] Avramidis, I. Bewertung der Regeln für die rechnerischen Exzentrizitäten in Erdbebenormen. Der Bauingenieur 65 (1990), p. 254-256.

[7.27] Müller, F.P., Keintzel, E.: Erdbebensicherung von Hochbauten. 2nd ed. Berlin: Ernst & Sohn (Wiley), 1984.

References Chapter 8

[8.1] Structural Engineers Association of California (SEAOC): „Vision 2000: Performance Based Seismic Engineering of Buildings" (1995)

[8.2] Linde, P. Numerical Modelling and Capacity Design of Earthquake Resistant Reinforced Concrete Walls. Report no. 200, Institut für Baustatik und Konstruktion, ETH Zürich, Birkhäuser Verlag, 1993.

[8.3] Hanskötter, U. Strategien zur Minimierung des numerischen Aufwands von Schädigungsanalysen seismisch erregter, räumlicher Hochbaukonstruktionen mit gemischten Aussteifungssystemen aus Stahlbeton. Report no. 94-11, Ruhr-University Bochum, KIB-SFB 151, September 1994.

[8.4] Kabeyasawa, H. Shiohara, H. und Otani, S. U.S.-Japan Cooperative Research on R/C Full-Scale Building Test Part 5: Discussion on Dynamic Response System. Proceedings 8th WCEE, Vol. 6, San Francisco 1984.

[8.5] Meyer, I.F.: Ein werkstoffgerechtes Schädigungsmodell und Stababschnittselement für Stahlbeton unter zyklischer nichtlinearer Beanspruchung. Technisch-wiss. Mitteilungen des Instituts für konstruktiven Ingenieurbau, No. 88-4, Ruhr-University Bochum, 1988

[8.6] Garstka, B.: Untersuchungen zum Trag- und Schädigungsverhalten stabförmiger Stahlbetonbauteile mit Berücksichtigung des Schubeinflusses bei zyklischer nichtlinearer Beanspruchung. Technisch-wiss. Mitteilungen des Instituts für konstruktiven Ingenieurbau, No. 93-2, Ruhr-University Bochum, 1993

[8.7] Park, Y.J., Ang, A.H-S.: Mechanistic Seismic Damage Model for Reinforced Concrete. Journal of Structural Engineering, ASCE, Vol. 111, No. 4, 722-739, 1985

[8.8] Linde, P. Analytical Modelling Methods for R/C Shear Walls. Report TVBK-1005, Department of Structural Engineering, Lund Institute of Technology, Sweden, 1989.

[8.9] Valles, R.E., Reinhorn, A.M., Kunnath, S.K., Li, C., Madan, A. IDARC2D version 4.0: A Computer Program for the Inelastic Damage Analysis of Buildings. Technical Report NCEER-96-0010, NCEER, Department of Civil Engineering, State University of New York at Buffalo, 1996.

[8.10] Kappos, A.J., Xenos, A.: A Reassessment of Ductility and Energy-Based Seismic Damage Indices for Reinforced Concrete Structures. Proceedings of the European Conference on Structural Dynamics Eurodyn' 96, Florence, 1996.

[8.11] Paulay, T., Bachmann, H., Moser, K. Erdbebenbemessung von Stahlbetonhochbauten. Basel: Birkhäuser Verlag 1994.

[8.12] Schauerte, J.: Raffineriedruckbehälter unter Erdbebenlast - Vergleichende Untersuchung numerischer Modellbildung unterschiedlicher Komplexität. Diploma thesis at the Chair of Structural Statics and Dynamics of the RWTH Aachen, March 1998

[8.13] Weitkemper, U.: Zur numerischen Untersuchung seismisch erregter Hochbauten mit Aussteifungssystemen aus Stahlbetonwandscheiben. Ph.D. thesis, RWTH Aachen, Chair of Structural Statics and Dynamics, Aachen 1999

References Chapter 9

[9.1] Housner, G.W.: The dynamic behaviour of water tanks. Bulletin of the Seismological Society of America 53 (1963), p. 381-387.

[9.2] API STANDARD 650, Welded Steel Tanks for Oil Storage, Manufacturing, Distribution and Marketing Department, ANSI, Ninth Edition, 1993, Appendix E.

[9.3] Fischer, F.D.: Ein Vorschlag zur erdbebensicheren Bemessung von flüssigkeitsgefüllten zylindrischen Tankbauwerken. Der Stahlbau 50 (1981), p. 13-20.

[9.4] Müller, F.P.; Keintzel, E.: Erdbebensicherung von Hochbauten. Berlin: Ernst & Sohn, (Wiley) 1984, p. 213-218.

[9.5] Hampe, E.: Bemerkungen zum Tragverhalten von Behälterbauwerken unter außergewöhnlichen dynamischen Einwirkungen. Beton- und Stahlbetonbau 80 (1985), p. 123-128.

[9.6] Wilson, E.L.; Khalvati, M.: Finite Elements for the Dynamic Analysis of Fluid-Solid Systems. International Journal for Numerical Methods in Engineering, 1983, p. 1657-1668.

[9.7] Kim, Y.S., Yun, C.B.: A Spurious Free Four-Node Displacement-Based Element for Fluid-Structure Interaction Analysis. Engineering Structures 19 (1997), p. 665-678.

[9.8] Nuclear Reactors and Earthquakes. TID7024, U.S.Atomic Energy Commission, 1963, p. 183-209.

[9.9] Epstein, H.I.: Seismic Design of Liquid Storage Tanks. ASCE, Journal of the Structural Division 102 (1976), p. 1659-1673.

[9.10] Housner, G.W.: Design Spectrum. In: Wiegel, R.L.: Earthquake Engineering. Prentice-Hall Inc., 1970, p. 93-106.

[9.11] Fischer, F.D.: Dynamic Fluid Effects in Liquid-Filled Flexible Cylindrical Tanks. Earthquake Engineering and Structural Dynamics 7 (1979), p. 587-601.

[9.12] Eibl, J.; Stempniewski, L.: Flüssigkeitsbehälter unter äußerem Explosionsdruck. Reports of the Institute for Reinforced Concrete and Material Technology, Karlsruhe Technical University, Karlsruhe, 1987.

[9.13] Stempniewski, L.: Flüssigkeitsgefüllte Stahlbetonbehälter unter Erdbebeneinwirkung. Reports of the Institute for Reinforced Concrete and Material Technology, Karlsruhe Technical University, 1990.

[9.14] Belytschko, T.; Flanagan, D.P.; Kennedy, L.M.: Finite Element Methods with User-Controlled Meshes for Fluid-Structure Interaction. Computer Methods in Applied Mechanics and Engineering, 1982, p. 669-688.

[9.15] Oden, J.T.; Zienkiewicz, O.C.; Gallagher, R.H.; Taylor, C.: Finite Elements in Fluids, Vol. I and II, New York: Wiley, 1975

[9.16] Zienkiewicz, O.C.; Bettes, P.: Fluid-Strukture Dynamic Interaction and Wave Forces. International Journal for Numerical Methods in Engineering, 1987, p. 1-16

Computer programs

1. System requirements

- IBM PC (or 100% compatible) with Windows 95, Windows 98 or Windows NT operating system
- Minimum 2 MB core memory
- Minimum 12 MB free memory available on the hard disk
- Graphics with minimum 800 x 600 pixels, better 1024 x 768 pixels

2. Disclaimer of Warranty

No warranties, express or implied, are made that the programs contained in this volume are free of error, or that they will meet your requirements for any particular application. They should not be relied upon for solving a problem the incorrect solution of which could result in injury to a person or loss of property. If you do use the programs in such a manner, it is at your own risk. The author and publisher disclaim all liability for direct or consequential damages resulting from your use of the programs.

3. License Information

This software contains files to help you utilize the models described in the accompanying book. By opening the package, you are agreeing to be bound by the following agreement:

This software product is protected by copyright and all rights are reserved by the author and Ernst & Sohn. You are licensed to use this software on a single computer. Copying the software to another medium or format for use on a single computer does not violate the Copyright Law. Copying the software for any other purpose is a violation of the Copyright Law.

Program descriptions

<table>
<tr><th colspan="2">AREMOM 1</th></tr>
<tr><td colspan="2">Evaluation of the cross sectional properties of polygonal cross-sections (Section 2.4)</td></tr>
<tr><td colspan="2">Input file:</td></tr>
<tr><td>File name</td><td></td></tr>
<tr><td>KOORD.ARE</td><td>1st row:
Number of cross section vertices, with vertex no. N being the same as vertex no. 1.
2nd to (N+1)st row:
The x- and y- Co-ordinates of the N vertices, two numbers per row.

The sequence of the vertices should be defined so that the cross section will lie on the left hand side of the straight line connecting the vertices n and n+1.</td></tr>
<tr><td colspan="2">Output file:</td></tr>
<tr><td>File name</td><td></td></tr>
<tr><td>MOMENT.ARE</td><td>The file contains the values for the area, the first moments, the second moments, the position of the centroid and the principal axes and principal second moments.</td></tr>
</table>

<table>
<tr><th colspan="2">BODMOM 2</th></tr>
<tr><td colspan="2">Evaluation of second order mass moments for arbitrary polyhedra (Section 2.4)</td></tr>
<tr><td colspan="2">Input file:</td></tr>
<tr><td>File name</td><td></td></tr>
<tr><td>KOORD.BOD</td><td>1st row:
Number m of all planes forming the body surface.
2nd to (m+1)th row:
Number n of the vertices of each of the m plane surfaces, one number per row.

The next rows contain the (x,y,z)- Co-ordinates of the n vertices for all m plane surfaces, one point (three coordinate values) per row.</td></tr>
<tr><td colspan="2">Output file:</td></tr>
<tr><td>File name</td><td></td></tr>
<tr><td>MOMENT.BOD</td><td>It contains the volume, the first moments, the Co-ordinates of the center of gravity and the second order mass moments. The vertex Co-ordinates are also printed for verification.</td></tr>
</table>

FFT1	3
Transformation of a time series from the time into the frequency domain (Section 2.6)	

Interactive in-/output:

Screen prompt	Notes
Number NPKT of data points in the time series (max. 8192)? NANZ, desired number as a power of 2 with NANZ.GE.NPKT (max. 8192)? The power of 2 corresponding to NANZ (max. 13)? Time step DT ?	The time series containing NPKT values is padded by (NANZ-NPKT) zeros; the actual time series to be transformed thus consists of NANZ values at a constant time step DT.

Input file:

File name	
TIMSER.DAT	NPKT rows containing the time t and the corresponding ordinate f(t) in a 2E14.7 format.

Output file:

File name	
OMCOF.DAT	OMCOF.DAT contains the computed (NANZ/2) complex FOURIER coefficients, with the circular frequencies in the first, the real part of the coefficient in the second and its imaginary part in the third column (3E14.7 format).
OMQU.DAT	OMQU.DAT contains the circular frequencies in the first column and the square value of the FOURIER coefficients in the second column (2E14.7 format).

FFT2	4
Transformation from the frequency domain back into the time domain (Section 2.6)	

Interactive in-/output:

Screen prompt	Notes
Number NANZ of time series data points (must be a power of 2, max. 8192 values)? Power of 2 corresponding to NANZ (max. 13)? Circular frequency step DOM = 2*pi/(NANZ*DT)?	(NANZ/2) complex values are read from the file OMCOF.DAT. They are the FOURIER coefficients at a frequency step of $\Delta\omega = \frac{2\pi}{NANZ \cdot DT}$

Input file:

File name	
OMCOF.DAT	OMCOF.DAT contains (NANZ/2) complex FOURIER coefficients, with the circular frequencies in the first, the real part of the coefficient in the second and its imaginary part in the third column (3E14.7 format).

Output file:

File name	
RESULT.DAT	Contains the corresponding time series (NANZ values), with the times in the first and the ordinates in the second column (2E14.7 format).

AUTKOR		5
Evaluation of the autocorrelation function of a time series (Section 2.6)		
Interactive in-/output:		
Screen prompt	Notes	
Number N of data points in the time series ? Time step DT? Ordinates of the autocorrelation function to be computed ?	The mean value, the variance, and the standard deviation of the series are also output on screen.	
Input file:		
File name		
TIMSER.DAT	N rows containing [t, F(t)] in a 2E14.7 format.	
Output file:		
File name		
KORREL.DAT	The autocorrelation function with the correlation length as a multiple of the time step DT in the first column and the ordinates in the second column (2E14.7 format).	

LININT		6
Linear interpolation of a polygonal function f(t) (Section 2.6)		
Interactive in-/output:		
Screen prompt	Notes	
Number NPKT of points defining the function? Constant interpolation time step? Number NANZ of ordinates to be computed?	The first point of the interpolated function is identical to the first point given in the file FKT.	
Input file:		
File name		
FKT	NPKT rows containing the [t, f(t)] co-ordinates of the points defining the function (assumed to be linear between these points).	
Output file:		
File name		
FKTINT	NANZ rows in a 2E14.7 format, with times in the first and function ordinates in the second column.	

DUHAMI	7
Evaluation of the particular solution for the damped SDOF system (Section 3.2)	

Interactive in-/output:

Screen prompt	Notes
Number NANZ of load function ordinates? Natural period T of the SDOF system? Damping ratio of the SDOF system? Constant time step DT? Load factor?	If the ordinates of the load function (right-hand side) have not already been divided by the mass m of the SDOF system, (1/m) should be entered as load factor. The calculated absolute maximum displacement is also output on screen.

Input file:

File name	
RHS	NANZ rows with [t, f(t)] of the load function (right-hand side of the differential equation). The time step DT (2E14.7 format) is constant. If the ordinates of the load function have not already been divided by the mass of the SDOF system, (1/m) should be entered as load factor.

Output file:

File name	
DUHOUT	Time and displacement values in a 2E14.7 format.

LEINM	8
Numerical integration of the equation of motion for the SDOF system by NEWMARK's method (Section 3.2)	

Interactive in-/output:

Screen prompt	Notes
Data of the SDOF system: Circular natural frequency? Damping ratio? Displacement for t=0? Velocity for t=0? Data of the load function: Number NANZ of time steps? Constant time step?	
Acceleration of gravity g in the units employed ? (e.g. 9.81)	This is required for outputting accelerations in units of g.
Load factor?	If the ordinates of the load function have not already been divided by the mass of the SDOF system, (1/m) should be entered as load factor. The displacement, velocity and acceleration maxima with the corresponding times are also output on screen.

Input file:

File name	
RHS	NANZ rows with [t, f(t)] of the load function (right-hand side of the differential equation). The time step DT (2E14.7 format) is constant. If the ordinates of the load function have not already been divided by the mass of the SDOF system, (1/m) should be entered as load factor.

Output file:

File name	
THNEW	The time and the calculated values for displacement, velocity and acceleration (the latter in units of g) are given in a 4E14.7 format.

<table>
<tr><th colspan="2">FILTER</th><th>9</th></tr>
<tr><td colspan="3">Filtering of a time series in the frequency domain (high-, low- and band-pass filter) (Section 3.5)</td></tr>
<tr><th colspan="3">Interactive in-/output:</th></tr>
<tr><th colspan="2">Screen prompt</th><th>Notes</th></tr>
<tr><td colspan="2">Number NANZ of the time series ordinates (max. 4096)?
Next higher exponent of 2 ? (e.g. 12 for 4096 points)

Constant time step?</td><td>Because of the FFT algorithm employed, the series is padded with zeros until its number of points is equal to a power of 2.</td></tr>
<tr><td colspan="2">High-pass filtering desired? Y/N
Circular corner frequency of the high-pass filter?

Low-pass filtering desired? Y/N
Circular frequency of the KANAI-TAJIMI filter?
Damping ratio of the KANAI-TAJIMI filter?

Band-pass filtering desired? Y/N
Circular frequencies OMH and OMT of the band-pass filter?</td><td>The filter parameters for the desired filter are entered.</td></tr>
<tr><th colspan="3">Input file:</th></tr>
<tr><th>File name</th><td colspan="2"></td></tr>
<tr><td>TIMSER.DAT</td><td colspan="2">The time series F(t) to be filtered, NANZ rows containing [t, F(t)] in a 2E14.7 format.</td></tr>
<tr><th colspan="3">Output file:</th></tr>
<tr><th>File name</th><td colspan="2"></td></tr>
<tr><td>FILT.DAT</td><td colspan="2">The resulting filtered time series, with times in the first and ordinates in the second column (2E14.7 format).</td></tr>
</table>

NLM	10
Time history analysis for a non-linear SDOF system; elastoplastic, bilinear or UMEMURA-type constitutive laws possible (Section 3.7)	

Interactive in-/output:

Screen prompt	Notes
Data of the nonlinear SDOF system: Bilinear constitutive law? Y/N Elastoplastic constitutive law? Y/N UMEMURA constitutive law? Y/N	
Strain hardening ratio p as proportion pK of the initial stiffness?	Only required for the bilinear constitutive law.
Circular natural frequency? Damping ratio?	
Mass of the SDOF system?	Required for the evaluation of the restoring force time history.
Initial conditions: Displacement for t=0? Velocity for t=0?	
Max. elastic displacement?	This is the limit displacement for the validity of HOOKE's law.
Load function data: Number NANZ of time steps? Constant time step?	
Acceleration of gravity g in the units employed? (e.g. 9.81)	This is required for outputting accelerations in units of g.
Load factor?	If the ordinates of the load function have not already been divided by the mass of the SDOF system, (1/m) should be entered as load factor. The displacement, velocity and acceleration maxima with the corresponding times are also output on screen.

Input files:

File name	
RHS	NANZ rows with [t, f(t)] of the load function (right-hand side of the differential equation). The time step DT (2E14.7 format) is constant. If the ordinates of the load function have not already been divided by the mass of the SDOF system, (1/m) should be entered as load factor.

Output files:

File name	
THNLM	The output file THNLM contains, in five columns (5E14.7 format), the times, the calculated values for displacement, velocity, and acceleration (the latter in units of g) and the restoring force $F_R(t)$.

RAHMEN	11
Evaluation of the internal forces and displacements in a plane structural frame with an arbitrary number of elastic supports due to static nodal forces and moments (Section 4.2)	

Interactive in-/output:

Screen prompt	Notes
Number NDOF of the system degrees of freedom ? Number NELEM of elements? Number NFED of elastic support matrices?	Each elastic support matrix couples a number of active kinematic degrees of freedom with one another and/or with the ground (degree of freedom 0).
Dimensions of the square elastic support matrices (NFED numbers) ?	The dimension of each elastic support matrix is equal to the number of the degrees of freedom it connects.

Input files:

File name	
ERAHM	The first NELEM rows contain EI, ℓ , EA and α for each beam element, in free format. EI is its constant bending stiffness (e.g. in kNm^2), ℓ is the element length (e.g. in m), EA is the axial stiffness (e.g.. in kN) and α is the angle between the global x-axis and the element axis (in degrees, positive counter-clockwise). The next NELEM rows contain the incidence vectors for all elements in free format, indicating the 6 global degrees of freedom corresponding to the 6 local degrees of freedom (u_1, w_1, φ_1, u_2, w_2, φ_2) of each element. Next, the NDOF load components (nodal forces and nodal moments) corresponding to the active kinematic degrees of freedom are entered, in free format.
INZFED	NFED rows, one for each elastic support matrix, containing the numbers of the system degrees of freedom which are constrained by this elastic support (free format).
FEDMAT	The coefficients of all NFED elastic support matrices, in free format.

Output files:

File name	
ARAHM	The output file ARAHM contains the computed displacements in all NDOF system degrees of freedom followed by the deformations and end forces/moments of each element. All end forces/moments refer to the global (x,z) coordinate system and are presented in the following sequence for both element ends: Horizontal (displacement or force) component, vertical (displacement or force) component, rotation or bending moment.

<table>
<tr><th colspan="2">VOLLST</th><th>12</th></tr>
<tr><td colspan="3">Evaluation of the complete (not condensed) stiffness matrix and the consistent mass matrix for a plane frame structure (Section 4.2)</td></tr>
<tr><th colspan="3">Interactive in-/output:</th></tr>
<tr><th colspan="2">Screen prompt</th><th>Notes</th></tr>
<tr><td colspan="2">Number NDOF of the system degrees of freedom ?
Number NELEM of elements?
Number NFED of elastic support matrices?</td><td>Each elastic support matrix couples a number of active kinematic degrees of freedom with one another and/or with the ground (degree of freedom 0).</td></tr>
<tr><td colspan="2">Dimensions of the square elastic support matrices (NFED numbers) ?</td><td>The order of each elastic support matrix is equal to the number of the degrees of freedom it connects.</td></tr>
<tr><th colspan="3">Input files:</th></tr>
<tr><th>File name</th><th colspan="2"></th></tr>
<tr><td>ERAHM</td><td colspan="2">The first NELEM rows contain EI, ℓ, EA and α for each beam element, in free format. EI is its constant bending stiffness (e.g. in kNm^2), ℓ is the element length (e.g. in m),EA is the axial stiffness (e.g.. in kN) and α is the angle between the global x-axis and the element axis (in degrees, positive counter-clockwise).
The next NELEM rows contain the incidence vectors for all elements in free format, indicating the 6 global degrees of freedom corresponding to the 6 local degrees of freedom (u_1, w_1, φ_1, u_2, w_2, φ_2) of each element.
Next, the NDOF load components (nodal forces and nodal moments) corresponding to the active kinematic degrees of freedom are entered, in free format.</td></tr>
<tr><td>INZFED</td><td colspan="2">NFED rows, one for each elastic support matrix, containing the numbers of the system degrees of freedom which are constrained by this elastic support (free format).</td></tr>
<tr><td>FEDMAT</td><td colspan="2">The coefficients of all NFED elastic support matrices, in free format.</td></tr>
<tr><td>EMAS</td><td colspan="2">NELEM values of the (constant) mass per unit length for all elements (e.g. in t/m) in free format.</td></tr>
<tr><th colspan="3">Output files:</th></tr>
<tr><th>File name</th><th colspan="2"></th></tr>
<tr><td>KVOLL</td><td colspan="2">Complete square stiffness matrix (NDOF * NDOF values in 6E14.7 format).</td></tr>
<tr><td>MVOLL</td><td colspan="2">Consistent mass matrix (NDOF * NDOF values in 6E14.7 format).</td></tr>
</table>

KONDEN	13
Performs a static condensation for a plane frame (Section 4.3)	

Interactive in-/output:

Screen prompt	Notes
Number NDOF of the system degrees of freedom?	
Number NDU of the master degrees of freedom?	
Number NELEM of elements?	
Number NFED of elastic support matrices?	Each elastic support matrix couples a number of active kinematic degrees of freedom with one another and/or with the ground (degree of freedom 0).
Dimensions of the square elastic support matrices (NFED numbers) ?	The dimension of each elastic support matrix is equal to the number of the degrees of freedom it connects.

Input files:

File name	
EKOND	The first NELEM rows contain EI, ℓ, EA and α for each beam element, in free format. EI is its constant bending stiffness (e.g. in kNm^2), ℓ is the element length (e.g. in m),EA is the axial stiffness (e.g.. in kN) and α is the angle between the global x-axis and the element axis (in degrees, positive counter-clockwise). The next NELEM rows contain the incidence vectors for all elements in free format, indicating the 6 global degrees of freedom corresponding to the 6 local degrees of freedom (u_1, w_1, φ_1, u_2, w_2, φ_2) of each element. Next, the NDU numbers of the master kinematic degrees of freedom are entered, in free format.
INZFED	NFED rows, one for each elastic support matrix, containing the numbers of the system degrees of freedom which are constrained by this elastic support (free format).
FEDMAT	The coefficients of all NFED elastic support matrices, in free format.

Output files:

File name	
KMATR	The computed condensed stiffness matrix (NDU * NDU coefficients, 6E14.7 format).
AMAT	The matrix A (NDOF * NDU) serves for evaluating the displacements in all degrees of freedom from known displacements in the NDU master degrees of freedom.

JACOBI	14
Solution of the general linear eigenvalue problem by the JACOBI method (Section 4.5)	

Interactive in-/output:

Screen prompt	Notes
Number of equations?	Number NDU of the master degrees of freedom, this being the order of the stiffness and mass matrix.

Input files:

File name	
KMATR	The condensed (NDU *NDU) stiffness matrix in free format or in 6E14.7 format, as computed by the program KONDEN.
MDIAG	The diagonal of the mass matrix consisting of the NDU masses in the master degrees of freedom (free format).

Output files:

File name	
AUSJAC	All eigenvalues and eigenvectors, the latter normalized for unit modal masses.
OMEG	This file, which contains the natural frequencies ω_i of the system, serves as the input file for a series of other programs.
PHI	This file, which contains the eigenvectors $\underline{\Phi}_i$ of the system, serves as the input file for a series of other programs.

INTERP	15
Linear interpolation of a time function as in LININT, here carried out for several components of a load vector which have different amplitudes but the same time function (Section 4.5)	

Interactive in-/output:

Screen prompt	Notes
Number NPKT of points defining the function? Constant interpolation step? Number NANZ of ordinates to be computed? Number of rows of the load vector ? Should the time be written in the first column of the output file (Y/N) ?	

Input files:

File name	
FKT	NPKT rows containing the [t, f(t)] co-ordinates of the points defining the time function which is assumed to be linear between these points.
AMPL	NDU amplitudes of the load vector components in free format.

Output file:

File name	
LASTV	If the time points are to be output, LASTV contains (NDU+1) columns in (E14.7, 30E12.4) format, with the time points in the first column and the components of the load vector in the next NDU columns. If no time points are to be output, LASTV consists of NDU columns with NANZ rows, giving the time histories of the NDU load vector components in free format.

MODAL	16
Modal analysis of MDOF systems after solving the eigenvalue problem by direct integration of the uncoupled modes (Section 4.5)	

Interactive in-/output:

Screen prompt	Notes
Number NDU of the master degrees of freedom ? Number NDPHI of the slave degrees of freedom ? Number NMOD of modes to be considered?	The NMOD circular natural frequencies and mode shapes are read from the files OMEG and PHI.
Number NT of time steps (= ordinates of the load vector) ? Constant time step DT?	
Damping of mode no. ... ?	Must be entered for each of the NMOD modes considered.

Input files:

File name	
MDIAG	The diagonal of the mass matrix consisting of the NDU masses in the master degrees of freedom, in free format.
OMEG	The eigenvalues ω_i of the system; this file is created by the JACOBI program
PHI	The mode shapes $\underline{\Phi}_i$ of the system; this file is created by the JACOBI program
V0	NDU values in free format indicating the displacements in the master degrees of freedom at time t=0.
VP0	NDU values in free format indicating the velocities in the master degrees of freedom at time t=0.
LASTV	Contains NT*NDU values corresponding to the NDU components of the load vector at all NT time points in free format. This file may be created by the INTERP program (without writing the time points in the output file).
AMAT	The matrix A (NDOF * NDU) serves for evaluating the displacements in all degrees of freedom from known displacements in the NDU master degrees of freedom.

Output files:

File name	
THIS.MOD	THIS.MOD contains the time histories of displacements in the NDU master degrees of freedom (F7.4, 30E14.6 format), with the time points in the first column.
THISDU	THISDU contains the displacements in the NDU master degrees of freedom in all NT time points in free format, without the time points. This file serves as input for the program INTFOR.
THISDG	THISDG contains the displacements in all (NDU+NDPHI) degrees of freedom in free format, without the corresponding time points.

<table>
<tr><th colspan="2">INTFOR</th><th>17</th></tr>
<tr><td colspan="3">Evaluation of internal forces and displacements in plane frames from known displacements in the master degrees of freedom (Section 4.5)</td></tr>
<tr><th colspan="3">Interactive in-/output:</th></tr>
<tr><th colspan="2">Screen prompt</th><th>Notes</th></tr>
<tr><td colspan="2">Number NDU of the master degrees of freedom?
Number NELEM of elements?
Number NDPHI of the slave degrees of freedom?
Number NT of time steps?
Constant time step DT?
For which time step should the state variables be computed?</td><td>Enter time step number.</td></tr>
<tr><td colspan="2">For which internal force component should the extreme values (max/min) be calculated?
Enter 1 for horizontal component
Enter 2 for vertical component
Enter 3 for bending moment</td><td>Evaluation of the positive and negative maximum values of the desired end force component in the entire frame. The remaining force components occuring at that time and the time points are also output in the file MAXMIN.</td></tr>
<tr><td colspan="2">Should a state variable time history be output into the file THHVM ?
Enter 1 for yes, 0 for no.
Corresponding element number?
End 1 or end 2 of the element?
Displacement (enter 1) or internal force (enter 2)?
Horizontal (1), vertical (2) or rotation (3)?
Horizontal (1), vertical (2) or moment (3)?</td><td>If „Yes“, the pertinent variable is specified through the following queries.

For displacement time histories.
For internal force time histories.</td></tr>
<tr><th colspan="3">Input files:</th></tr>
<tr><th>File name</th><th colspan="2"></th></tr>
<tr><td>EKOND</td><td colspan="2">The first NELEM rows contain EI, ℓ , EA and α for each beam element, in free format.
EI is its constant bending stiffness (e.g. in kNm^2), ℓ is the element length (e.g. in m),EA is the axial stiffness (e.g.. in kN) and α is the angle between the global x-axis and the element axis (in degrees, positive counter-clockwise).
The next NELEM rows contain the incidence vectors for all elements in free format, indicating the 6 global degrees of freedom corresponding to the 6 local degrees of freedom (u_1, w_1, φ_1, u_2, w_2, φ_2) of each element.
Next, the NDU numbers of the master kinematic degrees of freedom are entered, in free format.</td></tr>
<tr><td>THISDU</td><td colspan="2">THISDU contains the displacements in the NDU master degrees of freedom in all NT time points in free format, without the time points. It is created by the programs MODAL or NEWMAR.</td></tr>
<tr><td>AMAT</td><td colspan="2">The matrix A (NDOF * NDU) serves for evaluating the displacements in all degrees of freedom from known displacements in the NDU master degrees of freedom.</td></tr>
<tr><th colspan="3">Output files:</th></tr>
<tr><th>File name</th><th colspan="2"></th></tr>
<tr><td>FORSTA</td><td colspan="2">The displacements and end forces/moments at the given time point, sorted by element number. All state variables refer to the global (x,z)-coordinate system and are arranged in the following sequence: (Horizontal component, vertical component, rotation or bending moment) at end 1 and end 2 of the beam element.</td></tr>
<tr><td>THHVM</td><td colspan="2">Time history of an internal force or displacement component, with the times in the first column and the ordinates in the second column (2E14.7 format).</td></tr>
<tr><td>MAXMIN</td><td colspan="2">This file contains, for each element, the evaluated maximum and minimum values of the internal force (H, V or bending moment) at both element ends, its time of occurrence and the other internal force components acting simultaneously.</td></tr>
</table>

EIGVOL — 18

Evaluation of the first natural period and the corresponding mode shape for a frame without previous condensation to master degrees of freedom (Section 4.6)

Interactive in-/output:

Screen prompt	Notes
Number N of system degrees of freedom ? Number of iterations?	N is the number of the system degrees of freedom A number in the order of 10 to 20 is recommended. The calculated first natural period and the corresponding natural frequency are also output on screen.

Input files:

File name	
KVOLL MVOLL	Both files are created by the VOLLST program.

Output file:

File name	
AUSVOL	This file contains the first natural period and the components of the first mode shape.

ALFBET — 19

Evaluation of α and β in the RAYLEIGH damping expression $\underline{C} = \alpha\underline{M} + \beta\underline{K}$ and determination of the damping ratio D for a series of natural periods (Section 4.7)

Interactive in-/output:

Screen prompt	Notes
Type of damping: For mass and stiffness proportional damping (general case), enter 1 for IKN. For mass proportional damping enter 2 for IKN. For stiffness proportional damping enter 3 for IKN. IKN=?	
Enter the first natural period T1 (T1>T2): Corresponding damping ratio? Enter the second natural period T2 (T2<T1): Corresponding damping ratio?	For IKN equal to 1
Enter the natural period (in s): Corresponding damping ratio?	For IKN equal to 2 or 3
Evaluation of the damping ratio D for a series of periods (NPER values, starting at TANF with an increment DPER): Starting value TANF? Period increment DPER? Number of periods?	After evaluating α and/or β the damping is established for all natural periods.

Output file:

File name	
APERD	NPER rows containing the periods and the computed damping ratios in 2E14.7 format.

CRAY — 20

Evaluation of $\underline{C}=\alpha\underline{M}+\beta\underline{K}$, $\underline{C}=\alpha\underline{M}$ or $\underline{C}=\beta\underline{K}$ (Section 4.7)

Interactive in-/output:

Screen prompt	Notes
Dimension NDU of the square matrices M, C, K? Type of the desired damping matrix C: C = alfa * M + beta * K: IKN=1 C = alfa * M : IKN=2 C = beta * K : IKN=3 IKN=?	
Enter the first natural period T1 (T1>T2): Corresponding damping ratio?	For IKN = 1 (general RAYLEIGH damping)
Enter the second natural period T2 (T2<T1): Corresponding damping ratio?	
Enter the natural period (in s): Corresponding damping ratio?	For IKN = 2 or 3 The evaluated values α, β are output on screen.

Input files:

File name	
MDIAG	The diagonal of the mass matrix (NDU coefficients in free format).
KMATR	Contains the condensed stiffness matrix of the system (NDU*NDU values) as created by the program KONDEN.

Output file:

File name	
CMATR	Damping matrix, NDU*NDU coefficients in free format.

CMOD — 21

Evaluation of the damping matrix $\underline{C}$ by means of a complete modal approach (Section 4.7)

Interactive in-/output:

Screen prompt	Notes
Dimension NDU of the square matrices M, C, K ? Number NMOD of modes to be considered?	

Input files:

File name	
MDIAG	The diagonal of the mass matrix (NDU values in free format).
OMEG	The eigenvalues ω_i of the system; this file is created by the JACOBI program
PHI	The mode shapes $\underline{\Phi}_i$ of the system; this file is created by the JACOBI program
DAEM	This file contains the NMOD damping ratios for the NMOD modes, in free format.

Output file:

File name	
CMATR	Resulting damping matrix (NDU*NDU values in free format).

FTCF		22
Evaluation of the $\underline{CC} = \underline{\Phi}^T \underline{C} \underline{\Phi}$ matrix (Section 4.7)		
Interactive in-/output:		
Screen prompt	Notes	
Dimension NDU of the square matrix C? Number NMOD of modes to be considered?		

Input files:

File name	
CMATR	Damping matrix, e.g. as created by the programs CMOD or CRAY.
PHI	The mode shapes $\underline{\Phi}_i$ of the system; this file is created by the JACOBI program

Output file:

File name	
CCMAT	The matrix $\underline{CC} = \underline{\Phi}^T \underline{C} \underline{\Phi}$, NMOD*NMOD coefficients, in free format.

CALLG	23
Setting up $\underline{C}$ by assembling the stiffness proportional damping matrices $\beta_i \underline{K}_i$ of all elements (Section 4.7)	

Interactive in-/output:

Screen prompt	Notes
Number NDOF of the system degrees of freedom?	
Number NELEM of elements?	
Number NFED of additional explicit damping matrices?	Each additional explicit damping matrix couples a number of active kinematic degrees of freedom with one another and/or with the ground (degree of freedom 0).
Dimensions of all explicit damping matrices (NFED numbers)?	If NFED > 0: Each damping matrix dimension is equal to the number of coupled degrees of freedom.

Input files:

File name	
EKOND	The first NELEM rows contain EI, ℓ, EA and α for each beam element, in free format. EI is its constant bending stiffness (e.g. in kNm^2), ℓ is the element length (e.g. in m),EA is the axial stiffness (e.g.. in kN) and α is the angle between the global x-axis and the element axis (in degrees, positive counter-clockwise). The next NELEM rows contain the incidence vectors for all elements in free format indicating the 6 global degrees of freedom corresponding to the 6 local degrees of freedom (u_1, w_1, φ_1, u_2, w_2, φ_2) of each element.
INZFED	NFED rows, one for each explicit additional damping matrix, containing the numbers of the system degrees of freedom which are coupled by this matrix (free format).
FEDCMT	The coefficients of all NFED explicit additional damping matrices, in free format.
BETA	NELEM coefficients $\beta = D \cdot T/\pi$ (D = damping ratio, T = period in s), to be used as factors of the element stiffness matrices for determination of the stiffness proportional damping matrices, in free format.

Output file:

File name	
CVOLL	Resulting damping matrix in 6E14.5 format.

NEWMAR	24
Solution of the coupled differential equation system by implicit direct integration (NEWMARK's method, Section 4.8)	

Interactive in-/output:

Screen prompt	Notes
Number NDU of the master degrees of freedom ? Number NT of time steps (= ordinates of the load vector) ? Constant time step DT? Should displacements, velocities or accelerations be written in the output file THIS.NEW? Enter IOUT=1 for displacements Enter IOUT=2 for velocities Enter IOUT=3 for accelerations IOUT = ?	
The acceleration will be output in your units (e.g. in m/s**2), not in g!	This message will appear only if IOUT=3.
	The maximum absolute value of the displacement will be printed on screen together with the time it occurred, the time step number and the corresponding degree of freedom.

Input files:

File name	Notes
KMATR	Contains the condensed stiffness matrix of the system (NDU*NDU values) as created by the program KONDEN.
MDIAG	The diagonal of the mass matrix (NDU coefficients in free format).
CMATR	(NDU*NDU) damping matrix, e.g. as created by CRAY or CMOD.
V0	NDU values in free format indicating the displacements in the master degrees of freedom at time t=0.
VP0	NDU values in free format indicating the velocities in the master degrees of freedom at time t=0.
LASTV	Contains NT*NDU values corresponding to the NDU components of the load vector at all NT time points in free format. This file may be created by the INTERP program (without writing the time points in the output file).

Output files:

File name	
THIS.NEW	This file contains the time points in the first column, and either the displacements, the velocities or the accelerations in the NDU master degrees of freedom in the following columns (F7.4, 30E14.6 format).
THISDU	THISDU contains the displacements in the NDU master degrees of freedom in all NT time points in free format, without the time points. This file serves as input for the program INTFOR.

<table>
<tr><td colspan="2">EULBER</td><td>25</td></tr>
<tr><td colspan="3">Evaluation of the end forces/displacements for plane frames under stationary-harmonic excitation (circular frequency ω) according to the EULER/BERNOULLI (1st order) theory (Section 5.7)</td></tr>
<tr><td colspan="3">Interactive in-/output:</td></tr>
<tr><td colspan="2">Screen prompt</td><td>Notes</td></tr>
<tr><td colspan="2">Number NDOF of the system degrees of freedom?
Number NELEM of elements?
Circular excitation frequency ?</td><td>

ω, in rad/s, is the same for all load components.</td></tr>
<tr><td colspan="3">Input files:</td></tr>
<tr><td>File name</td><td colspan="2"></td></tr>
<tr><td>EEBN</td><td colspan="2">The first NELEM rows contain EI, ℓ, EA, m and α for each beam element, in free format. EI is its constant bending stiffness (e.g. in kNm^2), ℓ is the element length (e.g. in m), EA is the axial stiffness (e.g.. in kN), m is the (constant) mass per unit length (e.g. in t/m) and α is the angle between the global x-axis and the element axis (in degrees, positive counter-clockwise).
The next NELEM rows contain the incidence vectors for all elements in free format, indicating the 6 global degrees of freedom corresponding to the 6 local degrees of freedom (u_1, w_1, φ_1, u_2, w_2, φ_2) of each element.
Next, the NDOF load components (nodal forces and nodal moments) corresponding to the active kinematic degrees of freedom are entered, in free format. These components are the amplitudes F_i of harmonic loads $F_i \sin \omega t$.</td></tr>
<tr><td colspan="3">Output file:</td></tr>
<tr><td>File name</td><td colspan="2"></td></tr>
<tr><td>AEBN</td><td colspan="2">After an echo print of the input data, displacements and element internal forces are written into the output file. All end forces/displacements refer to the global (x,z)-coordinate system and are given in the sequence (Horizontal component, vertical component, rotation or bending moment) at each member end.</td></tr>
</table>

EUBER2	26
Evaluation of the end forces/displacements for plane frames under stationary-harmonic excitation (circular frequency ω) according to the EULER/BERNOULLI (2nd order) theory (Section 5.7)	

Interactive in-/output:

Screen prompt	Notes
Number NDOF of the system degrees of freedom? Number NELEM of elements ? Circular excitation frequency ?	ω, in rad/s, is the same for all load components.

Input file:

File name	Notes
EEB2	The first NELEM rows contain the six values EI, ℓ , EA, m, α and D for each beam element, in free format. EI is its constant bending stiffness (e.g. in kNm^2), ℓ is the element length (e.g. in m), EA is the axial stiffness (e.g.. in kN), m is the (constant) mass per unit length (e.g. in t/m), α is the angle between the global x-axis and the element axis (in degrees, positive counter-clockwise) and D (e.g. in kN) is the constant member axial force (always assumed to be a compressive force). The next NELEM rows contain the incidence vectors for all elements in free format, indicating the 6 global degrees of freedom corresponding to the 6 local degrees of freedom (u_1, w_1, φ_1, u_2, w_2, φ_2) of each element. Next, the NDOF load components (nodal forces and nodal moments) corresponding to the active kinematic degrees of freedom are entered, in free format. These components are the amplitudes F_i of harmonic loads $F_i \sin \omega t$.

Output file:

File name	
AEB2	After an echo print of the input data, displacements and element internal forces are written into the output file. All end forces/displacements refer to the global (x,z)-coordinate system and are given in the sequence (horizontal component, vertical component, rotation or bending moment) at each member end.

<table>
<tr><th colspan="2">TIMOSH</th><th>27</th></tr>
<tr><td colspan="3">Evaluation of the end forces/displacements for plane frames under stationary-harmonic excitation (circular frequency ω) according to the TIMOSHENKO theory (Section 5.7)</td></tr>
<tr><th colspan="3">Interactive in-/output:</th></tr>
<tr><th colspan="2">Screen prompt</th><th>Notes</th></tr>
<tr><td colspan="2">Number NDOF of the system degrees of freedom?
Number NELEM of elements?
Circular excitation frequency ?</td><td>

ω, in rad/s, is the same for all load components.</td></tr>
<tr><th colspan="3">Input file:</th></tr>
<tr><th>File name</th><th colspan="2"></th></tr>
<tr><td>ETIM</td><td colspan="2">NELEM rows containing EI, ℓ, EA, m, α, i and GA_s for each member in free format.
EI is the constant bending stiffness (e.g. in kNm^2), ℓ is the element length (e.g. in m), EA is the axial stiffness (e.g.. in kN), m is the (constant) mass per unit length (e.g. in t/m), α is the angle between the global x-axis and the element axis (in degrees, positive counter-clockwise), i is the radius of gyration (e.g. in m) and GA_s is the shear stiffness (e.g. in kN).
The next NELEM rows contain the incidence vector for all elements in free format, indicating the 6 global degrees of freedom corresponding to the 6 local degrees of freedom (u_1, w_1, φ_1, u_2, w_2, φ_2) for each element.
Next, the NDOF load components (nodal forces and nodal moments) corresponding to the active kinematic degrees of freedom are entered, in free format. These components are the amplitudes F_i of harmonic loads $F_i \sin \omega t$.</td></tr>
<tr><th colspan="3">Output file:</th></tr>
<tr><th>File name</th><th colspan="2"></th></tr>
<tr><td>ATIM</td><td colspan="2">After an echo print of the input data, displacements and element internal forces are written into the output file. All end forces/displacements refer to the global (x,z)-coordinate system and are given in the sequence (Horizontal component, vertical component, rotation or bending moment) at each member end.</td></tr>
</table>

EUBFRQ	28
Evaluation of the natural frequencies for plane frames according to the 1st order EULER/BERNOULLI theory (Section 5.7)	

Interactive in-/output:

Screen prompt	Eingabe/Notes
Number NDOF of the system degrees of freedom?	
Number NELEM of elements ?	
No. of the indicative degree of freedom?	The reciprocal value of the square root of the displacement in this indicative degree of freedom (due to a unit harmonic load) is evaluated as a function of the circular frequency (NOM ω-values) and output in the file AEBNFR.
Initial value for the circular frequency?	
Circular frequency increment?	
Number NOM of values to be computed?	The computed values are also output on screen.

Input file:

File name	
EEBN	The first NELEM rows contain EI, ℓ , EA, m and α for each beam element, in free format. EI is its constant bending stiffness (e.g. in kNm^2), ℓ is the element length (e.g. in m), EA is the axial stiffness (e.g.. in kN), m is the (constant) mass per unit length (e.g. in t/m) and α is the angle between the global x-axis and the element axis (in degrees, positive counter-clockwise). The next NELEM rows contain the incidence vectors for all elements in free format, indicating the 6 global degrees of freedom corresponding to the 6 local degrees of freedom (u_1, w_1, φ_1, u_2, w_2, φ_2) of each element.

Output file:

File name	
ECHO	Echo print of the input data.
AEBNFR	The circular frequencies and the reciprocals of the square root of the displacements in the indicative degree of freedom, in two columns (2E16.7 format). The roots (zero crossings) of this curve correspond to the circular natural frequencies of the system.

<table>
<tr><th colspan="2">EB2FRQ</th><th>29</th></tr>
<tr><td colspan="3">Evaluation of the natural frequencies for plane frames according to the 2nd order EULER/BERNOULLI theory (Section 5.7)</td></tr>
<tr><th colspan="3">Interactive in-/output:</th></tr>
<tr><th colspan="2">Screen prompt</th><th>Notes</th></tr>
<tr><td colspan="2">Number NDOF of the system degrees of freedom?
Number NELEM of elements ?
Factor for the beam axial forces?

No. of the indicative degree of freedom?

Initial value for the circular frequency?
Circular frequency increment ?
Number NOM of values to be computed ?</td><td>

The values in the 6th column of the input file EEB2 are multiplied by this factor.
The reciprocal value of the square root of the displacement in this indicative degree of freedom (due to a unit harmonic load) is evaluated as a function of the circular frequency (NOM ω-values) and output in the file AEB2FR.

The computed values are also output on screen.</td></tr>
<tr><th colspan="3">Input file:</th></tr>
<tr><th>File name</th><th colspan="2"></th></tr>
<tr><td>EEB2</td><td colspan="2">The first NELEM rows contain the six values EI, ℓ , EA, m, α and D for each beam element, in free format.
EI is its constant bending stiffness (e.g. in kNm^2), ℓ is the element length (e.g. in m), EA is the axial stiffness (e.g.. in kN), m is the (constant) mass per unit length (e.g. in t/m), α is the angle between the global x-axis and the element axis (in degrees, positive counter-clockwise) and D (e.g. in kN) is the constant member axial force (always assumed to be a compressive force).
The next NELEM rows contain the incidence vectors for all elements in free format, indicating the 6 global degrees of freedom corresponding to the 6 local degrees of freedom (u_1, w_1, φ_1, u_2, w_2, φ_2) of each element.</td></tr>
<tr><th colspan="3">Output file:</th></tr>
<tr><th>File name</th><th colspan="2"></th></tr>
<tr><td>ECHO
AEB2FR</td><td colspan="2">Echo print of the input data.
The circular frequencies and the reciprocals of the square root of the displacements in the indicative degree of freedom, in two columns (2E16.7 format). The roots (zero crossings) of this curve correspond to the circular natural frequencies of the system.</td></tr>
</table>

TIMFRQ	30
Evaluation of the natural frequencies for plane frames according to the TIMOSHENKO theory (Section 5.7)	

Interactive in-/output:

Screen prompt	Notes
Number NDOF of the system degrees of freedom?	
Number NELEM of elements ?	
No. of the indicative degree of freedom?	The reciprocal value of the square root of the displacement in this indicative degree of freedom (due to a unit harmonic load) is evaluated as a function of the circular frequency (NOM ω-values) and output in the file ATIMFR.
Initial value for the circular frequency?	
Circular frequency increment ?	
Number NOM of values to be computed?	The computed values are also output on screen.

Input file:

File name	
ETIM	NELEM rows containing EI, ℓ, EA, m, α, i and GA_s for each member in free format. EI is the constant bending stiffness (e.g. in kNm^2), ℓ is the element length (e.g. in m), EA is the axial stiffness (e.g.. in kN), m is the (constant) mass per unit length (e.g. in t/m), α is the angle between the global x-axis and the element axis (in degrees, positive counter-clockwise), i is the radius of gyration (e.g. in m) and GA_s is the shear stiffness (e.g. in kN). The next NELEM rows contain the incidence vector for all elements in free format, indicating the 6 global degrees of freedom corresponding to the 6 local degrees of freedom (u_1, w_1, φ_1, u_2, w_2, φ_2) for each element.

Output file:

File name	
ECHO	Echo print of the input data.
ATIMFR	The circular frequencies and the reciprocals of the square root of the displacements in the indicative degree of freedom, in two columns (2E16.7 format). The roots (zero crossings) of this curve correspond to the circular natural frequencies of the system.

GLOCKE	31
Solution of the nonlinear differential equation of a swinging church bell in order to evaluate the support reaction time histories (Section 6.1)	

Interactive in-/output:

Screen prompt	Notes
Maximum angle of rotation in degrees? Circular excitation frequency? Bell weight G in kN? Shape factor c = m*s**2/(Theta_s+m*s**2)? Distance between the bell centre of gravity and the axis of rotation in m?	The maximum values of the horizontal and vertical components of the support reaction are output on screen.

Output file:

File name	
AGLO	It contains the time points, the rotation angle, the horizontal and the vertical component of the bell support reaction (4E14.7 format) in four columns.

BASKOR	32
Linear base-line correction of an accelerogram and evaluation of the corresponding ground velocity and ground displacement time histories (Section 7.2)	

Interactive in-/output:

Screen prompt	Notes
Number NANZ of points in the accelerogram ? Constant time step DT? Factor FAKT to obtain acceleration ordinates in m/s**2 units ?	The ordinates of the record in ACC are multiplied by FAKT (e.g. 9.81 if they are originally in g) in order to obtain m/s^2-units. This factor can also be used to scale the accelerogram. The factors c_1 and c_2 of the linear base-line correction are also output on screen.

Input file:

File name	
ACC	Contains the accelerogram in 2E14.7 format, with time points in the first column and ordinates in the second column (NANZ rows).

Output files:

File name	
ACC.KOR VEL.KOR DIS.KOR VEL.UNK DIS.UNK	Corrected acceleration time history, ordinates in m/s^2. Corrected velocity time history, ordinates in m/s. Corrected displacement time history, ordinates in m. Uncorrected velocity time history. Uncorrected displacement time history. All output files consist of NANZ rows in 2E14.7 format, with time points in the first and the ordinates in the second column.

INTEG	33
Evaluation of the ground velocity and displacement time history for a given ground acceleration time history (accelerogram) (Section 7.2)	

Interactive in-/output:

Screen prompt	Notes
Number NANZ of points in the accelerogram ? Constant time step DT? Factor FAKT to obtain acceleration ordinates in m/s**2 units ?	The ordinates of the record in ACC are multiplied by FAKT (e.g. 9.81 if they are originally in g) in order to obtain m/s^2-units. This factor can also be used to scale the accelerogram.

Input file:

File name	
ACC	Contains the accelerogram in 2E14.7 format, with time points in the first and ordinates in the second column (NANZ rows).

Output file:

File name	
VEL	Velocity time history in m/s, with time points in the first column (NANZ rows in 2E14.7 format).
DIS	Displacement time history in m, with time points in the first column (NANZ rows in 2E14.7 format).

HUSID	34
Evaluation of the HUSID diagram, the ARIAS intensity and the strong motion duration of an accelerogram (Section 7.2)	

Interactive in-/output:

Screen prompt	Notes
Number NANZ of points in the accelerogram ? Constant time step DT? Factor FAKT to obtain acceleration ordinates in m/s**2 units ?	The ordinates of the record in ACC are multiplied by FAKT (e.g. 9.81 if they are originally in g) in order to obtain m/s^2-units. This factor can also be used to scale the accelerogram. The peak ground acceleration (PGA) in m/s^2, the ARIAS intensity (INT * π /2g) in m/s and the strong motion duration in s (time duration between 5% and 95% of the HUSID diagram) are output on screen.

Input file:

File name	
ACC	Contains the accelerogram in 2E14.7 format, with time points in the first and ordinates in the second column (NANZ rows).

Output file:

File name	
HUS	NANZ rows containing the normalized HUSID diagram (with time points in the first and ordinates in the second column) in 2E14.7 format.

SPECTR	35
Evaluation of response spectra of an accelerogram (absolute pseudo-acceleration, relative pseudo-velocity, relative displacement and energy spectra, Section 7.2)	

Interactive in-/output:

Screen prompt	Notes
Damping ratio D for the spectrum to be evaluated? Initial conditions: Displacement for t=0? Velocity for t=0? Number NANZ of points of the accelerogram? (<7000) Constant time step DT?	
Factor FAKT to obtain acceleration ordinates in m/s**2 units ?	The ordinates of the record in ACC are multiplied by FAKT (e.g. 9.81 if they are originally in g) in order to obtain m/s^2-units. This factor can also be used to scale the accelerogram.
Initial natural period? Period increment? Number of spectral ordinates to be computed ? (<200)	The spectral intensity according to HOUSNER (in cm) is additionally output on screen.

Input file:

File name	
ACC	Contains the accelerogram in 2E14.7 format, with time points in the first and ordinates in the second column (NANZ rows).

Output file:

File name	
SPECTR	Six columns in 6E14.7 format containing, respectively, the periods (s), the relative displacement spectrum (cm), the relative pseudo-velocity spectrum (cm/s), the absolute pseudo-acceleration spectrum (g) and the absolute and relative energy spectra (m^2/s^2).

NLSPEC	36

Evaluation of inelastic response spectra of an accelerogram for a given target maximum ductility μ (Section 7.2)

Interactive in-/output:

Screen prompt	Notes
Target ductility?	
Input data for the nonlinear spring model:	
For bilinear law JKN = 1	
For elastic-perfectly plastic law JKN = 2	
For UMEMURA law JKN = 3	
JKN = ?	
Strain hardening ratio p as a proportion pK of the initial stiffness K ?	Only required for the bilinear law.
Spectrum damping ratio D ?	
Initial conditions: Displacement for t=0? Velocity for t=0?	
Number NANZ of points of the accelerogram? (<7000)	
Constant time step DT?	
Factor FAKT to obtain acceleration ordinates in m/s**2 units ?	The ordinates of the record in ACC are multiplied by FAKT (e.g. 9.81 if they are originally in g) in order to obtain m/s^2-units. This factor can also be used to scale the accelerogram.
Initial natural period ? Period increment? Number of ordinates to be computed ? (<200)	The period and the maximum ductility actually obtained are also output on screen; the latter may deviate slightly from the target ductility due to the finite number of iteration cycles.

Input file:

File name	
ACC	Contains the accelerogram in 2E14.7 format, with time points in the first and ordinates in the second column (NANZ rows).

Output file:

File name	
NLSPK	Five columns in 5E14.7 format containing the periods (s), the relative displacement spectrum (cm), the relative pseudo-velocity spectrum (cm/s), the absolute pseudo-acceleration spectrum (g) and the ductility actually present, which may deviate slightly from the target ductility.

LINLOG		37
Transformation of spectra which are partly linear in the log-log diagram into a linear form (Section 7.3)		
Interactive in-/output:		
Screen prompt	Notes	
Number N of (T,Sv)-pairs in (s, cm/s) describing the spectrum?	Each vertex point in the log-log diagram is given by its period and relative pseudo-velocity ordinate.	

Input file:	
File name	
ESLOG	N rows, each containing the two values T and S_v of each spectrum vertex in free format.
Output file:	
File name	
ASLIN	ASLIN consists of 4 columns, containing, respectively, the periods in s, the relative displacement spectrum in cm, the relative pseudo-velocity in cm/s and the absolute pseudo-acceleration in g.

SYNTH	38
Generation of spectrum compatible accelerograms (Section 7.4)	
Input file:	
File name	
ESYN	This file contains the following data in free format: An arbitrary integer number IY (IY < 1024), Number NK of (T, S_v)-pairs to be read, defining the desired spectrum, Number N of accelerogram ordinates to be generated, with the constant time step fixed at 0.01 s, Number of the time step at the end of the increasing part of the trapezoidal envelope function (Fig. 7.4-1), Number of the time step at the beginning of the decreasing part of the trapezoidal envelope function (Fig. 7.4-1), Number of iteration cycles, usually 5 to 15, Periods TANF and TEND, between which the target spectrum is to be approximated, Damping ratio of the desired spectrum, The NK pairs (T, S_v) defining the target spectrum (T in s and S_v in cm/s); only one pair per row
Output files:	
File name	
KONTRL ASYN	Echo print of the input data. The generated accelerogram in 2E14.7 format, with time points in the first and accelerations (in m/s^2) in the second column (N rows).

MDA2DE	39
Response spectrum modal analysis of plane frame structures (Section 7.5)	

Interactive in-/output:

Screen prompt	Notes
Number NDU of the master degrees of freedom? Number NMOD of modes to be considered? Factor for the mass matrix?	The discrete mass values in MDIAG are multiplied by this factor.
Which type of spectral ordinate would you like to enter? Displacement Sd (m): JKN = 1 Velocity Sv (m/s): JKN = 2 Acceleration Sa (g): JKN = 3 JKN = ?	The following data must be entered for each of the NMOD modes:
Spectral displacement in m? Spectral velocity in m/s? Spectral acceleration in g?	The input depends on the selected type of spectral ordinate. For each mode the modal excitation factor, the effective mass and the base shear force are output on screen. In addition, the total seismically active mass and the mass actually considered are output on screen; the latter should be at least 90 % of the former.

Input files:

File name	
MDIAG	The diagonal of the mass matrix consisting of the NDU masses in the master degrees of freedom, in free format.
OMEG	The eigenvalues ω_i of the system; this file is created by the JACOBI program
PHI	The mode shapes $\underline{\Phi}_i$ of the system; this file is created by the JACOBI program
RVEKT	NDU values in free format representing the displacements in each master degree of freedom for a unit base displacement in the direction of the seismic excitation.

Output file:

File name	
STERS2	STERS2 contains the modal displacements and the equivalent static forces in the master degrees of freedom for all NMOD modes considered.

<table>
<tr><th colspan="2">MODBEN</th><th>40</th></tr>
<tr><td colspan="3">Seismic analysis of plane frame structures by direct integration of the decoupled equations of motion (Section 7.5)</td></tr>
<tr><th colspan="3">Interactive in-/output:</th></tr>
<tr><th>Screen prompt</th><th colspan="2">Notes</th></tr>
<tr><td>Number NDU of the master degrees of freedom?
Number NDPHI of the slave degrees of freedom?
Number NMOD of modes to be considered?
Number NT of time steps?
Constant time step DT ?
Factor for the accelerogram?

Enter damping ratio of mode no. ...</td><td colspan="2">

The resulting dimension of the acceleration multiplied by this factor should be (L/T^2), e.g. in m/s^2 units.

Must be entered for each of the NMOD modes. The evaluated modal excitation factor for each mode is output on screen.

The (absolute) maximum displacement value and the corresponding time point are also output on screen.</td></tr>
</table>

<table>
<tr><th colspan="2">Input files:</th></tr>
<tr><th>File name</th><th></th></tr>
<tr><td>MDIAG</td><td>The diagonal of the mass matrix consisting of the NDU masses in the master degrees of freedom, in free format.</td></tr>
<tr><td>OMEG</td><td>The eigenvalues ω_i of the system; this file is created by the JACOBI program</td></tr>
<tr><td>PHI</td><td>The mode shapes $\underline{\Phi}_i$ of the system; this file is created by the JACOBI program</td></tr>
<tr><td>V0</td><td>NDU values in free format indicating the displacements in the master degrees of freedom at time t=0.</td></tr>
<tr><td>VP0</td><td>NDU values in free format indicating the velocities in the master degrees of freedom at time t=0.</td></tr>
<tr><td>ACC</td><td>Contains the accelerogram in 2E14.7 format, with time points in the first and ordinates in the second column (NT rows).</td></tr>
<tr><td>AMAT</td><td>The matrix A (NDOF * NDU values) is required for the evaluation of the displacements in all degrees of freedom from known displacements in the master degrees of freedom.</td></tr>
<tr><td>RVEKT</td><td>NDU values in free format representing the displacements in each master degree of freedom for a unit base displacement in the direction of the seismic excitation.</td></tr>
</table>

<table>
<tr><th colspan="2">Output files:</th></tr>
<tr><th>File name</th><th></th></tr>
<tr><td>THIS.MOD</td><td>THIS.MOD contains the displacement time histories in the NDU master degrees of freedom, with time points in the first column and displacements in the other NDU columns (F7.4, 30E14.7 format)</td></tr>
<tr><td>THISDG</td><td>The displacements in all (NDU + NDPHI) degrees of freedom for all NT time points, in free format.</td></tr>
<tr><td>THISDU</td><td>The displacements in the NDU master degrees of freedom for all NT time points, in free format.</td></tr>
</table>

NEWBEN	41
Seismic analysis of plane structures by direct integration of the coupled equations of motion (Section 7.5)	

Interactive in-/output:

Screen prompt	Notes
Number NDU of the master degrees of freedom? Number NT of time steps? Constant time step DT? Factor for the accelerogram?	The resulting dimension of the acceleration multiplied by this factor should be (L/T^2), e.g. in m/s^2 units.
Should displacements, velocities or accelerations be output in THIS.NEW? For displacements: IOUT = 1 For velocities: IOUT = 2 For accelerations: IOUT = 3 IOUT = ?	The (absolute) maximum displacement, the corresponding degree of freedom number (FRH), the corresponding time and the time step number are output on screen.

Input files:

File name	Notes
MDIAG	The diagonal of the mass matrix consisting of the NDU masses in the master degrees of freedom, in free format.
KMATR	The (NDU*NDU) stiffness matrix, e.g. as generated by the KONDEN program.
CMATR	The (NDU*NDU) damping matrix, e.g. as generated by CRAY or CMOD
V0	NDU values in free format indicating the displacements in the master degrees of freedom at time t=0.
VP0	NDU values in free format indicating the velocities in the master degrees of freedom at time t=0.
ACC	Contains the accelerogram in 2E14.7 format, with time points in the first and ordinates in the second column (NT rows).
RVEKT	NDU values in free format representing the displacements in each master degree of freedom for a unit base displacement in the direction of the seismic excitation.

Output files:

File name	
KONTRL	Echo print of the input data.
THIS.NEW	Either the displacements, velocities or the accelerations of the NDU degrees of freedom at each time point with the time points in the first column (F7.4, 30E14.7 format).
THISDU	The displacements in the NDU master degrees of freedom for all NT time points in free format.

TRA3D		42
Spatial stiffness matrix for a single shear wall (Section 7.6)		
Interactive in-/output:		
Screen prompt	Notes	
Number of stories N? Input file name for the (N,N) lateral shear wall stiffness matrix? Output file name for the spatial (3N,3N) shear wall stiffness matrix? Shear wall angle (in deg, positive counter-clockwise)? Shear wall distance from the mass centre?	 This file can be created by the KONDEN or LATWND program.	
Input file:		
File name		
Optional	Input file name, containing the (N,N) in-plane lateral shear wall stiffness matrix	
Output file:		
File name	Notes	
Optional	Output file containing the spatial (3N,3N) shear wall stiffness matrix in the global coordinate system, in free format.	

MATSUM		43
Summation of a series of single spatial shear wall stiffness matrices (Section 7.6)		
Interactive in-/output:		
Screen prompt	Notes	
Number of stories N? File name for the first spatial (3N,3N) stiffness matrix? File name for the next spatial (3N,3N) stiffness matrix to be added? Add another spatial stiffness matrix? Y/N		
Input files:		
File name		
Optional	The file names of the spatial stiffness matrices created by TRA3D	
Output file:		
File name		
KMAT3D	The spatial stiffness matrix of the N-story structure in the 3*N degrees of freedom (two mutually orthogonal horizontal displacements and the rotation around the vertical axis for each story) as sum of the single shear wall spatial stiffness matrices, in free format.	

MDA3DE	44
Response spectrum modal analysis of 3D structures (Section 7.6)	

Interactive in-/output:

Screen prompt	Notes
Number of stories N?	
Maximum number NMOD of modes to be considered?	Corresponds to the number of eigenvalues and mode shapes contained in OMEG and PHI
Ground motion direction angle (in deg) ?	Angle α between the x-axis and the ground excitation direction, positive counter-clockwise.
	The following data refer to each of the NMOD modes:
Mode no. ... Natural period: ... s Mode participating factor: ...	For each mode the current mode number, the period and the calculated seismic excitation factor are shown on screen. Based on this information the following question can be answered:
Should this mode be considered, its equivalent static forces calculated and output in a separate file? Y/N	
Total effective mass percentage so far including this mode: ...%	Appears only if the question is answered affirmatively
Output file name for the equivalent static loads of this mode?	Desired file name, the output will be in free format
Spectral acceleration for this mode in g?	To be extracted by the user from the pertinent response spectrum.
	For each mode considered, the equivalent static loads in the x, y, and θ directions, the corresponding sums over all modes, and the percentage of effective total mass are shown on screen.

Input files:

File name	
MDIAG	The diagonal of the mass matrix consisting of the NDU masses in the master degrees of freedom in free format.
OMEG	The eigenvalues ω_i of the system; this file is created by the JACOBI program
PHI	The mode shapes $\underline{\Phi}_i$ of the system; this file is created by the JACOBI program

Output files:

File name	
STERS3 Optional	STERS3 contains the equivalent static loads for each mode and also some further data. Each of these files contains the load vector of the equivalent static loads in the pertinent mode (3*N coefficients, in free format).

DISP3D		45
Evaluates the displacements for each set of 3*N equivalent static loads of an N story pseudo-spatial building model, (Section 7.6)		
Interactive in-/output:		
Screen prompt	Notes	
Number of stories N ? Input file name containing the load vector? Output file name for the 3*N displacements?	 Name of the load vector file (containing 3*N values), e.g. as created by the MDA3DE program. The calculated displacements will be written in this file in free format (for further use in the FORWND program).	
Input files:		
File name		
KMAT3D Optional	Global stiffness matrix of the structure. created by MATSUM. Load vector file name.	
Output file:		
File name		
Optional	File containing the evaluated 3*N displacements in free format.	

FORWND		46
Evaluates the story shear forces for a shear wall from known global displacements of the structure (Section 7.6)		
Interactive in-/output:		
Screen prompt	Notes	
Number of stories N? Input file name for the (N,N) lateral shear wall stiffness matrix? Input file name containing the 3*N displacements of the structure? Output file name for the N horizontal story shear forces? Wall angle (in deg, positive counter-clockwise)? Distance of the wall from the mass centre?	 This file may have been created by KONDEN or LATWND. e.g. as created by DISP3D This file can be used as input for performing a static structural analysis (e.g. by programs like RAHMEN or STARAH). This is the angle between the x-axis and the longitudinal direction of the shear wall.	
Input files:		
File name		
Optional Optional	The N*N lateral stiffness matrix of the shear wall. The file containing the 3*N displacements of the structure.	
Output files:		
File name		
Optional	File containing the N story shear forces for the shear wall considered, in free format.	

LATWND	47
Evaluates the lateral stiffness matrix of a shear wall without openings using rectangular plane elements (Section 7.6)	

Interactive in-/output:

Screen prompt	Notes
Number of stories N?	
Number of rectangular elements in each story?	The height of all elements is equal to the story height.
Element thickness?	
Elastic modulus?	
POISSON ratio?	
Element height (= story height)?	
Element width?	Is equal to the shear wall width divided by the number of elements in each story.

Output file:

File name	
LATWD	Lateral N*N shear wall stiffness matrix in free format.

STAKON	48
Evaluates the lateral stiffness matrix of plane frames with rigid zones at element ends (Section 7.6)	

Interactive in-/output:

Screen prompt	Notes
Number NDOF of the system degrees of freedom?	
Number NDU of master degrees of freedom?	Typically, the horizontal story displacements are the master degrees of freedom (NDU values for an NDU story building).
Number NELRIE of beam elements with rigid end zones?	
Number NELSTU of beam elements without rigid end zones?	
Number NFED of elastic support matrices?	Each elastic support matrix couples a number of active kinematic degrees of freedom with one another and/or with the ground (degree of freedom 0)
Dimensions of the square elastic support matrices (NFED numbers) ?	The dimension of each elastic support matrix is equal to the number of the degrees of freedom it connects.

Input file:

File name	Notes
ESTAK	NELRIE rows containing the values of EI, ℓ, EA, α, AL and BL for each element with rigid end zones, in free format. EI is the constant bending stiffness in the middle part of the beam (e.g. in kNm^2), ℓ is the element length (e.g. in m), EA is the axial stiffness (e.g.. in kN); α is the angle between the global x-axis and the element axis (in degrees, positive counter-clockwise); AL is the rigid zone length at end 1 (e.g. in m) and BL is the rigid zone length at end 2 (e.g. in m). The next NELRIE rows contain the incidence vectors for all elements in free format, indicating the 6 global degrees of freedom corresponding to the 6 local degrees of freedom (u_1, w_1, φ_1, u_2, w_2, φ_2) for each element. The next NELSTU rows contain EI, ℓ, EA and α (as defined above) for the elements without rigid end zones, followed by NELSTU rows containing the corresponding incidence vectors. Next, the NDU numbers of the master kinematic degrees of freedom are entered, in free format.
INZFED	NFED rows, one for each elastic support matrix, containing the numbers of the system degrees of freedom which are constrained by this elastic support (free format).
FEDMAT	The coefficients of all NFED elastic support matrices, in free format.

Output file:

File name	Notes
ASTAK	The computed condensed stiffness matrix (NDU * NDU coefficients, 6E14.7 format).
AMAT	The matrix A (NDOF * NDU) serves for evaluating the displacements in all degrees of freedom from known displacements in the NDU master degrees of freedom.

STARAH	49
Evaluation of the internal forces and displacements for a frame system with beam/column rigid end zones (Section 7.6)	

Interactive in-/output:

Screen prompt	Notes
Number NDOF of the system degrees of freedom? Number NELRIE of beam elements with rigid end zones? Number NELSTU of beam elements without rigid end zones? Number NFED of elastic support matrices?	Each elastic support matrix couples a number of active kinematic degrees of freedom with one another and/or with the ground (degree of freedom 0).
Dimensions of the square elastic support matrices (NFED numbers) ?	The dimension of each elastic support matrix is equal to the number of the degrees of freedom it connects.

Input file:

File name	
ESTAR	NELRIE rows containing the values of EI, ℓ, EA, α, AL and BL for each element with rigid end zones, in free format. EI is the constant bending stiffness in the middle part of the beam (e.g. in kNm^2), ℓ is the element length (e.g. in m), EA is the axial stiffness (e.g.. in kN); α is the angle between the global x-axis and the element axis (in degrees, positive counter-clockwise); AL is the rigid zone length at end 1 (e.g. in m) and BL is the rigid zone length at end 2 (e.g. in m). The next NELRIE rows contain the incidence vectors for all elements in free format, indicating the 6 global degrees of freedom corresponding to the 6 local degrees of freedom (u_1, w_1, φ_1, u_2, w_2, φ_2) for each element. The next NELSTU rows contain EI, ℓ, EA and α (as defined above) for the elements without rigid end zones, followed by NELSTU rows containing the corresponding incidence vectors. Next, the NDOF load components (nodal forces and nodal moments) corresponding to the active kinematic degrees of freedom are entered, in free format.
INZFED	NFED rows, one for each elastic support matrix, containing the numbers of the system degrees of freedom which are constrained by this elastic support (free format).
FEDMAT	The coefficients of all NFED elastic support matrices, in free format.

Output file:

File name	
ASTAR	Contains the computed displacements in all NDOF system degrees of freedom followed by the deformations and end forces/moments of each element. All end forces/moments refer to the global (x,z) coordinate system and are presented in the following sequence for both element ends: Horizontal (displacement or force) component, vertical (displacement or force) component, rotation or bending moment.

UFORM		50
Format transformation of a file containing a time series with a constant time step.		
Interactive in-/output:		
Screen prompt	Notes	
Input file name: Output file name: Number of points in the time series: Input format: Factor to multiply the ordinates with: Print time points? Yes=1, No=0 : Time step: Output format:	 e.g. 14X, E14.7 The ordinates of the time series will be multiplied by this factor. e.g. 2F10.4	
Input file:		
File name		
Optional	Contains the original time series.	
Output file:		
File name		
Optional	Contains the time series in the desired format.	

EQSOLV		51
Solving an algebraic system of linear equations		
Interactive in-/output:		
Screen prompt	Notes	
Number of equations N?	The solution vector is also output on screen.	
Input file:		
File name		
KOEFMAT RSEITE	The N*N coefficient matrix, entered columnwise in free format. The load vector, N values in free format.	
Output file:		
File name		
ERGVEKT	The solution vector, N values in free format.	

MATINV		52
Inversion of a square matrix		
Interactive in-/output:		
Screen prompt	Notes	
Number of equations N ?		
Input file:		
File name		
MAT	The N*N matrix to be inverted, entered columnwise in free format.	
Output file:		
File name		
MATINV	The inverse matrix , columnwise in free format.	

Index